AF335292

Quantum Transport in Semiconductors

Quantum Transport in Semiconductors

Edited by

David K. Ferry

Arizona State University
Tempe, Arizona

and

Carlo Jacoboni

Università di Modena
Modena, Italy

Plenum Press • **New York and London**

Library of Congress Cataloging-in-Publication Data

Quantum transport in semiconductors / edited by David K. Ferry and
  Carlo Jacoboni.
       p.   cm. -- (Physics of solids and liquids)
    Includes bibliographical references and index.
    ISBN 0-306-43853-4
    1. Semiconductors. 2. Quantum theory.   I. Ferry, David K.
II. Jacoboni, C. III. Series.
QC611.Q36 1992
537.6'225--dc20   91-36620
                                                                  CIP

ISBN 0-306-43853-4

© 1992 Plenum Press, New York
A Division of Plenum Publishing Corporation
233 Spring Street, New York, N.Y. 10013

Printed in the United States of America

Contributors

N. d'Ambrumenil, Department of Physics, University of Warwick, Coventry CV4 7AL, United Kingdom

David K. Ferry, Center for Solid State Electronics Research, College of Engineering and Applied Sciences, Arizona State University, Tempe, Arizona 85287, USA

Gerald J. Iafrate, U.S. Army Research Office, Research Triangle Park, North Carolina 27709, USA

Carlo Jacoboni, Dipartimento di Fisica, Università di Modena, 41100 Modena, Italy

Antti-Pekka Jauho, Physics Laboratory, H. C. Ørsted Institute, University of Copenhagen, DK2100 Copenhagen Ø, Denmark. Present address: Nordita (Nordisk Institut for Teoretisk Fysik), DK2100 Copenhagen Ø, Denmark

M. Jonson, Solid State Division, Oak Ridge National Laboratory, Oak Ridge, Tennessee 37381-6024, USA. Present address: Institute of Theoretical Physics, Chalmers University of Technology, S-41296 Göteborg, Sweden

N. C. Kluksdahl, Department of Electrical Engineering, Arizona State University, Tempe, Arizona 85287, USA. Present address: Ford Aerospace, Houston, Texas 77058, USA

A. M. Kriman, Department of Electrical Engineering, Arizona State University, Tempe, Arizona 85287, USA. Present address: Department of Electrical and Computer Engineering, State University of New York at Buffalo, Buffalo, New York 14620, USA

Gerald D. Mahan, Department of Physics, University of Tennessee, Knoxville, Tennessee 37830, USA, and Solid State Division, Oak Ridge National Laboratory, Oak Ridge, Tennessee 37381-6024, USA

Lino Reggiani, Departimento di Fisica e Centro Interuniversitario di Struttura della Materia, Università di Modena, 41100 Modena, Italy

C. Ringhofer, Department of Mathematics, Arizona State University, Tempe, Arizona 85287, USA

K. K. Thornber, NEC Research Institute, Princeton, New Jersey 08540, USA

Preface

The majority of the chapters in this volume represent a series of lectures that were given at a workshop on quantum transport in ultrasmall electron devices, held at San Miniato, Italy, in March 1987. These have, of course, been extended and updated during the period that has elapsed since the workshop was held, and have been supplemented with additional chapters devoted to the tunneling process in semiconductor quantum-well structures. The aim of this work is to review and present the current understanding in *nonequilibrium* quantum transport appropriate to semiconductors. Generally, the field of interest can be categorized as that appropriate to inhomogeneous transport in strong applied fields. These fields are most likely to be strongly varying in both space and time.

Most of the literature on quantum transport in semiconductors (or in metallic systems, for that matter) is restricted to the equilibrium approach, in which spectral densities are maintained as semiclassical energy-conserving delta functions, or perhaps incorporating some form of collision broadening through a Lorentzian shape, and the distribution functions are kept in the equilibrium Fermi–Dirac form. The most familiar field of nonequilibrium transport, at least for the semiconductor world, is that of hot carriers in semiconductors. Here, the major problem is actually determining the form of the distribution function that arises in the high fields, and this is generally done by solving the Boltzmann transport equation with the scattering processes introduced through the Fermi golden rule and an energy-conserving delta function for the spectral density. When one then moves to nonequilibrium quantum transport, the situation is complicated by the fact that now the spectral density deviates from the simple forms and must also be determined along with the actual distribution function. The existence of a far-from-equilibrium steady state in the strong applied fields depends upon achieving a balance between the driving forces and the dissipative forces (as well as the boundary condition), and this usually entails a self-consistent approach to determining the spectral densities and the distribution function. As in the semi-classical regime, many approaches and approximations have been taken to try to achieve this end. In this volume we review a number of these different approaches. This is still a rapidly evolving field though, and no single approach has yet to begin to provide the required degree of understanding that one would desire. It is hoped that this volume will stimulate others to take up the field and that, through much more work, understanding will be achieved.

The editors of this volume would like to thank Dr. Larry Cooper of the Office of Naval Research for making the original workshop happen and for providing continuing support for the work leading to this volume. In addition, many people helped throughout the preparation of the final chapters. Of particular note, though, was the work of Drs. Rosella Brunetti, Lino Reggiani, and Al Kriman, and we extend our thanks to them.

David K. Ferry
Carlo Jacoboni

Contents

4. Quantum Transport in Solids: The Density Matrix
Gerald J. Iafrate

5. The Quantum Hall and Fractional Quantum Hall Effects
N. d'Ambrumenil

6. Green's Function Methods: Quantum Boltzmann Equation for Linear Transport

Gerald D. Mahan

7. Green's Function Methods: Nonequilibrium, High-Field Transport

Antti-Pekka Jauho

8. Numerical Techniques for Quantum Transport and Their Inclusion in Device Modeling

Lino Reggiani

9. Wave Packet Studies of Tunneling through Time-Modulated Semiconductor Heterostructures
Antti-Pekka Jauho

10. Tunneling Times in Quantum Mechanical Tunneling
M. Jonson

11. Wigner Function Modeling of the Resonant Tunneling Diode
A. M. Kriman, N. C. Kluksdahl, David K. Ferry, and C. Ringhofer

Introduction

Gerald J. Iafrate

Classical Boltzmann transport has been the mainstay of semiconductor technology from the early development of solid state physics. But in the last two decades, as semiconductor technology has continued to puruse the down scaling of IC device dimensions into the submicron (less than 10,000 Å) and ultrasubmicron (less than 1000 Å) regions, many new and interesting questions have emerged concerning the physics of small dimensions,[1,2] specifically with regard to the need for a quantum development of transport in the solid state. Table I highlights some of the basic physical effects that are viewed as important in ultrasmall electronics research. I want to focus attention on the issues of quantum transport in semiconductors.

Semiconductor transport in the ultrasubmicron region indeed approaches the regime of quantum transport. This is suggested by two trends: within the effective-mass approximation, the thermal de Broglie wavelength[3] for electrons in semiconductors is of the order of ultrasubmicron dimensions, thereby encroaching on the "physical optics" limit of wave mechanics; in addition, the time of flight for electrons traversing ultrasubmicron distances with velocities of about 10^7 cm/sec (typical velocities for electrons in semiconductors) is in the 10^{-15} to 10^{-12} sec region—a time scale which equals, if it is not less than, momentum and energy relaxation times in semiconductors, and precludes the validity of the golden rule. This leads to the need for transient, nonequilibrium solid state quantum transport considerations.

Whereas classical transport physics is based on the concept of a probability distribution function which is defined over the phase space of the system, the concept of a phase-space distribution function in the quantum

---

Gerald J. Iafrate • U.S. Army Research Office, Research Triangle Park, North Carolina 27709, USA.

TABLE I. Important Effects in Ultrasmall Electronics

Transport Effects
Drift
- Velocity overshoot
- Ballistic transport
- Oxide polar phonons
  decreasing channel mobility
- Hot-electron effects (scattering
  in high electric fields,
  injection into oxides)
- Hot-phonon effects

Diffusion
- Hot-electron diffusion
  (invalidation of Einstein relation)
- Anisotropy of diffusion
- Diffusion and reduced
  dimensionality

Size Effects
- Spatial quantization (one- and two-dimensional electrons)
- Quantum resonances—surface plasmon, phonons
- Interfaces, surfaces, metal boundaries (influence of these boundaries on important semiconductor parameters)

Environmental Effects
- Low-level radiation effects (α-particles from IC packages, cosmic rays)
- Synergetic effects
- Remote polar scattering (see Drift)
- Parasitic and interconnect factors; model contacts

Generation-Recombination Effects
- Hot-electron thermionic emissions
- G-R noise for nonstationary transport (Langevin)
- Impact ionization effects

Solid State Physics/Electronics
- Nonlinear response theory
- Reexamine effective-mass theory
- Statistical mechanics of finite Fermi systems
- Electron–phonon interactions with confined phonons
- Quantum transport: density matrix and Wigner formalism
- Interface physics modeling
- Low-dimensional effects

Materials/Fabrication
- Selective area epitaxy (MBE/MOCVD)
- RIE/plasma etching with ultrahigh anisotropy
- Ultrafine lithography, including sensitive resists
- Epitaxy of heterojunction and inverted heterojunction structures
- Abrupt modulation doping

Physical/Analytical Probes for Ultrasmall Structures
- Raman spectroscopy
- UV-XPS
- ESR/NMR
- Modulation spectroscopy
- I-V modulation techniques
- DLTS
- Photoluminescence/laser action
- Auger techniques
- Femtosecond/picosecond spectroscopy
- Magnetotransport
- Hot-electron spectrometry

formulation of transport is difficult, inasmuch as the noncommutation of
the position and momentum operators (the Heisenberg uncertainty principle) precludes the precise specification of a point in phase space. However,
within the formulation of quantum mechanics, various formalisms based
on density matrices, Wigner functions, Feynman path integrals, and Green's

functions have been developed. These embrace the quantum nature of transport; moreover, in recent years, each technique has been utilized to address key aspects of quantum transport in semiconductors.

Many of the issues relevant to quantum transport in semiconductors are highlighted in Table II. At present, there is no unifying, user-friendly approach to quantum transport in semiconductors. Density matrices, and the associated Wigner function approach, Green's functions, and Feynman path integrals all have their application and computational strengths and weaknesses, and all are equivalent representations of the quantum nature of transport. In the chapters to follow, each of these formulations for treating quantum transport in the solid state will be reviewed and its current status discussed.

In order to elucidate the limitations of the Boltzmann transport theory and to describe adequately the fundamental physical processes necessary to understand transport in solids, especially with regard to submicron and ultrasubmicron electronics, the basic assumptions and issues relevant to the Boltzmann transport equation (BTE) are reviewed.

The transport properties of carriers in bulk crystalline solids are derived from a probability distribution function $f(\mathbf{r}, \mathbf{p}, t)$. The function $f(\mathbf{r}, \mathbf{p}, t)$ is defined such that $f(\mathbf{r}, \mathbf{p}, t)\, d\mathbf{r}$ is the probability of finding an electron with quasi momentum $\mathbf{p} = h\mathbf{k}$ in a volume $d\mathbf{r}$ centered at $\mathbf{r}$ at time t. For a homogeneous solid in thermodynamic equilibrium, $f(\mathbf{r}, \mathbf{p}, t)$ is $f_0(\varepsilon(\mathbf{p}))$, where f_0 is the Fermi–Dirac distribution function at temperature T. The distribution function is found by solving the Boltzmann transport equation

$$\frac{\partial f}{\partial t} + \mathbf{v} \cdot \nabla_r f + \mathbf{F} \cdot \nabla f = \left(\frac{\partial f}{\partial t}\right)_{\text{coll}} \tag{1}$$

where

$$\mathbf{v} = \nabla_p \varepsilon(\mathbf{p})$$

and $\mathbf{F}$ is the force on the electron (or hole).

TABLE II. Issues Relevant to Quantum Transport in Semiconductors

- Strong electric and magnetic fields
- Periodic crystal potential; band-structure effects
- Electron–electron scattering
- Pauli exclusion principle
- Inhomogeneous electric fields
- Dissipation; charge extraction-injection
- Band-engineered barriers, quantum wells
- Dynamical screening
- Tunneling
- Transient effects
- Scattering from defects, phonons
- Temperature dependence
- Many-body effects

When an electric field is present, the force is given by $\mathbf{F} = q\mathbf{E}$, where q is the charge on the particle and $\mathbf{E}$ is the field at the point $\mathbf{r}$ at time t. Here

$$\left(\frac{\partial f}{\partial t}\right)_{\text{coll}} = \sum_{\mathbf{p}'} \{f(\mathbf{r}, \mathbf{p}', t)[1 - f(\mathbf{r}, \mathbf{p}, t)]W(\mathbf{p}', \mathbf{p})$$

$$- f(\mathbf{r}, \mathbf{p}, t)[1 - f(\mathbf{r}, \mathbf{p}', t)]W(\mathbf{p}, \mathbf{p}')\} \qquad (2)$$

where $W(\mathbf{p}, \mathbf{p}')$ is the transition rate from state $\mathbf{p}$ and $\mathbf{p}'$ and depends on the details of the scattering mechanisms,[4] e.g., electron–phonon scattering, electron-impurity scattering, etc.

Using the principle of microscopic reversibility, it is easy to show that, when $f = f_0(\varepsilon)$, the Fermi–Dirac distribution is $(\partial f/\partial t)_{\text{coll}} = 0$, thus ensuring that, for a homogeneous solid in thermodynamic equilibrium (i.e., where all terms on the left-hand side of (1) are zero), the solution of the Boltzmann transport equation reduces to the usual distribution function.[4] If we are dealing with electrons (holes) obeying nondegenerate statistics (i.e., $f(\mathbf{r}, \mathbf{p}, t) \ll 1$), then the terms in brackets in the collision term can be replaced by unity, thus significantly simplifying the scattering term.

In presenting an elementary derivation of the Boltzmann equation for solids, we consider a group of electrons in the volume element $d\mathbf{r}\, d\mathbf{p}$ in phase space (the number of such electrons for a given spin direction is $f(\mathbf{r}, \mathbf{p}, t)\, d\mathbf{r}\, d\mathbf{p}/(2\pi h)^3$. At a time δt later, with no scattering, these electrons move to positions $\mathbf{r} + \delta\mathbf{r}, \mathbf{p} + \delta\mathbf{p}$ in a volume $d\mathbf{r}\, d\mathbf{p}$. For a short enough time, δt, the volume element in phase space will not change ($d\mathbf{r}\, d\mathbf{p} = d\mathbf{r}\, d\mathbf{p}'$) since all the electrons in the original volume element have essentially the same position and quasimomentum, and therefore the changes in $\delta\mathbf{r}$ and $\delta\mathbf{p}$ for all electrons in the element are the same to the lowest order. However, if collisions are allowed at the rate of $(\delta f/\delta t)_{\text{coll}}$, then the distribution function evolves as

$$f(\mathbf{r} + \delta\mathbf{r}, \mathbf{p} + \delta\mathbf{p}, t + \delta t) = f(\mathbf{r}, \mathbf{p}, t) + \delta t\left(\frac{\partial f}{\partial t}\right)_{\text{coll}} \qquad (3)$$

Expanding the left-hand side of (3), dividing by δt, and using $\delta r/\delta t = \mathbf{\upsilon}$, one obtains

$$\frac{\delta f}{\delta t} + \mathbf{\upsilon} \cdot \nabla_r f + \mathbf{p}\nabla_{\mathbf{p}} f = \left(\frac{\delta f}{\delta t}\right)_{\text{coll}} \qquad (4)$$

If the group velocity of electrons, with energy $\varepsilon = \varepsilon(\mathbf{p})$, is

$$\mathbf{\upsilon} = \mathbf{\nabla}_p \in (\mathbf{p}) \qquad (5)$$

and, according to the usual theorem on quasiclassical dynamics of electron wave packets in external fields,[5]

$$\mathbf{p} = \hbar\mathbf{k} = \mathbf{F}(\mathbf{r}, \mathbf{p}, t) \qquad (6)$$

where $\mathbf{F}$ is the external force on the electron, then the left-hand side of (4) is identical to the BTE. The right-hand side of (1) and (2) can be easily obtained by noting that $(\delta f/\delta t)_{\text{coll}}$ should be merely the rate at which electrons from all other states ($\mathbf{p}'$) are scatttered into state $\mathbf{p}$, minus the rate at which electrons in state $\mathbf{p}$ are scattered into other states ($\mathbf{p}'$), all of which are evaluated at $\mathbf{r}$ and t. If $W(\mathbf{p}, \mathbf{p}')$ is the transition rate from a filled state $\mathbf{p}$ to an empty state $\mathbf{p}$, then one expects the transition rate from a state with fractional occupancy, $f(\mathbf{r}, \mathbf{p}', t)$, to a state of fractional occupancy, $f(\mathbf{r}, \mathbf{p}, t)$ to be given by the first term in (2), with the factor $1 - f(\mathbf{r}, \mathbf{p}, t)$ giving the fraction of states which are empty. This satisfies the exclusion principle that transitions to filled states are forbidden. Similarly, the second term gives the scattering of state $\mathbf{p}$ to $\mathbf{p}'$.

Despite the apparent simplicity of the above derivation, it has significant deficiencies. First, because of the uncertainty principle, the function $f(\mathbf{r}, \mathbf{p}, t)$ does not have a precise meaning as a probability function in both $\mathbf{r}$ and $\mathbf{p}$. In fact, if wave packets are formed, then $\Delta x \Delta p \sim \hbar$, and if the uncertainty in $\Delta \mathbf{p}$ is to be only a small fraction of $\mathbf{p}$ (so that we can describe a state with a well-defined $\mathbf{p}$), then the spread in the wave packet in space Δx must be many electron wavelengths long. This is a particularly serious restriction in semiconductors, where the thermal de Broglie wavelength can be quite large. Thus, the BTE cannot be expected to give a correct description of the spatial variation of the distribution if it changes significantly over several wavelengths of a typical carrier, as can be expected to occur if the force $\mathbf{F}$ has such a spatial variation.

Second, the picture employed treats the electrons as essentially free particles only, which are occasionally scattered by phonons, impurities, imperfections, and the like, but, otherwise, are not affected between collisions. However, the electrons interacting with additional potentials will alter their $\varepsilon(\mathbf{p})$ function and, hence, their velocity.

Third, the use of (6) is justified only if $\mathbf{F}$ is essentially constant over the width of the electron wave packet, which can be large for semiconductors, as indicated above. The violation of the effective-mass assumption also implies that interband and nonparabolicity effects must be considered if the field is turned on too rapidly, and this can be important for superlattices having very small band gaps.

Fourth, the assumption that the scattering takes place locally in space and time is incorrect in that the scattering potentials are extended in space and take a finite amount of time to complete. During the scattering process, an electron will have its energy changed because it is being accelerated by the force $\mathbf{F}$. Thus, the assumption that the effects of the field and the effects of collisions can be treated independently is not entirely accurate; such an assumption can be expected to break down when the energy change due to the field acting over the collision time τ_c (i.e., $Fv\tau_c$) is of the order of ε, the average energy of an electron.

Fifth, the transition rate is generally calculated by assuming it arises from an incoherent sum of single scattering events. However, if the scatterers are dense (i.e., more than one within a de Broglie wavelength), multiple scattering effects are possible.

Sixth, the electron–electron interaction can become significant for dense systems, such as quantum-confining wells.

Seventh, the assumption that $f(\mathbf{r}, \mathbf{p}, t)$ can be interpreted as a probability density in both position and momentum is really a Thomas–Fermi approximation. In fact, if we consider the time-dependent case, and let $\mathbf{F}$ correspond to the force on the electron due to a static impurity plus the self-consistent field due to the redistribution of electrons about the impurity, the solution for the potential, when Poisson's equation is employed, is just the usual Thomas–Fermi equation for the potential. However, it is well known that the solutions to this equation lead to grossly inaccurate results for the critical doping density which gives rise to the semiconductor–metal (Mott) transition in many valley semiconductors. Moreover, when dielectric screening (which, more properly, takes into consideration the wave mechanical nature of the electrons) is employed, the results for the critical density for the transition are in good agreement with experiment.[6] Furthermore, it is found that an understanding of the low-temperature conductivity of degenerately doped semiconductors, with highly anisotropic band structures (i.e., silicon and germanium), can be understood only through the use of dielectric screening, as the use of Thomas–Fermi screening significantly underestimates the scattering.[7]

Hence, even if there were no experimental manifestations for the breakdown of the predictions of the BTE, as discussed in the last point, the first six considerations suggest that it would be desirable to derive a transport equation from more general quantum mechanical considerations.

ACKNOWLEDGMENT

The author acknowledges invaluable assistance from C. S. Kavina for the preparation of the manuscript.

REFERENCES

1. G. J. IAFRATE, in: *Gallium Arsenide Technology* (D. K. Ferry, ed.), Ch. 12, H. W. Sams, Indianapolis (1985).
2. *Hot Carriers in Semiconductors* (J. Shah and G. J. Iafrate, eds.), Special Issue: *Solid State Electronics* **31**, No. 3–4 (1988).
3. G. J. IAFRATE, in: *The Physics of Submicron Semiconductor Devices* (H. L. Grubin, D. K. Ferry, and C. Jaboboni, eds.), Plenum, New York (1988).

4. W. JONES and N. MARCH, *Theoretical Solid State Physics*, Vol. 2, Ch. 6, Wiley, New York (1973).
5. A. HAUG, *Theoretical Solid-State Physics*, Vol. 1, p. 96 Pergamon Press, New York (1978).
6. J. B. KRIEGER and M. NIGHTINGALE, *Phys. Rev. B* **4**, 1266 (1971).
7. J. B. KRIEGER *et al.*, *Phys. Rev. B* **3**, 1262 (1971); **5**, 1499 (1972); **8**, 2780 (1973).

Chapter 1

Principles of Quantum Transport

Carlo Jacoboni

1.1. INTRODUCTION

In introducing the subject of quantum transport, it is useful to reconsider some of the basic assumptions that are more or less explicitly made in the semiclassical theory of electron transport in semiconductors.

(a) Apart from isolated scattering events, electrons are described by single particle states. To neglect many-body effects, such as collision broadening, means to consider low coupling. If τ is the mean time between collisions, the order of magnitude of the collision broadening $\Delta\varepsilon$ can be estimated from the uncertainty relation as

$$\Delta\varepsilon \approx \hbar/\tau \tag{1.1.1}$$

Thus, if we require, for example, $\Delta\varepsilon$ to be much less than the energy of an optical phonon, of the order of 50 meV, we need

$$\tau \gg 10^{-14}\,\mathrm{s} \tag{1.1.2}$$

(b) During the free flight between two successive collisions, electrons are considered as classical particles with well-defined positions and momenta. For this approximation to be acceptable, we need to be able to conceive wave packets with a momentum uncertainty Δp much less than their average momentum p and at the same time a position uncertainty Δx much less than the mean free path l:

$$\Delta p \ll p, \qquad \Delta x \ll l \tag{1.1.3}$$

Carlo Jacoboni • Dipartimento di Fisica, Università di Modena, 41100 Modena, Italy

Quantum Transport in Semiconductors, edited by David K. Ferry and Carlo Jacoboni. Plenum Press, New York, 1991.

From the uncertainty relations we then have

$$\hbar \sim \Delta p\, \Delta x \ll pl \sim \frac{p^2}{m}\tau \sim KT\tau \qquad (1.1.4)$$

or

$$\tau \gg \hbar/KT \sim 10^{-13} \div 10^{-14}\,\mathrm{s} \qquad (1.1.5)$$

at ordinary temperatures, where K is the Boltzmann constant and T is the temperature.

(c) In the classical treatment of the Boltzmann equation for electron transport in semiconductors, collisions are in general assumed to be instantaneous in both time and space. Since the interaction between the particles and the scattering agents has some time duration, this assumption is not correct in general, not even in classical theory. In the case of weak coupling, when scattering events are sufficiently rare, the duration of a collision may be negligible with respect to the time of free flights between successive collisions and the assumption may be reasonable. In order to estimate the requirement for such a condition to be fulfilled, a collision duration has to be estimated. This estimate is somewhat arbitrary and depends on the particular scattering mechanism considered.

If we consider the case of phonon scattering, which is of most interest to us, the duration of a collision, which is the time necessary for creating the interference between the initial state and the final state of the electron–phonon system, can be estimated as the time necessary for the electron to sense the periodicity of the phonon field. Because of a Doppler shift, this can be given by the inverse of the phonon frequency ω_p for slow electrons or large phonon q's or the phonon wavelength divided by the electron velocity for fast electrons or small q's. If we take into account that the phonon energies are of the order of the thermal energy and that the phonon q's are of the order of the electron k's, we obtain the estimate

$$\frac{1}{\omega_p} \sim \frac{\hbar}{KT} \qquad (1.1.6)$$

$$\frac{\lambda}{v} \sim \frac{m}{q\hbar k} \sim \frac{m}{\hbar k^2} \sim \frac{\hbar}{KT} \qquad (1.1.7)$$

so that, in any case,

$$\tau_{\mathrm{c}} \sim \hbar/KT \qquad (1.1.8)$$

Thus the requirement that this time is much shorter than the time between collisions coincides in this case with the above requirement of equation (1.1.5) in (b).

During the collision time a constant applied electric field acts on the initial and final electron wave vectors, changing them linearly with time. However, since the energy is not a linear function of k, the energy difference between the two states varies with time, and the interference effects that generate the transition are disturbed by the field. Such a process may modify the transition rates, an effect called the *intracollision field effect*. In order to estimate this effect, we may compare the energy $\Delta\varepsilon$ imparted to the electron by the field with the energy difference $\Delta\varepsilon_t$ of the transition.

$$\frac{\Delta\varepsilon}{\Delta\varepsilon_t} \sim v\tau_c \frac{eE}{\hbar\omega_q} \sim \frac{m}{\hbar k^2}\frac{eE}{\hbar\omega_q} \sim 1 \qquad (1.1.9)$$

for $E \sim 10^5$ V/cm. Even though these estimates are very rough, they seem to indicate that the intracollision field effect can be quite important under hot-electron conditions.

From the above discussion it is clear that the critical parameter for evaluating the applicability of the classical transport theory is (as physically plausible) the time τ between collisions in the classical theory itself. Now, in present-day technology, devices can be as short as 0.1 μm. If we apply a voltage of a few volts, we obtain fields of the order of 10^5 V/cm. Leaving aside problems related to transient transport or field inhomogeneities, the energy reached by the electrons at such fields is of the order of 1 eV. At these energies the average time between collisions may be of the order of 10^{-14} s, so that most of the classical approximations fail.

To the above discussion, we add the consideration that the geometrical dimensions of the physical system under consideration become comparable to several lengths characteristic of electron quantum physics, such as the de Broglie wavelength, or the radius of the electron orbits around impurities.

The natural consequence of everything said above is that a quantum treatment of charge transport in microelectronic devices is needed today in order to trust the physical understanding that comes from our theoretical framework.

1.2. THE GENERAL PROBLEM

The general problem under investigation is that of a quantum system described by some time-independent Hamiltonian $H^{(0)}$ to which we apply an external perturbation $H^{(\text{ext})}$

The system under investigation is often divided into two parts. One "small" part is the actual system of interest, for example the conduction electrons inside the crystal. The "large" part is a system in thermal contact with the small system and provides the necessary dissipation to maintain stationarity (for example the phonon bath and the heat bath). "Small" and "large" refer to the fact that under the effect of the external perturbation the small system can get out of equilibrium while usually the large part is considered to have such a large heat capacity (so many degrees of freedom) that it is not driven out of equilibrium by the perturbation. Hot-phonon effects are not considered here. If such a phenomenon is to be considered, however, phonons may be included in the small subsystem, and the large subsystem is the heat bath in contact with the boundary of the crystal.

The response of the small subsystem to the external perturbation is the goal of our theory. If the perturbation is so weak that a linear approximation can be good enough, the linear response theory provides very general and elegant results, but it has been well known for more than 30 years, that in electronic devices we often pass the limits of the linear response theory and we have to look for solutions to the nonlinear transport (hot-electron) problem.

In the following sections we give the basic definitions and outline the major techniques for the solution of the general transport problem in quantum physics. The ideas introduced here will then be developed at length later in this book.

1.3. QUANTUM DYNAMICS AND REPRESENTATIONS

Given a system characterized by the Hamiltonian H, its time evolution is described by the evolution operator $U_H(t, t_0)$ that verifies the following equation and initial conditions:

$$i\hbar \frac{d}{dt} U_H(t, t_0) = H U_H(t, t_0) \tag{1.3.1}$$

$$U_H(t_0, t_0) = 1 \tag{1.3.2}$$

If H does not depend explicitly on time, the solution of equations (1.3.1) and (1.3.2) is simply

$$U_H(t, t_0) = e^{-i(H/\hbar)(t-t_0)} \tag{1.3.3}$$

If, on the other hand, H depends upon time, a formal solution can be found by generalizing equation (1.3.3) in the following way: first equation (1.3.1) can be formally integrated to yield the integral equation

$$U_H(t, t_0) = 1 + \int_{t_0}^{t} dt' \frac{H(t')}{i\hbar} U_H(t', t_0) \tag{1.3.4}$$

Then, this last equation is inserted into itself iteratively; a series is obtained whose general term has the form

$$\int_{t_0}^{t} dt_1 \frac{H(t_1)}{i\hbar} \int_{t_0}^{t_1} dt_2 \frac{H(t_2)}{i\hbar} \cdots \int_{t_0}^{t_{n-1}} dt_n \frac{H(t_n)}{i\hbar} \tag{1.3.5}$$

In general, $H(t_i)$ does not commute with $H(t_j)$ so that the order of the operators H in the integrals is crucial. Let us then define a time-ordering operator T (it is actually a superoperator since it is applied to operators to yield new operators) such that, once applied to a product of operators evaluated at different times, it orders them in such a way that the earliest times appear on the right. If we apply T to the products in the terms of the series, we can formally change the orders of the Hamiltonians at will so that all the integrals can be extended up to t and, at the same time, divide by the multiplicity factor $n!$. The result is the formal extension of equation (1.3.3),

$$U_H(t, t_0) = T \exp\left[-\frac{i}{\hbar} \int_{t_0}^{t} H(t') \, dt' \right] \tag{1.3.6}$$

which is simply a compact notation for the series expansion discussed above. This expression reduces immediately to equation (1.3.3) if H does not depend on time.

In the Schrödinger representation the time evolution of the system is carried by the state vectors:

$$|\Phi_{\mathscr{S}}(t)\rangle = U_H(t, t_0)|\Phi_{\mathscr{S}}(t_0)\rangle \tag{1.3.7}$$

and physical quantities not depending explicitly on time are represented by constant operators.

In the Heisenberg representation the time evolution of the system is carried by the observables:

$$A_{\mathscr{H}}(t) = U_H^{\dagger}(t, t_0) A_{\mathscr{H}}(t_0) U_H(t, t_0) \tag{1.3.8}$$

while state vectors are constant. Note that in equation (1.3.8) the label $\mathscr{H}$ in A stands for Heisenberg, while the label H in U stands for the Hamiltonian.

We move from Schrödinger to Heisenberg representation by applying the unitary transformation $U_H^{\dagger}(t, t_0)$.

In the interaction representation both state vectors and observables carry part of the time evolution. This representation is often used when the Hamiltonian can be separated into an unperturbed part H_0, in general, time-independent and whose solutions are known, and a time-dependent perturbation H':

$$H(t) = H_0 + H'(t) \tag{1.3.9}$$

The interaction representation is obtained from the Schrödinger representation by means of the unitary transformation $U_{H_0}^{\dagger}(t, t_0)$:

$$|\Phi_{\mathscr{I}}(t)\rangle = U_{H_0}^{\dagger}(t, t_0)|\Phi_{\mathscr{S}}(t)\rangle \tag{1.3.10}$$

or from the Heisenberg representation by means of the unitary transformation

$$V(t, t_0) = U^\dagger_{H_0}(t, t_0) U_H(t, t_0) \qquad (1.3.11)$$

Since in the Heisenberg representation the state vector is constant, V is also the evolution operator for a state vector in the interaction representation. Thus, in such a representation, observables evolve according to the unperturbed dynamics:

$$A_\mathscr{I}(t) = U^\dagger_{H_0}(t, t_0) A_\mathscr{I}(t_0) U_{H_0}(t, t_0) \qquad (1.3.12)$$

while state vectors evolve according to the "partial" evolution:

$$|\Phi_\mathscr{I}(t)\rangle = V(t, t_0)|\Phi_\mathscr{I}(t_0)\rangle \qquad (1.3.13)$$

In order to better understand the meaning of the operator V, let us find its dynamical equation

$$i\hbar \frac{d}{dt} V(t, t_0) = i\hbar \frac{d}{dt}(U^\dagger_{H_0} U_H) = U^\dagger_{H_0}(-H_0 + H)U_H = H'_\mathscr{I} V \qquad (1.3.14)$$

Thus, V is the evolution operator for the perturbation Hamiltonian H' in the interaction representation: $V = U_{H'_\mathscr{I}}$. Equation (1.3.13), therefore, indicates that in the interaction representation the state vector evolves according to the perturbation Hamiltonian as the observables evolve according to the unperturbed Hamiltonian. In other words, in this representation the known unperturbed dynamics is assigned to the observables and taken away from the equation of motion for the state vectors.

With the same argument followed for equation (1.3.6) we obtain

$$V(t, t_0) = T \exp\left[-\frac{i}{\hbar} \int_{t_0}^{t} H'_\mathscr{I}(t')\, dt'\right] \qquad (1.3.15)$$

1.4. THE DENSITY MATRIX

Given that the system of interest is represented in some representation by the state vector $|\Phi\rangle$, the density matrix is defined, in the same representation, as the projection operator

$$\rho = |\Phi\rangle\langle\Phi| \qquad (1.4.1)$$

In a given basis, the density matrix operator has the following matrix elements

$$\rho_{nm} = \langle n|\Phi\rangle\langle\Phi|m\rangle = c_n c^*_m \qquad (1.4.2)$$

where c_n are the coefficients for the development of $|\Phi\rangle$ in the given basis.

For a physical quantity represented by the observable A, its mean value in the state $|\Phi\rangle$ is then given by

$$\langle A \rangle = \langle \Phi | A | \Phi \rangle = \langle \Phi | m \rangle \langle m | A | n \rangle \langle n | \Phi \rangle \tag{1.4.3}$$

(a sum over repeated indexes is implied), or

$$\langle A \rangle = \rho_{nm} A_{mn} = \mathrm{Tr}(\rho A) \tag{1.4.4}$$

The concept of density matrix is extended to ensemble statistics. By using the overbar for ensemble averages, we define for an ensemble of N systems

$$\rho = \overline{|\Phi\rangle\langle\Phi|} = \frac{1}{N} \sum_{\text{ens.}} |\Phi\rangle\langle\Phi| = \sum_{\Phi} p_\Phi |\Phi\rangle\langle\Phi| \tag{1.4.5}$$

where p_Φ is the probability of finding a system of the ensemble in state $|\Phi\rangle$. In this case

$$\overline{\langle A \rangle} = \overline{\langle \Phi | A | \Phi \rangle} = \overline{\langle \Phi | m \rangle \langle m | A | n \rangle \langle n | \Phi \rangle} = A_{mn} \overline{\langle n | \Phi \rangle \langle \Phi | m \rangle} \tag{1.4.6}$$

Thus, again

$$\overline{\langle A \rangle} = \mathrm{Tr}(\rho A) \tag{1.4.7}$$

Let us now consider the equation of motion of the density matrix in the different representations considered in the previous section. The time evolution of ρ is obtained immediately by its definition with the time evolutions of the state vectors in equations (1.3.7) and (1.3.13). The result in the Schrödinger representation is

$$\rho_{\mathscr{S}}(t) = U_H(t, t_0) \rho_{\mathscr{S}}(t_0) U_H^\dagger(t, t_0) \tag{1.4.8}$$

and in the interaction representation:

$$\rho_{\mathscr{I}}(t) = V(t, t_0) \rho_{\mathscr{I}}(t_0) V^\dagger(t, t_0) \tag{1.4.9}$$

By differentiating with respect to time, we obtain immediately the Liouville equations for the density matrix in the two representations:

$$i\hbar \frac{d\rho_{\mathscr{S}}}{dt} = [H, \rho_{\mathscr{S}}] \tag{1.4.10}$$

$$i\hbar \frac{d\rho_{\mathscr{I}}}{dt} = [H'_{\mathscr{I}}, \rho_{\mathscr{I}}] \tag{1.4.11}$$

In the second form, only the perturbation Hamiltonian appears in the commutator.

In the Heisenberg representation the density matrix is constant in time, since the state vectors are constant, equal to its initial value in the Schrödinger representation.

The time evolution of an average quantity A, which is always given by equation (1.4.4), is due to the time evolution of ρ in the Schrödinger

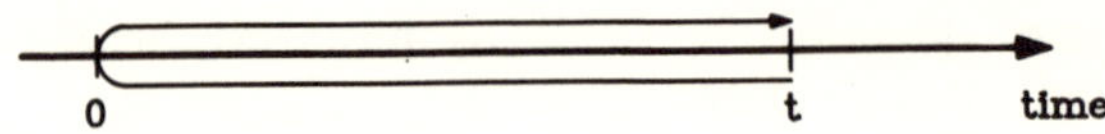

FIGURE 1. Closed-loop contour for time integration.

representation and by the time evolution of A in the Heisenberg representation. In the interaction representation, A carries the time evolution due to the known, unperturbed Hamiltonian H_0, and ρ the time evolution due to the perturbation Hamiltonian H'.

Thus, in the Schrödinger or interaction representation, the solution of the Liouville equation would provide an answer to the transport problem. The formal solutions given in equations (1.4.8) and (1.4.9), however, do not constitute real solutions since the evolution operators are not written in explicit terms. If we substitute the expression in equation (1.3.15) for the evolution operator V, we obtain

$$\rho_{\mathscr{I}}(t) = T \exp\left[-\frac{i}{\hbar}\int_{t_0}^{t} H'_{\mathscr{I}}(t')\,dt'\right]\rho_{\mathscr{I}}(t_0)\,\tilde{T}\exp\left[\frac{i}{\hbar}\int_{t_0}^{t} H'_{\mathscr{I}}(t')\,dt'\right] \quad (1.4.12)$$

where $\tilde{T}$ is the time antiordering operator. This expression can be written in a more compact form by considering a contour C of integration as in Figure 1.1, and the contour ordering operator T_c defined in such a way that operators which come first in the contour appear on the right. Then

$$\rho_{\mathscr{I}}(t) = T_c\left[\exp\left[-\frac{i}{\hbar}\int_{t_c} H'_{\mathscr{I}}(t')\,dt'\right]\rho_{\mathscr{I}}(t_0)\right] \quad (1.4.13)$$

1.5. SECOND QUANTIZATION

The fundamental concepts of the second quantization formalism will be briefly summarized here for future use. For a general discussion of these concepts we refer to the general literature.

If, in some way, we succeeded in indentifying single particle states or elementary excitations labeled as $1, 2, \ldots$, we define a state of the system $|n_1, n_2, \ldots, n_i, \ldots\rangle$ as one with n_1 particles or excitations in state 1, n_2 in state 2, and so on. The state $|0000\ldots\rangle$ or simply $|0\rangle$ is called the vacuum state with no particles. Starting from the vacuum state, we may generate all the other states by successive applications of the creation operators, $a_i^\dagger$ for bosons, and $c_i^\dagger$ for fermions. These operators and their hermitian conjugate, the annihilation operators, are defined by

$$a_i|n_1, n_2, \ldots\rangle = \sqrt{n_i}\,|n_1, n_2, \ldots, n_i - 1, \ldots\rangle \quad (1.5.1)$$

$$a_i^\dagger|n_1, n_2, \ldots\rangle = \sqrt{n_i + 1}\,|n_1, n_2, \ldots, n_i + 1, \ldots\rangle \quad (1.5.2)$$

for bosons, and

$$c_i|n_1, n_2, \ldots, n_i, \ldots\rangle = \begin{cases} 0, & \text{if } n_i = 0 \\ |n_1, n_2, \ldots, 0, \ldots\rangle & \text{if } n_i = 1 \end{cases} \qquad (1.5.3)$$

$$c_i^\dagger|n_1, n_2, \ldots, n_i, \ldots\rangle = \begin{cases} |n_1, n_2, \ldots, 1, \ldots\rangle & \text{if } n_i = 0 \\ 0 & \text{if } n_i = 1 \end{cases} \qquad (1.5.4)$$

for fermions.

The commutation relations for these operators are

$$[a_\mu, a_\nu^\dagger] = \delta_{\mu\nu} \qquad (1.5.5)$$

$$[a_\mu, a_\nu] = [a_\mu^\dagger, a_\nu^\dagger] = 0 \qquad (1.5.6)$$

$$[c_\mu, c_\nu^\dagger]_+ = \delta_{\mu\nu} \qquad (1.5.7)$$

$$[c_\mu, c_\nu]_+ = [c_\mu^\dagger, c_\nu^\dagger]_+ = 0 \qquad (1.5.8)$$

The electron field operators $\Psi^\dagger(\mathbf{r})$ and $\Psi(\mathbf{r})$ are similar creation and annihilation operators for the position representation. In other words, the state $\Psi^\dagger(\mathbf{r})|\Phi\rangle$ is the state of the system obtained from the state $|\Phi\rangle$ by adding an electron in position $\mathbf{r}$. They again obey the commutation relations

$$[\Psi(\mathbf{r}), \Psi^\dagger(\mathbf{r}')]_+ = \delta(\mathbf{r} - \mathbf{r}') \qquad (1.5.9)$$

$$[\Psi(\mathbf{r}), \Psi(\mathbf{r}')]_+ = [\Psi^\dagger(\mathbf{r}), \Psi^\dagger(\mathbf{r}')]_+ = 0 \qquad (1.5.10)$$

In the Heisenberg or interaction representation $\Psi^\dagger$ and Ψ become time-dependent operators. At equal times, the commutation relations are still the same as in equations (1.5.9) and (1.5.10), but at different times this does not hold true anymore.

It may be useful to discuss the meaning of these field operators in the Heisenberg or interaction representation. We cannot say that $\Psi_{\mathcal{H}}^\dagger(\mathbf{r}, t)|\Phi_{\mathcal{H}}\rangle$ represents a new state at time t in the Heisenberg representation because in such a representation state vectors are time-independent. However, it is true that $\Psi_{\mathcal{H}}^\dagger(\mathbf{r}, t)$, applied to $|\Phi_{\mathcal{H}}\rangle$, gives a new state in Heisenberg representation with an extra particle in $\mathbf{r}$ at time t. In fact, given the Hamiltonian H, the vector $\Psi_{\mathcal{S}}^\dagger(\mathbf{r})|\Phi_{\mathcal{S}}(\mathbf{r}, t)\rangle$ does not evolve according to the dynamics of H since it represents a state evolving according to H plus an extra particle with constant position $\mathbf{r}$. If at time t_1, this state is taken as the initial condition and it is then left to evolve according to H, its state vector in Schrödinger representation is

$$\begin{aligned} |\Phi_{\mathcal{S}}'(t)\rangle &= U_H(t, t_1)\Psi_{\mathcal{S}}^\dagger(\mathbf{r})|\Phi_{\mathcal{S}}(\mathbf{r}, t_1)\rangle \\ &= U_H(t, t_0) U_H^\dagger(t_1, t_0)\Psi_{\mathcal{S}}^\dagger(\mathbf{r}) U_H(t_1, t_0)|\Phi_{\mathcal{S}}(t_0)\rangle \\ &= U_H(t, t_0)\Psi_{\mathcal{H}}^\dagger(\mathbf{r}, t_1)|\Phi_{\mathcal{H}}\rangle \end{aligned} \qquad (1.5.11)$$

However, we must have

$$|\Phi'_{\mathscr{S}}(t)\rangle = U_H(t, t_0)|\Phi'_{\mathscr{H}}\rangle$$

Thus,

$$\Psi^{\dagger}_{\mathscr{H}}(\mathbf{r}, t_1)|\Phi_{\mathscr{H}}\rangle = |\Phi'_{\mathscr{H}}\rangle \qquad (1.5.12)$$

From this equation we obtain the meaning of the field operator $\Psi^{\dagger}_{\mathscr{H}}(\mathbf{r}, t)$ in the Heisenberg representation: applied to a state $|\Phi_{\mathscr{H}}\rangle$, it yields a new state $|\Phi'_{\mathscr{H}}\rangle$ that has an extra particle, with respect to $|\Phi_{\mathscr{H}}\rangle$, which is found to be in $\mathbf{r}$ at time t.

In the same way we can show that, given a state represented by $|\Phi_{\mathscr{S}}(t)\rangle$ in the interaction representation, if we apply $\Psi^{\dagger}_{\mathscr{S}}(\mathbf{r}, t)$ to it, we obtain a new state vector representing, in the interaction representation, a state with an extra particle in $\mathbf{r}$ at time t.

When a field operator is written with a time argument without a representation label, it is assumed to be considered in the Heisenberg representation. When it is written without a time argument, it is assumed to be considered in the Schrödinger representation.

1.6. GREEN'S FUNCTIONS

In transport theory we study how particles move under the action of external forces. If we know that a system is formed by a particle that is in $\mathbf{r}_2$ at time t_2, in the Heisenberg representation this state is given by the vector $\Psi^{\dagger}(\mathbf{r}_2, t_2)|0_{\mathscr{H}}\rangle$, and we may ask ourselves what is the probability of finding the particle in $\mathbf{r}_1$ at t_1. The probability amplitude will be given by the scalar product

$$\langle 0_{\mathscr{H}}|\Psi(\mathbf{r}_1, t_1)\Psi^{\dagger}(\mathbf{r}_2, t_2)|0_{\mathscr{H}}\rangle \qquad (1.6.1)$$

This is the basic concept of the Green's function, defined, more precisely, by

$$G(\mathbf{r}_1, t_1, \mathbf{r}_2, t_2) = -i\langle\Phi_{\mathscr{H}}|\Psi(\mathbf{r}_1, t_1)\Psi^{\dagger}(\mathbf{r}_2, t_2)|\Phi_{\mathscr{H}}\rangle \qquad (1.6.2)$$

where the propagation of an extra particle is considered, in addition to a many-body state $|\Phi\rangle$. Actually this is one of the many Green's functions that will be introduced, discussed, and used later in this book.

In the Heisenberg representation G has its simplest form, since all the time information is carried by the field operators. By applying the necessary unitary transformation, the Green's function in equation (1.6.2) can be written in the other representations as

$$G(\mathbf{r}_1, t_1, \mathbf{r}_2, t_2) = -i\langle\Phi_{\mathscr{S}}(t_1)|\Psi(\mathbf{r}_1)U_H(t_1, t_2)\Psi^{\dagger}(\mathbf{r}_2)|\Phi_{\mathscr{S}}(t_2)\rangle \quad (1.6.3)$$

or

$$G(\mathbf{r}_1, t_1, \mathbf{r}_2, t_2) = -i\langle\Phi_\mathscr{I}(t_1)|\Psi_\mathscr{I}(\mathbf{r}_1, t_1)\,V(t_1, t_2)\Psi_\mathscr{I}^\dagger(\mathbf{r}_2, t_2)|\Phi_\mathscr{I}(t_2)\rangle \quad (1.6.4)$$

From the above discussion of the physical idea underlying the concept of the Green's function, it should be acceptable that its knowledge carries all the information necessary for the solution of a transport problem.

In order to establish a link between the density matrix and the Green's functions, let us assume that at time t_0 the state is given by a particle in $\mathbf{r}$, so that

$$\rho_\mathscr{S}(t_0) = |\mathbf{r}_\mathscr{S}(t_0)\rangle\langle\mathbf{r}_\mathscr{S}(t_0)| = \Psi^\dagger(\mathbf{r})|0\rangle\langle0|\Psi(\mathbf{r}) \quad (1.6.5)$$

The $(\mathbf{r}_1, \mathbf{r}_2)$ matrix element of the density matrix at time t is given by

$$\rho_\mathscr{S}(\mathbf{r}_1, \mathbf{r}_2, t) = \langle\mathbf{r}_1|\rho_\mathscr{S}(t)|\mathbf{r}_2\rangle = \langle\mathbf{r}_1|U_H(t, t_0)\rho_\mathscr{S}(t_0)U_H^\dagger(t, t_0)|\mathbf{r}_2\rangle$$

$$= \langle0|\Psi(\mathbf{r}_1)\,U_H(t, t_0)\Psi^\dagger(\mathbf{r})|0\rangle\langle0|\Psi(\mathbf{r})\,U_H^\dagger(t, t_0)\Psi^\dagger(\mathbf{r}_2)|0\rangle \quad (1.6.6)$$

that is, the product of the two Green's functions describing the propagation of the particle from $\mathbf{r}$ at t_0 to $\mathbf{r}_1$ and to $\mathbf{r}_2$ at t. In more complex many-body situations this connection still exists, based on the expansion of the density matrix implicit in equation (1.4.13), since both H' and $\rho(t_0)$ contain field operators, as will be seen later in the book.

As regards the derivation of an equation of motion for G, however, the situation is more complicated. In fact, since G is a function of the variables of one particle we cannot hope to write an equation of motion for G alone which also accounts for the interaction of the whole many-body system. In the equation of motion for G (the Dyson equation) another quantity usually appears, the self-energy Σ, that accounts for the interaction of a single particle with the rest of the many-body system, and that must be obtained by independent means. A similar situation appears in the semiclassical Boltzmann equation, where the distribution function $f(\mathbf{r}, \mathbf{v}, t)$ is a function of single particle variables, but in the equation the scattering integral contains the scattering probabilities that carry the information about the interaction of a particle with the rest of the system. The derivation of the equation of motion for G and Σ is usually performed in terms of the separation of the Hamiltonian shown in equation (1.3.9) and specialized for actual cases of interest. Examples will be seen in later chapters of this book.

1.7. WIGNER FUNCTIONS

The Wigner function was introduced as an extension of the classical distribution function. If we consider, for simplicity, a one-dimensional system with canonical variables q and p, it should be clear that we cannot strictly define a probability function $P(q, p)$ such that $P(q, p)\, dq\, dp$ is equal to the probability of finding the system in dq around q and dp around p, since this probability is ill-defined in quantum mechanics owing to the incompatibility of the two necessary measurements (or variables).

Since the function we are looking for should be the classical analog of the density matrix operator ρ, we may raise the more general problem of defining the function $\mathscr{A}(q, p)$, where q and p are c-numbers, which is the classical analog of a general quantum operator $A(q, p)$ function of the operators q and p. A requisite could be that the sum of all possible values is the same for both quantities:

$$\mathrm{Tr}\{A\} = \frac{1}{2\pi\hbar} \int\!\!\int \mathscr{A}(q, p)\, dq\, dp \qquad (1.7.1)$$

This, however, is not enough to determine $\mathscr{A}$ since only an integrated property of it is given. On the other hand, if we require that the property expressed by equation (1.7.1) holds when we also give different weights to the different points in phase space, then the equation can be inverted. Here we have an arbitrariness in how to choose these different weights. If Fourier functions are chosen, we obtain the Weyl correspondence rule.

Let us then require that, for any real ξ and η,

$$\mathrm{Tr}\{Ae^{i[\xi q/\hbar + \eta p/\hbar]}\} = \frac{1}{2\pi\hbar} \int\!\!\int \mathscr{A}(\alpha, \beta)e^{i\xi\alpha/\hbar}e^{i\eta\beta/\hbar}\, d\alpha\, d\beta \qquad (1.7.2)$$

Now the Fourier transform in the right-hand side can be inverted:

$$\mathscr{A}(\alpha, \beta) = \frac{1}{2\pi\hbar} \int e^{-i\xi\alpha/\hbar}\, d\xi \int e^{-i\eta\beta/\hbar}\, d\eta \int dx$$

$$\int dx'\langle x|A|x'\rangle\langle x'|e^{i[\xi q/\hbar + \eta p/\hbar]}|x\rangle \qquad (1.7.3)$$

By using

$$e^{i[\xi q + \eta p]} = e^{i\eta p/2}e^{i\xi q}e^{i\eta p/2} \qquad (1.7.4)$$

and

$$e^{i\eta p/\hbar}|x\rangle = |x - \eta\rangle \qquad (1.7.5)$$

we obtain, after straightforward calculations,

$$\mathscr{A}(\alpha, \beta) = \int e^{i\eta\beta/\hbar}\left\langle \alpha - \frac{\eta}{2}\right|A(q, p)\left|\alpha + \frac{\eta}{2}\right\rangle\, d\eta \qquad (1.7.6)$$

which is the Weyl correspondence rule.

The Wigner function is now defined as the Weyl transform of the density matrix operator (the factor $1/2\pi\hbar$ is added in order to use simple integration over the phase space):

$$P_W(q, p) = \frac{1}{2\pi\hbar} \int e^{i\eta p/\hbar} \left\langle q - \frac{\eta}{2} \middle| \psi \right\rangle \left\langle \psi \middle| q + \frac{\eta}{2} \right\rangle d\eta$$

or

$$P_W(q, p) = \frac{1}{2\pi\hbar} \int e^{i\eta p/\hbar} \psi^*\left(q + \frac{\eta}{2}\right) \psi\left(q - \frac{\eta}{2}\right) \tag{1.7.7}$$

Several properties of P_W echo its origin of classical analog of the density matrix. In particular,

$$\int P_W(q, p)\, dp = |\psi(q)|^2 \tag{1.7.8}$$

$$\int P_W(q, p)\, dq = |\phi(p)|^2 \tag{1.7.9}$$

Therefore, if $F(q)$ and $G(p)$ are functions only of q and p, respectively, then

$$\iint F(q) P_W(q, p)\, dq\, dp = \langle F \rangle \tag{1.7.10}$$

$$\iint G(p) P_W(q, p)\, dq\, dp = \langle G \rangle \tag{1.7.11}$$

More generally, it can be shown that if A is a general observable function of q and p, then

$$\langle A \rangle = \text{Tr}(\rho A) = \iint \mathcal{A}(q, p) P_W(q, p)\, dq\, dp \tag{1.7.12}$$

where $\mathcal{A}$ is the Weyl transform of A.

On the other hand, it is not possible to give P_W a simple probabilistic interpretation for the reasons indicated at the beginning of this section, and this is confirmed by the fact that P_W takes in general negative, as well as positive, values in the phase space for those situations in which quantum interference occurs.

As it will be seen later in the book, a generalization of P_W is introduced as a special Fourier transform of a Green function.

Concluding this brief introduction to the Wigner function, we note that if the interaction of one particle with the rest of the system can be described by a perturbation potential, an equation for the Wigner function can be written which reduces to the Boltzmann equation in the classical limit. Particular care must be taken in this case to form the initial condition for $P_W(q, p)$, since not all functions of q and p are eligible to be correct Wigner functions.

1.8. KINETIC EQUATIONS AND IRREVERSIBILITY

In the previous section we saw that a major problem of a many-body theory is the reduction of the whole complexity of the system to a simplified version of it, which is analyzed in terms of a reduced number of variables of interest. A typical example is that of the kinetic equation, which can be defined as an equation for a function of only a subset of the variables of the system.

In order to discuss further this reduction problem, let us consider again the example of electrons interacting with phonons in a crystal. The state vectors and therefore the density matrix will be a function of electron variables or quantum numbers x and phonon variables or quantum numbers ξ:

$$\rho = \rho(\mathbf{x}, \xi, \mathbf{x}', \xi') \tag{1.8.1}$$

If an observable $A^{(el)}$ acts only on the electron variables, its mean value is given by

$$\langle A^{(el)} \rangle = \mathrm{Tr}(\rho A) = \sum_{\mathbf{x}} \sum_{\xi} \rho(\mathbf{x}, \xi, \mathbf{x}', \xi) A^{(el)}(\mathbf{x}', \mathbf{x}) \tag{1.8.2}$$

or

$$\langle A^{(el)} \rangle = \mathrm{Tr}(\rho^{(el)} A^{(el)}) \tag{1.8.3}$$

where

$$\rho^{(el)}(\mathbf{x}, \mathbf{x}') = \sum_{\xi} \rho(\mathbf{x}, \xi, \mathbf{x}', \xi) = \mathrm{Tr}_{\xi}(\rho) \tag{1.8.4}$$

is the reduced electronic density matrix.

The reduced density matrix is often the only one we need for our transport problem. However if we try to take the trace over the phonon states of the Liouville equation (1.3.9), we obtain

$$i\hbar \frac{d\rho^{(el)}}{dt} = \mathrm{Tr}_{\xi}[H, \rho] \tag{1.8.5}$$

Since the Hamiltonian contains terms of interaction between electrons and phonons, the trace operation does not commute with H, and we do not obtain a simple closed equation for $\rho^{(el)}$.

The problem of obtaining a closed equation for the reduced system is one of the major problems of quantum transport theory. In this reduction process part of the information is lost, and when, in doing so, we substitute some detailed deterministic information with some average or probabilistic quantity, in general we introduce irreversibility in the description of the process. This fundamental step in transport theory is performed, for example, when we assume a random distribution of impurities and substitute their exact position with the probability for the electron to encounter an impurity in its path or when we substitute the exact phonon field (described by the dynamics of the large subsystem) with a probability of having the phonon system in given states according to the thermal distribution.

Chapter 2

The Kubo Formula and Linear Response

David K. Ferry

In much of semiconductor transport theory, it is our main aim to calculate the response of the distribution of electrons, within our device or bulk semiconductor sample, to an applied perturbation. While this perturbation is usually an electric field, it may also be a magnetic field, temperature gradient, density gradient, pressure, or any combination of these generalized forces. Energy from these forces is coupled to the electrons and subsequently decays to the lattice via interaction with the phonons. In small semiconductor devices, we would like to know the entire time dependence of the appropriate interactions, and usually we use some form of kinetic theory based upon the Boltzmann transport equation, which itself is not valid on the short time scales. The response is directly related to the nature of the scattering of the electrons by the lattice, and is often characterized in terms of relaxation times, such as the momentum relaxation time and the energy relaxation time. In turn, these averaged quantities are the macroscopic effects of the microscopic fluctuations introduced by the scattering processes themselves. As a consequence, it is possible to relate them directly to the averaged response to the spectrum of the fluctuations themselves—the traditional fluctuation-dissipation theorem.

While a kinetic theory is usually the basis of transport theory, there is an alternative approach tied rather directly to the use of Langevin equations to describe the transport of carriers. The latter has been known in statistical physics for rather a long time, but it is only in the past few decades that a

---

David K. Ferry ● Center for Solid State Electronics Research, College of Engineering and Applied Sciences, Arizona State University, Tempe, Arizona 85287, USA.

Quantum Transport in Semiconductors, edited by David K. Ferry and Carlo Jacoboni. Plenum Press, New York, 1991.

formal linear response theory, pioneered by Kubo,[1] has grown up to utilize this approach. In this chapter, it is our goal to introduce the Kubo formalism, to introduce the connection with Green's functions, to give a concrete example for electron transport, and to show the extension to quasilinear situations.

The approach to be used here is primarily that of linear response theory. In many cases, when we treat semiconductor transport, we are primarily dealing with concepts such as hot electrons and velocity saturation. These ideas do not arise from deviations from formal linear response theory, but from the idea that the system itself evolves to a far-from-equilibrium state for which the normal ideas of *stationarity* do not always hold. It is more important to incorporate changes arising from the evolution of the distribution function itself (energy relaxation) than to address higher-order terms in response theory. Accompanying this nonstationary behavior is the need to recognize that most of the response functions will be two-time functions, and the quasilinear treatment will take this into account. Nevertheless, our introductory treatment will primarily deal with these functions in a single time-variable formalism. The explicit two-time nature will be discussed at the end of the treatment in terms of a quasiequilibrium density operator.

2.1. LINEAR RESPONSE THEORY

In this treatment, we desire to find the response of the coupled electron–phonon system to a time-dependent perturbation by calculating to lowest order the change induced in the density matrix. Thus, we write the Hamiltonian as

$$H_{\text{total}}(t) = H_0 + H_1(t) \tag{2.1.1}$$

where H includes the electron, lattice, and interaction terms, and H_1 includes the perturbing potential. Here, we will be interested in the current response, and this is traditionally incorporated via the vector potential $\mathbf{A}(\mathbf{r}, t)$ as

$$H_1(t) = \int d^3r \, \mathbf{A}(\mathbf{r}, t) \cdot \mathbf{j}(\mathbf{r}) \tag{2.1.2}$$

Here, $\mathbf{j}(\mathbf{r})$ is the paramagnetic part of the symmetrized total current operator

$$\mathbf{J}(\mathbf{r}, t) = \frac{e}{2m} \sum_i \{[\mathbf{p}_i - e\mathbf{A}(\mathbf{r}, t)]\delta(\mathbf{r} - \mathbf{r}_i) + \delta(\mathbf{r} - \mathbf{r}_i)[\mathbf{p}_i - e\mathbf{A}(\mathbf{r}, t)]\}$$

$$= \mathbf{j}(\mathbf{r}) - \frac{ne^2}{m} \mathbf{A}(\mathbf{r}, t) \tag{2.1.3}$$

Here, $\mathbf{r}_i$ is the position of each individual electron contributing to the current. The second quantized form of the first term above is given simply by[2]

$$\mathbf{j}(\mathbf{r}) = \sum_{\mathbf{p,q}} e^{i\mathbf{q}\cdot\mathbf{r}} a^{+}_{\mathbf{p}+\hbar\mathbf{q}} a_{\mathbf{p}} \left(\mathbf{p} + \hbar\frac{\mathbf{q}}{2} \right) \tag{2.1.4}$$

The form of (2.1.4) is easy to understand in that $\mathbf{p}$ is the Fourier component that arises from transforming the $\mathbf{r}$ variable, and corresponds to the electron momentum, while $\mathbf{q}$ corresponds to the difference variable in the position of the creation and annihilation of the excitation.

To proceed, we can either go over to the interaction representation, keeping just the lowest-order terms in the exponential expansion for linear response, or we can expand the density matrix in the Liouville equation

$$i\hbar \frac{\partial \rho(t)}{\partial t} = [H(t), \rho(t)] \tag{2.1.5}$$

We chose the latter approach, and in the spirit of the linear response approach we write $\rho = \rho_0 + \delta\rho$, where ρ_0 describes the system prior to the application of the perturbation, and is thus a system in which the electrons and phonons are in equilibrium with each other. Then the linearized equation for $\delta\rho$ is (our derivation is standard and closely follows that of Ref. 3)

$$i\hbar \frac{\partial \delta(t)}{\partial t} = [H_0, \delta\rho(t)] - [\rho_0, H_1] \tag{2.1.6}$$

We may solve for $\delta\rho$ as

$$\delta\rho(t) = \frac{i}{\hbar} e^{-iHt/\hbar} \int_{-\infty}^{t} dt' \, [\rho_0, H_1(t')] \, e^{iHt/\hbar} \tag{2.1.7}$$

In (2.1.7), we now have an additional time variation in the term for H_1, which will result in having the current operators written in the interaction representation. Although we did not start with it, this representation follows immediately from solving (2.1.6).

We now proceed with solving for the expectation value of the current (2.1.3), which gives us

$$\langle \mathbf{J}(\mathbf{r}, t) \rangle = \langle \mathbf{j}(\mathbf{r}) \rangle_0 + \frac{i}{\hbar} \int_{-\infty}^{t} dt' \, \langle [H_1(t'), \mathbf{j}(\mathbf{r}, t)] \rangle_0$$

$$- \frac{ne^2}{m} \mathbf{A}(\mathbf{r}, t) \tag{2.1.8}$$

and the current operators in the brackets are now written in the equivalent interaction representation as

$$\mathbf{j}(\mathbf{r}, t) = \exp\left(\frac{i}{\hbar} Ht\right)\mathbf{j}(\mathbf{r})\exp\left(-\frac{i}{\hbar} Ht\right) \tag{2.1.9}$$

and this form follows from using the cyclic properties of the trace. The first term on the right-hand side of (2.1.8) may be ignored because we assume there is no current flow in the equilibrium state. Then, using the form of the perturbing potential from (2.1.2), we may write

$$\langle J_\alpha(\mathbf{r}, t)\rangle = \int_{-\infty}^{t} dt' \int d^3\mathbf{r}' \sum_\beta K_{\alpha\beta}(\mathbf{r} - \mathbf{r}', t - t')A_\beta(\mathbf{r}', t') \tag{2.1.10}$$

where we have introduced the linear response function $K(t)$:

$$K(\mathbf{r} - \mathbf{r}', t - t') = -\frac{i}{\hbar}\Theta(t - t')\langle[j_\alpha(\mathbf{r}, t), j_\beta(\mathbf{r}', t')]\rangle$$

$$-\frac{ne^2}{m}\delta_{\alpha\beta}\delta(\mathbf{r} - \mathbf{r}')\delta(t - t') \tag{2.1.11}$$

and Θ is the Heaviside function. In (2.1.10), we used the fact that A_β is a c-number to rearrange the various operators. The response function K describes a causal effect induced by the application of the perturbation. Thus, the range of integration in (2.1.10) summarizes the response obtained from the excitation at the time t' of application of the perturbation up to the observation time t. The first term on the right-hand side of (2.1.11) is the current–current response function, while the second term is a direct current response to the applied vector potential, which is effectively the displacement current.

We can go a step further and look at the Fourier transform of (2.1.10) in order to get the frequency-dependent conductivity. Here, we assume that the vector potential is given by the electric field, and

$$\mathbf{A}(\mathbf{r}, \omega) = -\frac{i}{\omega}\mathbf{E}(\mathbf{r}, \omega) \tag{2.1.12}$$

Then

$$\langle J_\alpha(\mathbf{r}, \omega)\rangle = -\frac{i}{\hbar\omega}\int d^3\mathbf{r}'\, K_{\alpha\beta}(\mathbf{r} - \mathbf{r}', \omega)E_\beta(\mathbf{r}, \omega) \tag{2.1.13}$$

and, if the electric field is homogeneous,

$$\sigma_{\alpha\beta}(\mathbf{r}, \omega) = \frac{1}{\hbar\omega}\int d^3\mathbf{r}'\,\langle[j_\alpha(\mathbf{r}, \omega), j_\beta(\mathbf{r}', 0)]\rangle$$

$$-i\frac{ne^2}{m\hbar\omega}\delta_{\alpha\beta} \tag{2.1.14}$$

The term in the ensemble average is a retarded two-particle Green's function and must have an imaginary part that cancels the last term in (2.1.14) in the steady state, since the conductivity must be entirely real at zero frequency. In inhomogeneous situations, however, we cannot separate out the electric field directly because the current involves a nonlocal spatial integration over $\mathbf{r}'$. Then the conductivity can only be achieved by taking a spatial Fourier transform to get $\sigma_{\alpha\beta}(\mathbf{q}, \omega)$ for each representative component. This latter transformation is also achieved by averaging the current over the volume $\mathbf{r}$.

One important consequence of the above discussion is that we have now defined the current response in terms of the fluctuations of the current itself. In essence, this is just the correlation function that appears on the right-hand sides of the various equations. The basis of this relationship is the fluctuation-dissipation theorem itself. The current that flows is determined directly by the dissipation arising in the presence of an applied force. On the other hand, the correlation function describes the fluctuations. The Kubo formula is no more than a direct statement of this important theorem.

2.2. THE ZERO-FREQUENCY FORM

The Kubo formula (2.1.14) is valid at finite (nonzero) temperatures and for transverse as well as longitudinal electric fields. At low frequencies, however, there is a problem. This problem arises from the fact that the vector potential $\mathbf{A}$ diverges as $\omega \to 0$. In order to get around this problem, we employ a "trick" as outlined by Mahan.[3] For this, we rewrite the commutator that appears in (2.1.7) using the following identity [we change t to $-t$ inside the integral in (2.1.7)]:

$$[F(-t), \rho_0] = i\hbar\rho_0 \int_0^\beta d\beta' \frac{\partial}{\partial t} F(-t - i\hbar\beta') \qquad (2.2.1)$$

The commutator is easily proved as follows:

$$\int_0^\beta d\beta' \frac{\partial F(-t - ih\beta')}{\partial t} = -\frac{i}{\hbar} \int_0^\beta d\beta' \frac{\partial F(-t - ih\beta')}{\partial \beta'}$$

$$= -\frac{i}{\hbar} [F(-t - i\hbar\beta) - F(-t)] \qquad (2.2.2)$$

With

$$\rho_0[F(-t - i\hbar\beta) - F(-t)] = e^{-\beta H}[e^{\beta H}F(-t)e^{-\beta H} - F(-t)]$$

$$= [F(-t), \rho_0] \qquad (2.2.3)$$

the identity is established. Then the average over the current becomes, after adjusting the limits of the integrations,

$$\langle \mathbf{J}(\mathbf{r}) \rangle = \int_0^\infty dt \int_0^\beta d\beta' \int d^3\mathbf{r}' \, \langle \mathbf{j}(\mathbf{r}', -t - i\hbar\beta') \cdot \mathbf{E}(\mathbf{r}')\mathbf{j}(\mathbf{r}) \rangle \qquad (2.2.4)$$

and[3]

$$\sigma_{\alpha\gamma}(\mathbf{r}, 0) = \int_0^\infty dt \int_0^\beta d\beta' \int d^3\mathbf{r}' \, \langle j_\gamma(\mathbf{r}', -t - ih\beta')j_\alpha(\mathbf{r}) \rangle \qquad (2.2.5)$$

which does not contain a frequency. However, it still contains a nonlocal integration over the spatial variables, and the form (2.2.5) assumes that the electric field is uniform throughout the sample.

2.3. RELAXATION AND GREEN'S FUNCTIONS

If we now look at a system that relaxes toward equilibrium, we can introduce another function, the relaxation function. This new function describes the change in the expectation value of a dynamical operator after the external perturbation has been turned off. An explicit form can be obtained by letting $\mathbf{A}(t) = \exp(\eta t)$ for $t \leq 0$ and zero for $t > 0$. The change in the operator $\mathbf{J}$ is then (we use the homogeneous form here for simplicity)

$$\langle J \rangle = \int_{-\infty}^0 dt' \, \exp(\eta t')K(t - t')$$

$$= \int_t^\infty dt' \, \exp[\eta(t - t')]K(t') \qquad (2.3.1)$$

By letting $\eta \to 0$, we can define the relaxation function as

$$R(t) = \int_t^\infty dt' \, K(t'), \qquad \frac{\partial R(t)}{\partial t} = -K(t) \qquad (2.3.2)$$

One can easily verify by direct differentiation that the relaxation function is a quantum correlation function, given by[4]

$$R(t) = \int_0^\beta d\beta' \, \langle J^+(-i\hbar\beta')P(t) \rangle - \beta\langle A^+ \rangle\langle P \rangle \qquad (2.3.3)$$

Here, P is any operator which we are trying to correlate with the current. It is a peculiarity of the semiclassical case that the functional forms that arise in the initial velocity response form are such that $R(t)$ and $K(t)$ differ only in the constant term (which is zero for the initially relaxed case anyway) and in the normalization. However, the use of $R(t)$, which is a proper correlation function, brings the semiclassical case into line with the more rigorous quantum treatment.

2.4. SOME EXAMPLES FOR THE CONDUCTIVITY

In this section, we want to examine the application of the Kubo formula for finding the conductivity in different cases. Although we examine the conductivity, the same approaches can be used to examine the response function for any physical process. The two examples chosen are the calculation of the conductivity in a metallic system where an expansion in energy eigenfunctions can be used, and the case of a strongly localized system where a site expansion is used. In these, we assume that we have a homogeneous system so that we can drop the integration over $\mathbf{r}$ (it actually gives a factor of V, where V is the volume) from the explicit variables.

2.4.1. The Metallic Conductivity

Equation (2.2.5) can now be rewritten for the steady-state, homogeneous conductivity σ as

$$\sigma_{\alpha\beta} = V \int_0^\infty dt \int_0^\beta d\beta' \, \mathrm{Tr}\{\rho_0 j_\beta(-t - i\hbar\beta')j_\alpha\} \qquad (2.4.1)$$

We can now expand the trace operation by inserting an intermediate set of wave functions. Here, we shall use the energy eigenfunctions. We also incorporate explicitly the Pauli exclusion principle, so that (2.4.1) becomes (by inserting the explicit temporal variation of the various operators and inserting a set of energy eigenvectors in order to compute the matrix elements)

$$\sigma_{\alpha\beta} = \frac{1}{V} \int_0^\infty dt \int_0^\beta d\beta'$$

$$\times \sum_{n,m} e^{-\beta' E_n} \langle n|j_\beta(-t - i\hbar\beta')|m\rangle [1 - e^{-\beta' E_m}]\langle m|j_\alpha|n\rangle \quad (2.4.2)$$

The first matrix element may be expanded as

$$\langle n|j_\beta(-t - i\hbar\beta')|m\rangle = \langle n|e^{-iH(t+i\hbar\beta')/\hbar} j_\beta e^{iH(t+i\hbar\beta')/h}|m\rangle$$

$$= \langle n|j_\beta|m\rangle \exp[-it(E_n - E_m)/\hbar]$$

$$\times \exp[\beta'(E_n - E_m)] \qquad (2.4.3)$$

The first exponential on the right-hand side of (2.4.3) can be integrated over t and yields $\hbar\pi\delta(E_n - E_m)$. The second exponential then can be integrated over β' to yield just a factor β. This combines with the $f[1 - f]$ [where f is the Maxwellian distribution in equation (2.4.2)] term to give the negative derivative of the distribution function, and the conductivity is then (from here on we work only with the diagonal part of the conductivity)

$$\sigma = \frac{\pi\hbar}{V}\sum_{n,m}\left(-\frac{\partial f}{\partial E}\right)\langle n|j|m\rangle\langle m|j|n\rangle\delta(E_n - E_m)$$

$$= \frac{\pi h}{V}\int dE'\int dE''\, N(E')N(E'')\left(-\frac{\partial f}{\partial E'}\right)|\langle n|j|m\rangle|^2\delta(E' - E'') \qquad (2.4.4)$$

where we have introduced a continuous energy variable in the second form. The energy derivative of the distribution function just gives an additional delta function at the Fermi surface for degenerate materials, and after introducing the current operators, (2.4.4) becomes

$$\sigma = \frac{\pi e^2\hbar^3}{Vm^2}[N(E)]^2|D_{nm}|^2 \qquad (2.4.5)$$

where E is understood to be the Fermi energy, and

$$D_{nm} = \int \phi_n^*\frac{\partial}{\partial x}\phi_m\, d\mathbf{r} \qquad (2.4.6)$$

is the reduced momentum matrix element.

The density of states is easily evaluated to be $Vmk/\pi^2\hbar^2$. The evaluation of the matrix elements is a little more complicated. For this, we follow an argument given by Mott.[5] We introduce an effective volume Ω which is defined by the mean free path λ through

$$\Omega = 4\pi\lambda^3/3 \qquad (2.4.7)$$

For plane-wave states, the integral produces

$$I = \begin{cases} k\Omega/V, & \text{for } |\mathbf{k} - \mathbf{k}'|\lambda < 1 \\ 0, & \text{otherwise} \end{cases} \qquad (2.4.8)$$

Then

$$|D_{nm}|^2 \simeq N_\Omega I^2\xi \qquad (2.4.9)$$

where N_Ω is the number of small cells $(= V/\Omega)$ and ξ is a factor due to the angular averaging $(k = k_F)$

$$\frac{1}{4\pi}2\pi\int_0^{1/k\lambda}\sin\theta\, d\theta \simeq \frac{1}{4k^2\lambda^2} \qquad (2.4.10)$$

Using (2.4.6)–(2.4.10) in (2.4.5), we finally find

$$\sigma = e^2k^2\lambda/3\pi^2\hbar \qquad (2.4.11)$$

This can be put into a more recognizable form by taking $\lambda = v\tau = (\hbar k/m)\tau$, where τ is evaluated at the Fermi surface. By using $n = k^3/3\pi^2$, we finally obtain

$$\sigma = ne^2\tau/m \tag{2.4.12}$$

which is the standard form for the conductivity.

2.4.2. Localized Conductivity in the Site Approximation

We now want to pursue another approach that has become useful in simulations of amorphous materials in which states are localized. To do this, we rewrite (2.4.4) as (we assume the derivative of the distribution also becomes a δ-function at $T = 0$, and we replace the current operator with the velocity operator while dropping the explicit notation of the matrix elements for the moment)

$$\sigma = \frac{2\pi e^2\hbar}{V}\,\mathrm{Tr}\{\delta(E - H)v\delta(E - H)v\} \tag{2.4.13}$$

where we have returned the two density-of-states factors to their delta function form under the trace operation (we have returned to the sum over energy form), and the factor of 2 is now the result of the sum over spin states. The reason for this is that we will ultimately expand not in an energy function basis but in a site basis. Thus, the delta functions will lead to Green's functions at local sites, and the velocity operators will couple one site to the next. The energy E in (2.4.13) is the Fermi energy, and we will find in some cases that the conductivity is quite sensitive to the value of this energy. While we could have used the finite-temperature Green's functions directly in the forms above, we work here with the zero-temperature forms. To proceed, we note that the delta functions are related to the imaginary part of the Green's functions where

$$G^r(E, \eta) = \frac{1}{E - H + i\eta} = P\frac{1}{E - H} - i\pi\delta(E - H) \tag{2.4.14}$$

is the retarded Green's function and P denotes the principal part. The advanced function is obtained by replacing η by $-\eta$. These can be used to introduce the spectral function

$$A(E) = i[G^r - G^a] \tag{2.4.15}$$

In the following, we denote the imaginary part of the Green's function by G'', and this function is one-half of the spectral function. Then (2.4.13) becomes

$$\sigma = \frac{2e^2\hbar}{\pi V}\,\mathrm{Tr}\{G''(E - H)vG''(E - H)v\}$$

$$= \frac{2e^2\hbar}{\pi V}\sum_{ijkl} G''_{ij}v_{jk}G''_{kl}v_{li} \tag{2.4.16}$$

Here, we assume a one-dimensional model so that the indices refer to states localized on a single site of an atomic chain. The velocity operator couples motion along this chain. If we were to add a second dimension, with the velocity still desired in this first direction, then the indices for this second dimension would only occur on the Green's functions, i.e.,

$$\sigma = \frac{2e^2\hbar}{\pi V}\sum_{rs}\sum_{ijkl} G''_{ij,rs} v_{jk} G''_{kl,sr} v_{li} \qquad (2.4.17)$$

where

$$G_{kl,sr} = G(\mathbf{r}_{ks}, \mathbf{r}_{lr}) \qquad (2.4.18)$$

and $\mathbf{r}_{ij}$ is the vector position to the site (i, j). We concentrate only on the one-dimensional case, but note that the two-dimensional case has been treated by Fisher and Lee.[6]

In one dimension, the velocity operator is primarily achieved by the hopping of an electron from one site to the next, since all states in one dimension are essentially localized.[7] We write this velocity operator as[8,9]

$$v = i\frac{a\Delta}{\hbar}\sum_n (|n\rangle\langle n+1| - |n+1\rangle\langle n|) \qquad (2.4.19)$$

where Δ is the coupling energy between neighboring sites. Using this result in (2.4.16) gives, with the volume $V = Na$,

$$\sigma(i\eta) = -\frac{2e^2 a\Delta^2}{\pi\hbar N}\sum_{ij} [G''_{i,j+1} - G''_{i,j-1}][G''_{j,i+1} - G''_{j,i-1}] \qquad (2.4.20)$$

and the limit as $\eta \to 0$ is taken.

The method of calculating the Green's functions is based upon the self-energy corrections imposed by the coupling energy Δ. We carry this out within the Anderson model,[10] in which the random potential is that of the site energy itself. Thus, we assume that the energy level corresponding to site i is ε_i, and that this energy is randomly distributed between $-W/2$ and $W/2$ with a uniform probability distribution. The sites are coupled in the tight-binding approximation with a nearest-neighbor interaction potential Δ, as introduced above for the velocity. From these energies, we can calculate the Green's functions and, more importantly, the self-energies. These are calculated by the iterative scheme[9]

$$\begin{aligned}
\Sigma^R_N = 0, \qquad \Sigma^R_j = \Delta^2/(E + i\eta - \varepsilon_{j+1} - \Sigma^R_{j+1}) \\
\Sigma^L_1 = 0, \qquad \Sigma^L_j = \Delta^2/(E + i\eta - \varepsilon_{j-1} - \Sigma^L_{j-1})
\end{aligned} \qquad (2.4.21)$$

Once the left- and right-propagated self-energies are calculated according to (2.4.21), the individual Green's functions can be built up from

$$G_{jk} = \frac{1}{E + i\eta - \varepsilon_j - \Sigma^R_j - \Sigma^L_j}\prod_{r=j}^{k-1}\frac{\Sigma^R_r}{\Delta} \qquad (2.4.22)$$

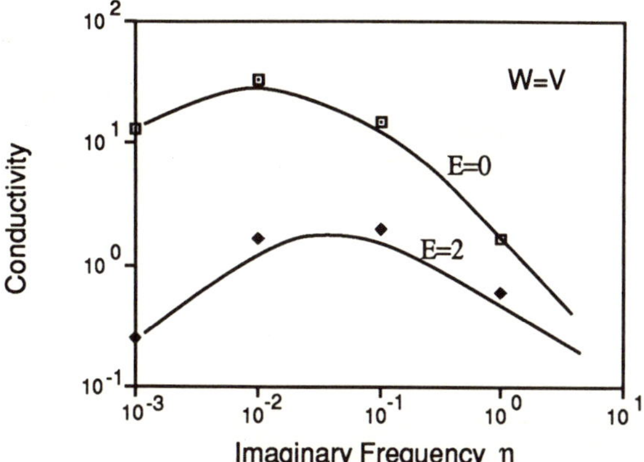

FIGURE 2.1. The conductivity for a fully localized one-dimensional model as a function of the pseudofrequency η. The two curves are for two different energies for the disorder ratio $W/V = 1$. Although η is added to the Green's functions as a convergence factor, it plays the role of a real frequency component. The figure demonstrates that the dc conductivity vanishes in this one dimensional system.

In Figure 2.1, we show the conductivity as a function of the "frequency" η for two different values of energy for the disorder $W/\Delta = 1$, as calculated from the above model.[11] In Figure 2.2, we show how σ varies with the disorder at a single value of η for a two-dimensional model. Thouless and Kirkpatrick[9] used this approach to verify that the conductivity in one dimension goes to zero as η and, in fact, does vanish at zero frequency for any nonzero level of disorder.

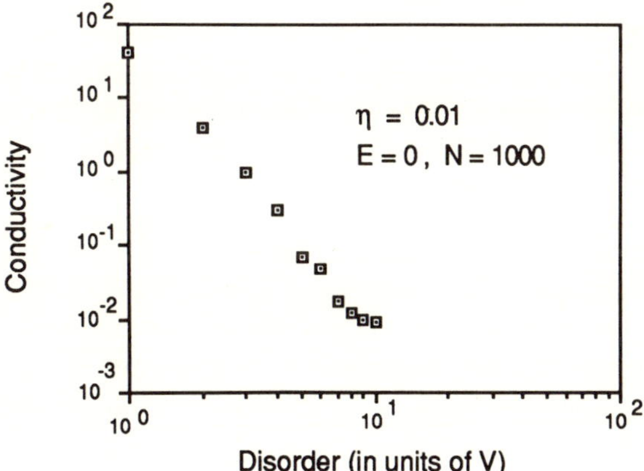

FIGURE 2.2. The conductivity in a two-dimensional Anderson model, as a function of the disorder ratio W/V.

2.5. EXTENSION TO TWO-TIME FUNCTIONS

In the far-from-equilibrium state that describes the hot-electron problem arising in semiconductor devices, the correlation functions, response
functions, and relaxation functions are proper two-time-variable functions.
One explicit reason for this is that the inverse temperature β is itself a
time-evolving function. Moreover, the ensemble used in the averages (the
density matrix) can be a time-evolving function also. In the above treatments,
we used ρ_0 as the equilibrium density matrix. This is a severe limitation on
the linear response formalism, because the deviation term can be quite
large. As a result, more modern approaches try to utilize as much of the
complete, time-evolving density matrix as possible. One common approach
is to introduce a *quasiequilibrium* density matrix which is parametrized[12-14]
in terms of the various quantities that are suitable for consideration as
integral invariants of the motion.[15] The parameters have the same form as
they would in a near-equilibrium situation, and the far-from-equilibrium
density matrix is assumed to have a form similar to the equilibrium density
matrix. However, the various parameters are assumed to evolve in time.
The philosophy is the same as that of the drifted-Maxwellian distribution
function in the Boltzmann equation approach to classical transport. The
crucial assumption is that some quite rapid scattering process is present
that forces the distribution function into a near equilibrium form.[16] Then
the linear response formalism is used to describe the deviations around this
quasiequilibrium density matrix, and a set of moment equations, such as
the retarded Langevin equation for momentum, is developed to describe
the temporal variation of the parameters. As a consequence, a family of
correlation functions is obtained describing the fluctuations affecting each
of the parameters. For example, the momentum is described by the velocity-
velocity correlation function, the temporal evolution of β_e (the inverse of
the electron temperature) is described by an energy-energy fluctuation
function (which normally vanishes in equilibrium due to the firm connection
of the system with the "bath"), etc. As in equilibrium, it is usually assumed
that the rapid scattering process is just carrier-carrier scattering,[16] which
requires that the carrier density be relatively high. The important point is
that if the intercarrier scattering is sufficiently rapid, then the evolution of
the quasiequilibrium statistical operator is independent of the initial distribution, and there should be a reduction in the number of parameters
necessary to describe the nonequilibrium response of the system. The
number of these parameters is just that which is introduced into the assumed
operator form.

Determination of the correlation functions is a complicated process,
especially in the two-time-variable form required in devices far from equilibrium. At least in the weak scattering limit, in which the electron-phonon

interaction is kept to second order in the matrix elements, these functions can be evaluated with the quasiequilibrium formulation.[13-14,17] The results are effectively the kernel of the Fermi golden rule for scattering induced relaxation of the variable as determined in semiclassical transport theory. For effects beyond this order, it is not clear if any simplification is possible, or whether the full complexity of the Green's function approach is necessary. However, even in this latter approach (which differs only in the fact that no reference is made to the Langevin equations), the use of the quasiequilibrium form of the density matrix is often employed. Yet, there is a crucial difference from the Fermi golden rule results, in that the presence of the high carrier density implies that the correlation functions must be calculated taking into account the fact that the carriers are strongly interacting with each other.

2.5.1. The Quasiequilibrium Statistical Operator

It is assumed that a time t_0 exists, beyond which correlations with lifetimes less than t_0 can be ignored and the state of the system may be described by a reduced set of macroscopic observables $Q_m(\mathbf{r}, t)$, which are the average values, taken over a quasiequilibrium ensemble, of the set of dynamical variables P_m. These dynamical variables and their conjugate forces $f_m(\mathbf{r}, t)$ may then be used to define a nonequilibrium statistical ensemble (there is a difference between the nonequilibrium and the quasiequilibrium density operators, which is discussed below). The approach is based upon Poincaré's theorem on integral invariants generalized to quantum systems.[12,15] Thus, the system density matrix should be constructable from the principal invariants of the system. We define a quasiequilibrium density matrix ρ_τ as a local, far-from-equilibrium density matrix—local in that it is in a local equilibrium with its surroundings and driving forces at time τ. This latter quantity serves as an evolving idealized "initial" condition for calculating deviations from it via the linearized response formulation. For all times $t > t_0$, the true nonequilibrium density matrix is a functional of the quasiequilibrium density operator ρ_τ.

It is now reasonable to use the quasiequilibrium statistical operator, which is smoothed in the microfluctuations and from the beginning time $t > t_0$ describes the slow irreversible evolution of the system, to obtain the balance equations necessary to describe the time evolution of the macroscopic observables $Q_m(\mathbf{r}, t)$. We consider the case of conduction electrons in a semiconductor under the influence of a strong electric field $\mathbf{F}(\mathbf{r})$. The Hamiltonian of the system can be written as

$$H = H_e + H_L + H_F + H_{eL} \tag{2.5.1}$$

where the terms on the right are the Hamiltonians of the electrons, the lattice, the field, and the electron-lattice interaction. Here,

$$H_F = -e \int \mathbf{F}(\mathbf{r}) \cdot d\mathbf{r} \qquad (2.5.2)$$

is the unscreened potential of the external field (although screening can be included, we do not do so here).

Within a time t_0, the energy and momentum which the carriers obtain from the field are redistributed among the ensemble due to the rapid carrier–carrier interactions. Thus, the quasiequilibrium statistical operator describes electrons which are in equilibrium with each other, but not with the lattice. On a time scale $t > t_0$, the system may be characterized by the average values of the set of operators

$$P_m = \{H_e, \mathbf{P}_e, N, \mathbf{P}_L, H_L + H_{eL}\} \qquad (2.5.3)$$

where $\mathbf{P}_e$ and $\mathbf{P}_L$ are the operators of total momentum of the electrons and the lattice, respectively, and N is the carrier number operator. The term H_e contains the carrier Hamiltonian, the coherent part of any interactions with the environment, and the carrier–carrier interaction, although we retain only the first of these in a renormalized effective mass treatment. However, it is assumed that this term is diagonal. While the set of parameters in (2.5.3) is a minimal set, we remark that there is no requirement to a priori assume that the lattice remains in equilibrium with the bath. These P_m satisfy the equations of motion

$$\frac{dP_m}{dt} = \frac{i}{\hbar}[H, P_m] \qquad (2.5.4)$$

We next introduce a set of time-dependent conjugate forces $f_m(\mathbf{r}, t)$, which are c-numbers. These are constructed to be thermodynamic conjugates of the individual Q_m as follows:[15]

$$f_m(\mathbf{r}, t) = \{-\beta_e, -\beta_e v_d, -\beta_e(\mu - mv_d^2), 0, -\beta_L\} \qquad (2.5.5)$$

Here, v_d is the drift velocity and μ is the quasi-Fermi energy. The force for the lattice momentum is zero since the lattice cannot have an average momentum.

The asymptotic, time-smoothed quantities $B_m(t)$ may be defined from the operators and their conjugate forces through[18]

$$B_m(t) = s \int d\mathbf{r} \int_{-\infty}^{0} dt'\, e^{st'} f_m(\mathbf{r}, t + t') P_m(\mathbf{r}, t') \qquad (2.5.6)$$

where

$$P_m(\mathbf{r}, t) = e^{-iHt/\hbar} P_m(\mathbf{r}) e^{iHt/\hbar} \qquad (2.5.7)$$

and we have introduced the operator *density* $P_m(\mathbf{r})$ rather than P_m in order to describe inhomogeneities in the electronic system.[18] In order to ensure that f_m and $Q_m = \langle P_m \rangle_t$ (the subscript t on the ensemble average denotes a time-varying average due to the time variation of the quasiequilibrium statistical operator) are thermodynamically conjugate, we require

$$\langle P_m \rangle_t = \langle P_m \rangle_{\tau(t)} \qquad (2.5.8)$$

where the latter denotes an average over the quasiequilibrium density operator.

We can now write the nonequilibrium density operator in the form of

$$\rho(t) = \exp\left\{-\Phi - \sum_m B_m(\mathbf{r}, t)\right\} \qquad (2.5.9)$$

where

$$\Phi = \ln \mathrm{Tr} \exp\left\{-\sum_m B_m(\mathbf{r}, t)\right\} \qquad (2.5.10)$$

The choice (2.5.9) for the nonequilibrium density operator ensures that this quantity reduces to the generalized Gibbsian canonical ensemble in the thermal equilibrium state and results in a positive entropy production in the nonequilibrium state. If we define the entropy operator as

$$S(t, 0) = \Phi_0 + \sum_m B_m(t) \qquad (2.5.11)$$

with Φ_0 suitably redefined, we can now write the nonequilibrium density operator as

$$\rho(t) = \exp\left\{-S(t, 0) - \int_{-\infty}^{0} dt'\, e^{st'} \dot{S}(t + t', t')\right\} \qquad (2.5.12)$$

where

$$\dot{S}(t + t', t') = \sum_m \int \big[\dot{P}_m(\mathbf{r}, t') f_m(\mathbf{r}, t + t')$$

$$+ \big[P_m(\mathbf{r}) - \langle P_m(\mathbf{r})\rangle_t\big] \dot{f}_m(\mathbf{r}, t + t')\big]\, d\mathbf{r} \qquad ((2.5.13)$$

Finally, we note that the limit $s \to 0^+$ is taken after the thermodynamic limit to ensure that ρ represents a retarded solution. By using (2.5.12) for the nonequilibrium density operator, we can then introduce the corresponding quasiequilibrium density operator as

$$\rho_\tau(t, 0) = \exp\{-S(t, 0)\} \qquad (2.5.14)$$

2.5.2. The Balance Equations

With the observables in (2.5.3), and the above definitions of the non-equilibrium density matrix and the quasiequilibrium density matrix, we can use the operator densities $P_m(\mathbf{r})$ defined through

$$P_m = \int P_m(\mathbf{r})\, d\mathbf{r} \qquad (2.5.15)$$

This allows us to look at inhomogeneous situations as well as homogeneous ones in the electronic systems of interest. Using standard field operator notation, we then have (P_L is not considered since its average must remain zero)[18]

$$N(\mathbf{r}) = \psi^+(\mathbf{r})\psi(\mathbf{r}) \qquad (2.5.16\text{a})$$

$$\mathbf{P}_e(\mathbf{r}) = \frac{\hbar}{2i}[\psi^+(\mathbf{r})\nabla\psi(\mathbf{r}) - \psi(\mathbf{r})\nabla\psi^+(\mathbf{r})] \qquad (2.5.16\text{b})$$

$$H_e(\mathbf{r}) = \frac{\hbar^2}{2m}\nabla\psi^+(\mathbf{r})\cdot\nabla\psi(\mathbf{r}) \qquad (2.5.16\text{c})$$

$$H_{eL}(\mathbf{r}) = \sum_{\mathbf{q}\lambda}[U_{\mathbf{q}\lambda}(\mathbf{r})b_{\mathbf{q}\lambda} + U_{-\mathbf{q}\lambda}b_{\mathbf{q}\lambda}^+]\psi^+(\mathbf{r})\psi(\mathbf{r}) \qquad (2.5.16\text{d})$$

$$H_L(\mathbf{r}) = \sum_{\mathbf{q}\lambda}\hbar\omega_{\mathbf{q}\lambda}b_{\mathbf{q}\lambda}^+ b_{\mathbf{q}\lambda} \qquad (2.5.16\text{e})$$

for the number, momentum, and energy density operator of the electrons, the electron–phonon interaction density, and the lattice, respectively. Here, ψ is the field operator for the electrons, $U_{\mathbf{q}\lambda}$ is the potential with component at wave number $\mathbf{q}$ and mode λ, and $b_{\mathbf{q}\lambda}^+$ is the creation operator for an excitation in this lattice mode. The conjugate forces were given in (2.5.5) and are described in terms of the parameters β_e, $\mathbf{v}_d$, μ, and β, which are the inverse electron temperature, the drift velocity, the quasi-Fermi energy, and the inverse lattice temperature, respectively. Usually the latter quantity is taken to be the bath temperature, but in nonequilibrium phonon situation this is not the case. Using the total Hamiltonian, one finds for the first three operators the following equations of motion:

$$\frac{dN(\mathbf{r}, t)}{dt} = -\frac{1}{m}\nabla\cdot\mathbf{P}_e(t) \qquad (2.5.17\text{a})$$

$$\frac{d\mathbf{P}_e(\mathbf{r}, t)}{dt} = -\mathbf{P}_e(\mathbf{r}, t)\cdot\nabla\mathbf{J}(\mathbf{r}, t)$$

$$-\nabla[H_F(\mathbf{r}, t)N(\mathbf{r}, t)] + \dot{\mathbf{P}}_{(L)}(\mathbf{r}, t), \qquad (2.5.17\text{b})$$

$$\frac{dH_e(\mathbf{r}, t)}{dt} = -\nabla\cdot\mathbf{J}_H(\mathbf{r}, t) - \frac{1}{m}\nabla\cdot[H_F(\mathbf{r}, t)\mathbf{P}_e(\mathbf{r}, t)]$$

$$+ \dot{H}_{e(L)}(\mathbf{r}, t) \qquad (2.5.17\text{c})$$

The divergence terms of the fluxes $\mathbf{J}_m(\mathbf{r}, t)$ are due to the commutators of the respective P_m with H_e.[12] In addition,

$$\dot{P}_{m(L)} = -\frac{i}{\hbar}[P_m(\mathbf{r}, t), H_{eL}] \tag{2.5.18}$$

gives the contributions due to the electron-phonon interaction.

Expanding the nonequilibrium density operator up to first order in the electron-phonon interaction and in the divergences of the fluxes, the balance equations have the general form

$$\frac{d\langle P_m(\mathbf{r})\rangle_t}{dt} = \sum_n \alpha_{mn}(\mathbf{r}, t)\langle P_n(\mathbf{r})\rangle_t - \nabla \cdot \langle \mathbf{J}_m(\mathbf{r})\rangle_t$$

$$-\frac{i}{\hbar}\int_{-\infty}^{0} dt'\, e^{st'}\Bigg\{-\langle[\nabla \cdot \mathbf{J}_m(\mathbf{r}), H_e(t')]\rangle_t$$

$$+\sum_k \int d\mathbf{r}'\, \langle[P_k(\mathbf{r}'), H_e(t')]\rangle_t \frac{\delta\langle \nabla \cdot \mathbf{J}_m(\mathbf{r})\rangle_t}{\delta\langle P_k(\mathbf{r}')\rangle_t}$$

$$+\langle[\dot{P}_{m(L)}(\mathbf{r}), H_{eL}(t')]\rangle_t\Bigg\} \tag{2.5.19}$$

All of the averages $\langle \cdot \rangle_t$ have to be taken with respect to the quasiequilibrium density operator $\rho_\tau(t)$, which is the evolving pseudoequilibrium function. In this case, the electron-phonon interaction has been eliminated from this form of the density matrix, and we are now in the equivalent of the equilibrium Kubo form discussed in previous sections, with the electron-phonon interaction buried in the various correlation functions that appear in (2.5.19). The term $\alpha_{mn}(\mathbf{r}, t)$ is proportional to ∇H_F and provides the driving effects of the external fields. The set of equations (2.5.19) for the various averages of the operators contain terms up to second order in $[P_m, H_e + H_{eL}]$ and are the sources of the various correlation functions that arise in the Kubo form. The thermodynamical averages of the fluxes can be expressed in terms of the sets $\{f_m(\mathbf{r}, t)\}$ and $\{\langle P_m(\mathbf{r})\rangle_t\}$. Some of these are

$$\langle \mathbf{P}_e(\mathbf{r})\rangle_t = m\langle N(\mathbf{r})\rangle_t \mathbf{v}_d(\mathbf{r}, t) \tag{2.5.20a}$$

$$\langle \mathbf{J}_H(\mathbf{r})\rangle_t = \left\{\tfrac{5}{3}\langle H_e(\mathbf{r})\rangle_t - \frac{\hbar^2}{12m}\nabla^2\langle N(\mathbf{r})\rangle_t - \tfrac{1}{3}\langle \mathbf{P}_e(\mathbf{r})\rangle \cdot \mathbf{v}_d(\mathbf{r}, t)\right\}\mathbf{v}_d(\mathbf{r}, t)$$

$$+\frac{\hbar^2}{4m}\nabla\langle N(\mathbf{r})\rangle_t \cdot \nabla\mathbf{v}_d(\mathbf{r}, t) \tag{2.5.20b}$$

$$\langle \mathbf{J}(\mathbf{r})\rangle_t = \left\{\tfrac{2}{3}\langle H_e(\mathbf{r})\rangle_t - \frac{\hbar^2}{12m}\langle N(\mathbf{r})\rangle_t - \tfrac{1}{3}\langle \mathbf{P}_e(\mathbf{r})\rangle_t \cdot \mathbf{v}_d(\mathbf{r}, t)\right\}\mathbf{v}_d(\mathbf{r}, t)$$

$$+\langle \mathbf{P}_e(\mathbf{r}, t)\rangle_t \tag{2.5.20c}$$

These averages describe a convective, dissipationless motion of the carriers in which the terms proportional to $\hbar^2$ represent the quantum corrections over similar classical results.

In order to determine the remaining expressions in (2.5.19) in terms of the set of operators $\{P_m(\mathbf{r})\}$ and conjugate forces $\{f_m(\mathbf{r}, t)\}$, one has to determine the time evolution of H_e and H_{eL}. Exploiting the assumption of modest inhomogeneities, one can split up the space coordinate $\mathbf{r}$ into $\mathbf{R} + \boldsymbol{\xi}$, where $\mathbf{R}$ labels small macroscopic cells within which the system is assumed to be almost homogeneous and $\boldsymbol{\xi}$ is the space vector within such a fictitious cell. We can then expand $H_F = H_F(\mathbf{R}, t) + \boldsymbol{\xi} \cdot \nabla H_F(\mathbf{R}, t)$ and define (Ω is the cell volume)

$$a(\mathbf{k}, \mathbf{R}) = \Omega^{-1/2} \int_\Omega d\boldsymbol{\xi} \, e^{-i\mathbf{k}\cdot\boldsymbol{\xi}} \psi(\mathbf{r}) \tag{2.5.21}$$

so that

$$a(\mathbf{k}, \mathbf{R}, t) = \exp[-\lambda(\mathbf{k}, \mathbf{R}, t)]a[\mathbf{k}_+(\mathbf{R}, t, 0)] \tag{2.5.22a}$$

with

$$\lambda(\mathbf{k}, \mathbf{R}, t) = \frac{i}{\hbar} \int_0^t d\tau \, \{H_F(\mathbf{R}, t) + E[\mathbf{k}_+(\mathbf{R}, t, \tau)]\} \tag{2.5.22b}$$

and

$$\mathbf{k}_\pm(\mathbf{R}, t, \tau) = \mathbf{k} \pm \frac{1}{\hbar} \int_\tau^t dt' \, \nabla H_F(\mathbf{R}, t') \tag{2.5.22c}$$

Here, the surface terms over the cell and H_{eL} have been neglected in the differential equation for $a(\mathbf{k}, \mathbf{R}, t)$. Neglecting inhomogeneities within the cells, one can then write the collision term in (2.5.19) as

$$-\frac{i}{\hbar} \int_{-\infty}^0 dt' \, e^{st'} \langle [\dot{P}_{m(L)}(\mathbf{r}, H_{eL}(t'))] \rangle_t$$

$$= \frac{2}{\Omega} \sum_{\mathbf{k}'\mathbf{k}''} \sum_{\mathbf{q}\lambda} [P_m(\mathbf{k}') - P_m(\mathbf{k}'')] |U_{\mathbf{q}\lambda}^{\mathbf{k}'\mathbf{k}''}|^2$$

$$\times \{(1 + N_{\mathbf{q}\lambda})f(\mathbf{k}', \mathbf{r}, t)[1 - f(\mathbf{k}'', \mathbf{r}, t)] - N_{\mathbf{q}\lambda}f(\mathbf{k}'', \mathbf{r}, t)[1 - f(\mathbf{k}', \mathbf{r}, t)]\}$$

$$\times \int_{-\infty}^0 dt' \, e^{(s+\Gamma)t'} \cos\{\omega_{\mathbf{q}\lambda}t' + \lambda[\mathbf{k}'_-(\mathbf{r}, t', 0), t', \mathbf{r}]$$

$$- \lambda[k''_-(\mathbf{r}, t', 0), t', \mathbf{r}]\} \tag{2.5.23}$$

A term $\exp(\Gamma t)$ has been inserted to include collision broadening of the carrier state explicitly. This effect and the intracollisional field effect have been studied elsewhere and will not be discussed here.[19,20] The distribution function introduced into (2.5.23) is defined as a quasiequilibrium Fermi–Dirac distribution, defined as[15]

$$f(\mathbf{k}, \mathbf{r}, t) \simeq f(\mathbf{k}, \mathbf{R}, t)$$

$$= [\exp\{\beta_e[E(\mathbf{k}) - \mathbf{v}_d \cdot \hbar\mathbf{k} - \mu(\mathbf{r}, t) + \tfrac{1}{2}mv_d^2(\mathbf{r}, t)]\} + 1]^{-1} \quad (2.5.24)$$

The influence of the inhomogeneities in the system on the collision terms is expressed in terms of space-dependent thermodynamical parameters in the carrier distribution function. For $E(\mathbf{k}) = \hbar^2 k^2 / 2m$, the commutator of an operator $A(\mathbf{r})$ with $H_e(t)$ in (2.5.23) can be written as

$$[A(\mathbf{r}), H_e(t)] = \{[A(\mathbf{r}), H_e] + \frac{i\hbar}{m} \int_0^t d\tau \, \nabla H_F(\mathbf{r}, \tau) \cdot \nabla A(\mathbf{r})\}e^{\Gamma t} \quad (2.5.25)$$

so that the averages can be easily expressed in terms of the various sets of parameters.

Although strong carrier–carrier scattering has been built into the approach implicitly through the selection of the set $\{P_m\}$, the equations given above do not include it explicitly (i.e., dynamically). However, consistent with the assumption of modest inhomogeneities in the system, one can include this interaction globally by adding the Poisson equation to the set (2.5.19) in the same spirit as has been done for the Vlasov equation classically.[21] This total approach gives a set of balance equations that are suitable for an inhomogeneous electric field, such as occurs within a device, provided that the assumptions built into the quasiequilibrium density operator are valid. The results deviate from those of the classical Boltzmann equation basically in that they show retardation terms and contain higher-order terms in the inhomogeneities, some of which are due to quantum effects. Moreover, the scattering terms themselves introduce spatial variations into the equations, an effect not normally considered in classical transport.

Although the early work of this approach treated the electric field response, excellent results have been obtained in the response of picosecond laser excitation studies as well.[21]

REFERENCES

1. R. KUBO, *J. Phys. Soc. Jpn.* **12**, 570 (1957).
2. S. DONIACH and E. H. SONDHEIMER, *Green's Functions for Solid State Physicists*, Benjamin/Cummings, Reading, MA (1974).
3. G. D. MAHAN, *Many-Particle Physics*, Plenum Press, New York (1981).
4. S. W. LOVESEY, *Condensed Matter Physics: Dynamic Correlations*, Benjamin/Cummings, Reading, MA (1980).

5. N. F. MOTT, *Phil. Mag.* **22**, 7 (1970).

6. D. S. FISHER and P. A. LEE, *Phys. Rev. B* **23**, 6851 (1981); *Phys. Rev. Lett.* **47**, 882 (1981).

7. E. ABRAHAMS, P. W. ANDERSON, D. C. LICCIARDELLO, and T. V. RAMAKRISHNAN, *Phys. Rev. Lett.* **42**, 673 (1979).

8. G. CZYCHOLL and B. KRAMER, *Sol. State Commun.* **32**, 945 (1979).

9. D. J. THOULESS and S. KIRKPATRICK, *J. Phys. C* **14**, 235 (1981).

10. P. W. ANDERSON, *Phys. Rev.* **109**, 1492 (1958).

11. R. MEZENNER and D. K. FERRY, unpublished.

12. D. N. ZUBAREV, *Usp. Fiz. Nauk* **71**, 71 (1960) [translation in *Sov. Phys. Uspekhi* **3**, 320 (1960)]; also, *Nonequilibrium Statistical Mechanics*, consultants Bureau, New York (1974).

13. V. P. KALASHNIKOV, *Physica* **48**, 93 (1970).

14. D. K. FERRY, *J. Physique* (*Colloq.*) **42**, C7-253 (1981).

15. U. FANO, *Rev. Mod. Phys.* **29**, 74 (1957).

16. N. N. BOGOLIUBOV, *Problemi dinam. teorii u stat. Fiz.*, Moscow (1946); also, in *Studies in Statistical Mechanics*, Vol. 1 (J. de Boer and G. E. Uhlenbeck, eds.), p. 11, North-Holland, Amsterdam (1962).

17. J. J. NIEZ and D. K. FERRY, *Phys. Rev. B* **28**, 1988 (1983).

18. W. PÖTZ and D. K. FERRY, in: *Proc. 17th Intern. Conf. Physics of Semiconductors* (J. D. Chadi and W. A. Harrison, eds.), p. 1329, Springer-Verlag, New York (1985).

19. D. K. FERRY and J. R. BARKER, *J. Phys. Chem. Solids* **41**, 1083 (1980).

20. A. P. JAUHO and J. W. WILKINS, *Phys. Rev. B* **29**, 1919 (1984).

21. A. R. VASCONCELLOS and R. LUZZI, *Phys. Rev. B* **27**, 3874 (1983).

Chapter 3

Path Integral Method: Use of Feynman Path Integrals in Quantum Transport Theory

K. K. Thornber

3.1. INTRODUCTION

The problems normally treated in quantum transport theory differ fundamentally from those of the usual transport theories. In the latter, one deals with well-defined probability distributions whose changes in space and time are governed by integral-differential equations involving complicated scattering rates. While sophisticated methods are often necessary to effect solutions, there is seldom any question as to the correctness of the form of the equations, and occasionally the individual scattering rates can be measured, or even calculated, independently.

In quantum transport theory conditions are very different.[1-14] Carriers can scatter so rapidly that the scattering process is no longer readily representable in terms of scattering rates alone. Instead, more details of the scattering amplitudes must be analyzed and included in the description of the carrier states themselves. High fields[7-14] complicate this description further. Potentials can change rapidly in space and time, rendering adiabatic techniques useless. Carriers can be confined to such a degree that their wave functions or density matrices are combinations of discrete and continuous components. And such confinement can often be externally altered and electronically controlled. As a result, new problems abound.

Most techniques of quantum transport theory fall into one of three broad categories: (1) quasiparticle methods, (2) operator-eigenfunction

---

K. K. Thornber ● NEC Research Institute, Princeton, New Jersey 08540, USA.

Quantum Transport in Semiconductors, edited by David K. Ferry and Carlo Jacoboni. Plenum Press, New York, 1991.

methods, and (3) path integral methods. In quasiparticle methods one attempts to retain as much of the classical formalism as possible in order to be able to express results in terms of carrier momentum or velocity, which is of greatest experimental interest. This approach also permits maximum flexibility; however, greatest care is necessary to ensure that all important effects are properly included. By contrast, operator-eigenfunction techniques adhere closely to the actual quantum states present in the structure of interest in the absence of scattering, and then introduce scattering. These techniques can obtain the greatest sensitivity to the confinement and lattice potentials, but are relatively inflexible in studying nonlinear, dynamical properties in the presence of full scattering.

Path integral methods[1-6] rely on an influence-functional technique in which the source of the dissipation, the carrier-phonon interaction, has been integrated over all phonon modes. This elimination (of the phonons) then leads to a model influence functional in which the phonon-scattering dissipation can be represented as an interaction with a collection of harmonic oscillator modes in which the translational invariance of the carrier is preserved. Confining potentials can be modeled by harmonic oscillator potentials coupled to the carriers so as to achieve representative level spacings. The resulting model is one in which constant, applied electric and magnetic fields, oscillatory or transient electric fields, screening, scattering and dissipation, confinement, background temperature, and initial conditions can be dealt with as readily as with a free particle. In other words, a basis already containing many interesting phenomena associated with quantum transport is achieved. This basis can serve as a starting point for more specific calculations. Choosing the model influence functional appropriately is a problem which is also addressed.

3.2. FORMULATION OF THE PROBLEM

The Hamiltonian of the system with which we shall be concerned can be written in the form[3-14]

$$
\begin{aligned}
H = {\tfrac{1}{2}}(p - A) \cdot m^{-1} \cdot (p - A) &- E_t \cdot x + V_m(x) + V_c(x) \\
&+ \sum_{n,k} \hbar \omega_{k,n} a_{k,n}^{\dagger} a_{k,n} \\
&+ \sum_{n,k} (C_{k,n} a_{k,n} \exp(ik \cdot x) + C_{k,n}^{*} a_{k,n}^{\dagger} \exp(-ik \cdot x)) \\
&+ \sum_{n,k} \hbar \Omega_{k,n} b_{k,n}^{\dagger} b_{k,n} \\
&+ \sum_{n,k} (S_{k,n} b_{k,n} \exp(ik \cdot x) + S_{k,n}^{*} b_{k,n}^{\dagger} \exp(-ik \cdot x)) \\
&+ \sum_{n,k} Q_{k,n} (C_{k,n} a_{k,n} + C_{k,n}^{*} a_{k,n}^{\dagger})(S_{k,n} b_{k,n} + S_{k,n}^{*} b_{k,n}^{\dagger})
\end{aligned}
\tag{3.2.1}
$$

These terms have the following significance. The first term is the kinetic energy of a charged carrier in a magnetic field $B = \nabla \times A$ and mass tensor m. The next three terms represent the applied electric field E_t, the electron potential of the medium (the atomic lattice) V_m, and the confinement potential V_c. The latter is often externally controllable, to some extent, electronically. The terms in $a_{k,n}$ and $a_{k,n}^{\dagger}$ are the energy and coupling to the carrier of the phonon modes, and those in $b_{k,n}$ and $b_{k,n}^{\dagger}$ are the energy and coupling of any screening modes such as plasmons. The final term is the (polarization) coupling between the phonon and the screening modes. Inclusion of screening in this discussion is somewhat academic since the phonon and screening modes can be decoupled into normal modes with modified frequencies and carrier-coupling parameters by standard methods. Henceforth, we shall express the resulting modes by a single notation, that of the fifth and sixth terms above.

The next step, the elimination of the phonon and screening modes from the problem, is by now a standard operator technique.[1–6,12] The underlying assumptions, however, should be reviewed. In order to calculate the expected carrier velocity, momentum distribution, scattering rates, etc., represented in terms of some operator $\hat{O}$, we must ascertain the density matrix of the entire system at the time of interest t_2:

$$\langle \hat{O} \rangle \equiv \mathrm{Tr}(\hat{O}\rho_{t_2})/\mathrm{Tr}(\rho_{t_2}) \qquad (3.2.2)$$

where ρ_t satisfies ($\hbar = 1$)

$$i(\partial\rho/\partial t) = [H, \rho] \qquad (3.2.3)$$

and hence

$$\rho_{t_2} = \exp\left(-i \int_{t_1}^{t_2} H_t \, dt\right) \rho_{t_1} \exp\left(i \int_{t_1}^{t_2} H_t \, dt\right) \qquad (3.2.4)$$

The choice of the *initial* density matrix ρ_{t_1} depends on the problem of interest. Normally the phonon modes will be of some (lattice) temperature, the screening modes at some probably higher temperature depending on the current density, and the carrier(s) in some electronic state(s) which we need not specify for now. More general starting density matrices can also be handled.[12] However, the essential point to be made is that (spatially) local changes in the occupancy of the modes due to net emissions and/or absorptions by the carrier(s) of interest are *included* in (3.2.4), from which these modes can be eliminated exactly.[7,8,12] No ad hoc modification is needed to include these changes no matter how strong the coupling. If carrier-medium-carrier effects are of interest, more carriers can be introduced into H and studied after integrating out the phonon and screening modes. To determine the occupancy, or, if applicable, the temperature, of these modes based on current densities, their lifetimes and other characteristics would have to be known and included in H as well. Though important for a complete analysis, such problems will not concern us further here.

Carrying out the indicated mode elimination in (3.2.4) and passing from operator to path integral representation,[1-14] we obtain in place of (3.2.2),

$$\iint D(x_t)D(x_t') \exp\left(i \int dt\, (f_t \cdot x_t - f_t' \cdot x_t')\right) \exp S_{m,c}^e[x_t, x_t'] \quad (3.2.5)$$

where the exact influence functional S^e is given by

$$S_{m,c}^e[x_t, x_t'] \equiv i \int dt\, [\tfrac{1}{2}\dot{x}_t \cdot m \cdot \dot{x}_t + \tfrac{1}{2}\dot{x}_t \cdot B \times x_t + E_t \cdot x_t - V_m(x_t) - V_c(x_t)]$$

$$- i \int dt\, [\tfrac{1}{2}\dot{x}_t' \cdot m \cdot \dot{x}_t' + \tfrac{1}{2}\dot{x}_t' \cdot B \times x_t' + E_t \cdot x_t' - V_m(x_t') - V_c(x_t')]$$

$$- \int dt \int^t dt' \sum_{n,k} (V_{x',x'} - V_{x,x'} + V_{x,x}^* - V_{x',x}^*) \quad (3.2.6)$$

$$V_{x,x'} \equiv |C_{k,n}|^2 T_{\omega_{k,n}}(t-t') \exp(-ik \cdot (x_t - x_{t'}')) \quad (3.2.7)$$

$$T_\omega(\tau) \equiv e^{i\omega\tau}(1 - e^{-\beta\omega})^{-1} + e^{-i\omega\tau}(e^{\beta\omega} - 1)^{-1} \quad (3.2.8)$$

$$\beta_{k,n} \equiv 1/\kappa T_{k,n} \quad (3.2.9)$$

In (3.2.9) $T_{k,n}$ is the temperature ascribed to the n, k mode prior to the introduction of the carriers.[12] The first and second bracketed terms in (3.2.6) represent the propagation of the density matrix of the carrier as influenced by the magnetic and electric fields and the medium (or lattice) and confining potentials. The third term, which couples the motion of the carrier with that of its past, includes all fluctuation, dissipation, and self-energy effects included by its interaction with the phonon and screening modes, both of which are included in (3.2.7), modified by their mutual interaction as described above. Finally, through the appropriate choice of f_t and f_t' in (3.2.5), the expected value of the actual carrier operator of interest can be represented.[1-6,12-14] (The influence functional $S_{m,c}^e$ includes both the medium and the confining potentials; S_m^e, only the former; S_c^e, only the latter; and S^e-neither.)

3.3. CONSERVATION LAWS AND CONSTANTS OF THE MOTION

Two of the most useful expressions in transport theory are equations expressing the conservation of momentum and conservation of energy.

These as well as most other relations of a similar nature can be derived in the path integral representation as follows.[5] Consider the identity

$$\iint D(x_t)D(x_t') \exp(S[x_t, x_t']) = \iint D(y_t)D(y_t') \exp(S[y_t, y_t']) \quad (3.3.1)$$

in which (x_t, x_t') and (y_t, y_t') are equivalent pairs of integration variables. $S[x_t, x_t']$ is any functional in $[x_t, x_t']$. If now we let $y_t = x_t + \lambda r_t$, where $\lambda \ll 1$, r_t is some function of time, and $D(x_t) = D(y_t)$. Inserting this into the left-hand side above and expanding to lowest order in λ, we obtain

$$0 = \int d\tau \iint D(x_t)D(x_t') \exp(S[x_t, x_t']) \delta S/\delta x_\tau \cdot r_\tau \quad (3.3.2)$$

If we let $r_\tau = \hat{n}\delta(\tau - t_2)$, then we recover the equations for the conservation of momentum in the direction $\hat{n}$ at time t_2. This is the same as replacing the path x_t by the path $x_t + \lambda\hat{n}\delta(\tau - t_2)$, which in classical mechanics leads to Lagrange's equations of motion. Letting $r_\tau = \hat{n}u(t_2 - \tau)$ leads to a constant of the motion analogous to momentum in the direction $\hat{n}$.

Time invariance lies at the root of energy conservation. Thus, replacing y_t by $x(t + \lambda\eta(t)) = x_t + \lambda\eta(t)\dot{x}_t$ implies that letting $r_\tau = \eta(\tau)\dot{x}_\tau$ in (3.3.2) will yield the equation for energy conservation at t_2 if $\eta(t) = \delta(t - t_2)$. For $\eta(t) = u(t_2 - t)$ one obtains the constant of the motion corresponding to energy. Equivalently, one could replace each path x_t in the original path integral by $x(t + \lambda\eta(t))$. The change of variables $\tau = t + \lambda\eta(t)$ can be used to avoid the functional differentiations; however, care must be taken to take any $x(t - t')$ into

$$x(t - t' + \lambda\eta(t - t')) = x(t + \lambda\eta(t) - t' - \lambda\eta(t') \\ + \lambda(\eta(t - t') - \eta(t) + \eta(t')))$$

$$dt = d\tau(1 - \lambda\dot{\eta}(\tau)) \quad \text{and} \quad d/dt = (1 + \lambda\dot{\eta}(\tau))\,d/d\tau$$

the latter two being valid to first order in λ.

First consider the following simple example. If the classical action S_c is given by

$$S_c = \int dt\, L(x_t, \dot{x}_t, t) \quad (3.3.3)$$

then the path variation $y_t = x_t + \lambda\eta(t)\dot{x}_t$, which results in

$$\delta S_c = \lambda \int dt\, \eta(t)\dot{x}_t \delta L/\delta x_t = 0 \quad (3.3.4)$$

would yield, upon letting $\tau = t + \lambda\eta(t)$,

$$S_c = \int d\tau\, (1 - \lambda\dot{\eta}(\tau))L(x_\tau, \dot{x}_\tau(1 + \lambda\dot{\eta}(\tau)), \tau - \lambda\eta(\tau)) \quad (3.3.5)$$

so that

$$\delta S_c = \lambda \int d\tau\, \eta(\tau) \left(\frac{dL}{d\tau} - \frac{d(\dot{x}_\tau\, \partial L/\partial \dot{x}_\tau)}{d\tau} - \frac{\partial L}{\partial \tau} \right) = 0 \qquad (3.3.6)$$

Hence

$$\frac{dH}{dt} = \frac{d(\dot{x}\, \partial L/\partial \dot{x})}{dt} - \frac{dL}{dt} = -\frac{\partial L}{\partial t} = \frac{\partial H}{\partial t} \qquad (3.3.7)$$

the usual expression for conservation of energy in classical mechanics. (Note that $\dot{x}\,\delta L/\delta x = dL/dx - d(\dot{x}\,\partial L/\partial \dot{x})/dt - \partial L/\partial t$.)

Applying the above method to our influence functional $S^e_{m,c}[x_t, x'_t]$, as expressed in (3.2.6)–(3.2.9), we obtain for the conservation of momentum the expression

$$\mathbf{E}_t + \langle \dot{\mathbf{x}}_t \rangle \times \mathbf{B} - m\langle \ddot{\mathbf{x}} \rangle_t = \sum_{n,k} k\langle R_{k,n} \rangle_t + \left\langle \frac{\partial V_m}{\partial \mathbf{x}} \right\rangle_t + \left\langle \frac{\partial V_c}{\partial \mathbf{x}} \right\rangle_t \qquad (3.3.8)$$

where

$$
\begin{aligned}
R_{k,n}(t) = |C_{k,n}|^2 \int^t dt' \ & \frac{\exp(-i\omega_{k,n}(t - t') + ik(x_t - x_{t'}))}{1 - \exp(-\beta\omega_{k,n})} \\
& - \frac{\exp(i\omega_{k,n}(t - t')) - ik\cdot(x_t - x_{t'})}{\exp(\beta\omega_{k,n}) - 1} \\
& + \frac{\exp(i\omega_{k,n}(t - t')) - ik\cdot(x_t - x'_{t'})}{1 - \exp(-\beta\omega_{k,n})} \\
& - \frac{\exp(-i\omega_{k,n}(t - t')) + ik\cdot(x_t - x'_{t'})}{\exp(\beta\omega_{k,n}) - 1}
\end{aligned}
\qquad (3.3.9)
$$

$$
\begin{aligned}
\left\langle \frac{\partial V_m}{\partial x} \right\rangle_t = {}& \iint D(x_\tau)D(x'_\tau)\exp(S^e_c[x_\tau, x'_\tau])\frac{\partial V_m}{\partial x_t} \\
& - i \int^t dt' \iint D(x_\tau)D(x'_\tau)\exp(S^e_{c,m(t,t')}[x_\tau, x'_\tau])\frac{\partial V_m}{\partial x_t} \\
& \cdot (V_m(x_{t'}) - V_m(x'_{t'}))
\end{aligned}
\qquad (3.3.10)
$$

and

$$
\begin{aligned}
\left\langle \frac{\partial V_c}{\partial x} \right\rangle_t = {}& \iint D(x_\tau)D(x'_\tau)\exp(S^e_m[x_\tau, x'_\tau])\frac{\partial V_c}{\partial x_t} \\
& - i \int^t dt' \iint D(x_\tau)D(x'_\tau)\exp(S^e_{m,c(t,t')}[x_\tau, x'_\tau])\frac{\partial V_c}{\partial x_t} \\
& \cdot (V_c(x_{t'}) - V_c(x'_{t'}))
\end{aligned}
\qquad (3.3.11)
$$

In $S^e_{c,m}$ in (3.3.10) [$S^e_{m,c}$ in (3.3.11)], $V_m(V_c)$ only acts between t' and t as indicated. Expression (3.3.11) will vanish for directions of interest which are perpendicular to the gradient of the confining potential. Expression (3.3.9) is the net rate of emission (emission less absorption) of coupled-phonon and screening modes, index n, and momentum transfer k.

To calculate the invariant corresponding to momentum, one integrates (3.3.8) between some initial time and some final time t, obtaining

$$m\langle\dot{x}\rangle_t + 2\langle A\rangle_t + \int^t dt'\left(\sum_{n,k} k\langle R_{k,n}\rangle_{t'} + \left\langle\frac{\partial V_m}{\partial x}\right\rangle_{t'} + \left\langle\frac{\partial V_c}{\partial x}\right\rangle_{t'} - E_{t'}\right)$$

$$= \text{constant of the motion} \tag{3.3.12}$$

Here A is the vector potential for the constant magnetic field B: $A = \frac{1}{2}B \times x$. For A independent of x, $\langle m\dot{x} + A\rangle$ is a constant of the motion for a free particle ($\langle p\rangle$). (Recall that the kinetic momentum is usually defined to be $m\dot{x} = p - A$.) However, for uniform magnetic fields, the invariant for a free particle is $\langle m\dot{x} + 2A\rangle$ as long as A is independent of time. The remaining terms in (3.3.12) express the residual between momentum gained from the electric field and that lost to the medium and to the Hall current.

The principal utility of quantities such as (3.3.12) is that constants of the motion represent invariants which can be used to label basis states in an expansion. For example, in writing out evolution equations (Liouville equations) for the density matrix of the system, it is customary to use as a basis the momentum state of either free or uniformly accelerating carriers corresponding to zero-order Hamiltonians for noninteracting particles. The interactions then drive the evolution of the system. Such an approach is adequate when indeed the state of the carrier is well-characterized by one of these basis states at any given time. However, when the scattering becomes sufficiently strong that the particles cannot be described, either theoretically or experimentally, as occupying a well-defined momentum state at any time (i.e., the lifetime in a given state is so small that the state is unacceptably broadened), then the utility of this basis and the applicability of expansions in such a basis are called into serious question. In the absence of dissipation the concept of quasiparticles is sometimes adequate to handle self-energy renormalizations. But in the presence of applied fields and dissipation, where energy and momentum are continually acquired from the applied fields and transferred to the medium at nonnegligible rates, quasiparticle bases are inadequate for the reasons given.

In sharp contrast to bases of particles or quasiparticles behaving ballistically between collisions, the representation developed below in the context of an approximate influence functional describes carriers whose motion mimics that of the carriers under the actual conditions of the problem. For particles in these states, constants of the motion such as (3.3.12) are preserved between scatterings characterized by scattering rates reflecting differences between the actual evolution of the system and its simulated evolution.

In this manner such invariants of an approximate influence functional can serve as an adequate basis for working out evolution equations.

The corresponding equation for conservation of energy and the related constant of the motion can be obtained from time invariance as indicated following equation (3.3.2). The result is

$$E_t \cdot \langle \dot{x} \rangle_t - \frac{1}{2} \frac{d \langle \dot{x} \cdot m \cdot \dot{x} \rangle_t}{dt}$$

$$= \sum_{n,k} \omega_{k,n} \langle R_{k,n} \rangle_t + \frac{d(\langle V_c(x_t) \rangle + \langle V_m(x_t) \rangle)}{dt}$$

$$+ \frac{d\left(\int^t dt' \langle V_{x_t,x'_{t'}}(t-t') - V^*_{x_t,x_{t'}}(t-t') \rangle \right)}{dt} \qquad (3.3.13)$$

Here $R_{k,n}(t)$ is as in (3.3.9), while $\langle V_c \rangle$ and $\langle V_m \rangle$ are analogous to expressions (3.3.10) and (3.3.11). Expression (3.3.13) illustrates how the energy absorbed from the applied electric field is distributed among the kinetic, lattice, and interaction energies and dissipated through the coupled phonon and screening modes. Two features of interest include the <u>absence</u> of a term in the magnetic field, since a static magnetic field cannot alter the energy of a charged particle, and the last term which expresses changes in the electron-phonon screening interactions through which the energy dissipation must flow. In the corresponding momentum equation (3.3.11) there was no term representing a transitional or intermediate state possessing momentum.

As with momentum, an integration over time of (3.3.13) will give us an energy-related invariant in the following form:

$$\tfrac{1}{2}\langle \dot{x} \cdot m \cdot \dot{x} \rangle_t + \langle V_m(x_t) \rangle + \langle V_c(x_t) \rangle$$

$$+ \int^t dt' \langle V_{x_t,x'_{t'}}(t-t') - V^*_{x_t,x_{t'}}(t-t') \rangle$$

$$+ \int^t dt' \left(\sum_{n,k} \omega_{k,n} \langle R_{k,n} \rangle_{t'} - E_{t'} \cdot \langle \dot{x} \rangle_{t'} \right)$$

$$= \text{constant of the motion} \qquad (3.3.14)$$

Here, as in (3.3.12), each term, each expectation value, is represented by its path integration expectation value indicated above. The primary purpose of expressions (3.3.8)–(3.3.14) lies not only in their implications for steady-state quantities and transient and ac response but in that such quantities evaluated for more general models can be used as bases for density matrices in place of free-particle energy and momenta. We turn now to the evaluation of the above expectation values.

3.4. APPROXIMATIONS FOR COMPUTATIONS

Unfortunately no means is known with which to carry out the final path integrals over the carrier coordinates in an exact, analytical manner. However, we can construct an influence functional $S^0_{m,c}$ similar to the $S^e_{m,c}$ of (3.2.6) which contains much of the physics of the original problem and which can be integrated[1–14] to completion for arbitrary f_t and f'_t. In order to carry out this program, it is important to realize that influence functionals involving only harmonic-oscillator, potential interactions can be worked out completely. Free-particle functionals are, of course, a special case; although algebraically simpler, it is only with considerable computation that equivalent effects can be realized with them. The path integral method is unique in its ability to handle analytically confinement and dissipation for arbitrary field strengths, coupling, temperature, and probe fields f_t and f'_t.

Returning to (3.2.6), we wish to modify the full (exact) $S^e_{m,c}$ to obtain an approximate $S^0_{m,c}$ which, nonetheless, contains sufficient content that many of the phenomena of interest are included as accurately as possible. The first three terms involving kinetic, magnetic, and electric energy can be retained in their present form. The band structure of the material, which arises from the lattice potential, is difficult to deal with in the path integral formalism apart from making m the effective-mass tensor m^* in the energy range of interest, and introducing an oscillator potential to model narrow, isolated bands. Where scattering is weak, this is clearly unsatisfactory. Where scattering is strong, however, the carriers' nominal de Brolie wavelengths greatly exceed their scattering "mean free paths," and hence the lattice potential becomes relatively unimportant. Consequently, band-structure phenomena will not be well modeled by this approach, while Bloch oscillations and poorly conducting materials can be treated much better.

The confinement potential is much easier to model. A harmonic oscillator potential readily confines a carrier to a sequence of well-defined energy bands. The spacing or density is readily adjusted to that of the problem of interest in the region of energy of interest. Furthermore, changes in the confinement potential brought about electronically in the structure of interest can be modeled and used to predict the effects of such modulation. Thus in $S^0_{m,c}$ we include terms

$$\frac{i}{2} \int dt\, (-x_t \cdot K \cdot x_t + x'_t \cdot K \cdot x'_t) \tag{3.4.1}$$

in place of

$$i \int dt \, (-V_c(x_t) + V_c(x'_t)) \tag{3.4.2}$$

Here $K = 0$ represents an unconfined carrier; K in one or in two dimensions, a carrier confined to two or to one dimension, respectively; and K in all three directions, a totally confined carrier with a very interesting ac response (once scattering is included).

Turning now to the scattering terms, we replace the last line of (3.2.6) by

$$+ \int dt \int^t dt' \, (V^0_{x',x'} - V^0_{x,x'} + V^{0*}_{x,x} - V^{0*}_{x',x}) \tag{3.4.3}$$

with

$$V^0_{x,x'} \equiv (x_t - x'_{t'}) G^*(t - t')(x_t - x'_{t'}) \tag{3.4.4}$$

where

$$G(\tau) = \int d\Omega \, G(\Omega) \exp(-i\Omega\tau) \tag{3.4.5}$$

Here $G(\Omega)$ represents the spectrum of the distribution of oscillators used in (3.4.3) and (3.4.4) to simulate the carrier-medium interaction fully expressed in (3.2.6)–(3.2.9). This modification can be motivated as follows. For strong scattering the interval between scattering events $(x_t - x_{t'})$ is expected to be small. Expanding to harmonic order (the first nonzero terms) yields (3.4.3) and (3.4.4) with $G^0(\Omega)$ specified in terms of (3.2.7). Later we note how $G^0(\Omega)$ can be obtained in a physically reasonable, self-consistent manner.

Summarizing, therefore, we have described an integrable influence functional $S^0_{m,c}[x_t, x'_t]$ which we can use as a basis for describing a number of interesting features of quantum transport. Written out in full, we have

$$S^0_{m,c} \equiv i \int dt \, (\tfrac{1}{2}\dot{x}_t \cdot m^* \cdot \dot{x}_t + \tfrac{1}{2}\dot{x}_t \cdot B \times x_t + E_t \cdot x_t - \tfrac{1}{2}x_t \cdot K \cdot x_t)$$

$$- i \int dt \, (\tfrac{1}{2}\dot{x}'_t \cdot m^* \cdot \dot{x}'_t + \tfrac{1}{2}\dot{x}'_t \cdot B \times x'_t + E_t \cdot x'_t - \tfrac{1}{2}x'_t \cdot K \cdot x'_t)$$

$$+ \int dt \int^t dt' \, (V^0_{x',x'} - V^0_{x,x'} + V^{0*}_{x,x} - V^{0*}_{x',x}) \tag{3.4.6}$$

where V^0 is written out in (3.4.3) and (3.4.4).

The next step is to carry out the path integration indicated in (3.2.5) with $S^e_{m,c}$ replaced[7-14] by $S^0_{m,c}$. The result is

$$\iint D(x_t)D(x'_t) \exp\left(i \int dt \, (f_t \cdot x_t - f'_t \cdot x'_t) \right) \exp(S^0_{m,c})$$

$$= \exp\left(\int dt \int^t dt' \, (f_t - f'_t)(L^*_{t-t'}f_t - L_{t-t'}f'_t) \right) \qquad (3.4.7)$$

where

$$L_\xi \equiv \int \frac{d\nu}{2\pi i} Z_\nu^{-1}(-i4\pi G_{-\nu})Z_\nu^{\dagger-1} \exp(-i\nu\xi) \qquad (3.4.8)$$

$$Z_\nu \equiv K - m^*(\nu + i\varepsilon)^2 - i(\nu + i\varepsilon)\varepsilon \cdot B$$

$$- 4(\nu + i\varepsilon)^2 \int d\Omega \, \frac{P}{\Omega} \frac{G(\Omega)}{\Omega^2 - (\nu + i\varepsilon)^2} \qquad (3.4.9)$$

or, equivalently,

$$Z_\nu = K - m^*(\nu + i\varepsilon)^2 - i(\nu + i\varepsilon)\varepsilon H$$

$$- 4 \int_0^\infty d\rho \, (1 - \exp(i\nu\rho) \, \mathrm{Im} \, G(\rho) \qquad (3.4.10)$$

and $\varepsilon \cdot B$ is defined by $A \cdot \varepsilon \cdot B \cdot C = A_i \varepsilon_{ijk} B_j C_k = A \cdot B \times C$. Z_ν plays an important role in establishing self-consistency relations. Expression (3.3.7) is the principal basis for evaluating expectation values and performing perturbation expansions of the form[14]

$$\iint D(x_t)D(x'_t) \exp\left(i \int dt \, (f_t \cdot x_t - f'_t \cdot x'_t) \right) \exp(S^0) \exp(S^e - S^0) \quad (3.4.11)$$

The next section will be devoted to applications of this result.

At this juncture two points are paramount. First, although we have attempted to model the actual problem as physically as possible, (3.4.7)–(3.4.10) as they stand should not be used to calculate expectation values of generic operators, such as momentum, which have <u>no</u> specific association with the problem at hand. To do so would be to express the quantity of interest only in terms of the model itself. Thus in zero order the kinetic momentum is

$$\langle p - A \rangle_\nu = \langle m\dot{x} \rangle_\nu = -i(\nu + i\varepsilon)mZ_\nu^{-1}e_\nu \qquad (3.4.12)$$

where e_ν is the not necessarily small, oscillatory part of the applied electric field. As seen from (3.3.9), Z_ν depends on the confinement potential strength K, on the effective-mass tensor m^*, and on the oscillator coupling spectrum $G(\Omega)$. Hence quantities like (3.4.12) are very sensitive to the particular approximation. On the other hand, quantities such as the rate of scattering by momentum k at time t (see (3.3.9)) involve the <u>full</u> scattering potential. When the rate $R_{k,t}$ is multiplied by the momentum change k and summed over k, one obtains in steady state the expression

$$E + v \times B = \sum_{n,k} |C_{k,n}|^2 k$$

$$\times \int d\rho \, T_{\omega_{k,n}}(\rho) \exp(-ik \cdot v\rho) \exp(-k \cdot \bar{L}_\rho \cdot k) \qquad (3.4.13)$$

where

$$L_\rho = L_\rho - L_{0^+} \qquad (3.4.14)$$

for the average drift velocity v versus applied dc field E, both perpendicular to the directions of confinement, in the presence of magnetic field B. In (3.4.13) all model dependence resides in the last term in a relatively insensitive manner.

The second paramount point is that quantities m, k and $G(\Omega)$ must be determined to complete our description of S^0. The band structure, if known, can provide m^*, and fitting to the confinement potential can yield K. Alternatively, m^* and K can be determined self-consistently along with $G(\Omega)$ utilizing the following general technique,[8-14] which includes the interrelationships between these parameters and the actual interactions expressed in (3.2.6)–(3.2.9).

3.5. SELF-CONSISTENCY FOR THE APPROXIMATE INFLUENCE FUNCTIONAL

The foregoing method of approximation relies on the choice of effective mass m^*, confinement coupling K, and frequency-dependent, oscillator strength $G(\Omega)$. Quite generally these model parameters depend on the (free) carrier mass, the applied fields E and B, and all interaction potentials. The following (very nonlinear) approach enables these parameters to be determined self-consistently.[8-14]

The first step is to translate the above equations to a frame of reference translating with the steady-state motion of the carriers as induced by the static portions of the applied E and B fields. In the path integral representation this is accomplished simply by letting $x_t = y_t + vt$, $x'_t = y'_t + vt$, where $v = v(E, B)$ as determined by (3.4.13). For transient problems other drift velocities, including time-dependent ones, could be more appropriate. The second step is to notice that the impedance associated with $S^0_{m,c}$ (3.4.6) is simply Z_v, (3.4.9) or (3.4.10). [See (3.4.12).] Moreover, the equation for the conservation of momentum (3.3.8) also yields an expression for the impedance in terms of the full scattering rates of the approximate influence function $S^0_{m,c}$. Requiring these impedances to be equal specifies the self-consistency.

Before working out the details of the above prescription, the following results should be summarized as motivation. There are many ways to calculate impedance, or its inverse, admittance. The methods outlined above are two. Other approaches include using perturbation expansions in the difference between the exact and the approximate influence functionals to include the actual interactions to first order in either the admittance or the impedance.[4] Alternatively, there is the self-energy admittance, which has the property of exactly satisfying the ground-state-energy-theorem sum rule,[15,16] and the free-energy-theorem sum rule,[10,13] for any oscillator distribution adjusted to minimize the free energy ($E = 0$, $B = 0$). Calculated exactly, of course, all would yield the same result. Calculated approximately, all would ordinarily be expected to yield very different results. However, when the oscillatory distribution $G(\Omega)$ which satisfies the above self-consistency is used to calculate each of these diverse impedances, they become identical analytically.[10,13] In addition, each step in the derivation of the free-energy sum rule is exactly satisfied. (The free energy is of the entire system.) Although this sum rule is satisfied by many distributions, only the dynamically self-consistent impedance leads to an exact frequency-by-frequency agreement when the fluctuation-dissipation theorem is calculated using the full impedance form derived from (3.3.8) for $E = 0$, $B = 0$. As before,[10] we believe this represents the <u>only</u> example yet known where a <u>non</u>trivial, approximate solution to a <u>non</u>trivial many-body problem exactly satisfies <u>non</u>trivial sum rules analytically. (Here "<u>non</u>trivial" refers to problem-specific solutions and sum rules in contrast to general conservation laws which practically any approximate solution would be expected to satisfy.)

Returning now to the self-consistency method, if we first ignore the gradient terms in V_m and V_c in the expression for the conservation of momentum (3.3.8)—in other words, if we first consider the problem of the carrier interacting only with the phonon-screening modes—then expanding (3.3.8) to lowest order in the probe field e_t and solving for the impedance Z_ν yields

$$Z_\nu = -m(\nu + i\varepsilon)^2 - i(\nu + i\varepsilon)\varepsilon B$$

$$-4(\nu + i\varepsilon)^2 \int_{-\infty}^{\infty} d\Omega \, \frac{P}{\Omega} \frac{G(\Omega)}{\Omega^2(\nu + i\varepsilon)^2} \tag{3.5.1}$$

where now

$$G(\Omega) = \tfrac{1}{2} \sum_{n,k} |C_{k,n}|^2 kk \int_{-\infty}^{\infty} \frac{d\xi}{2\pi} \exp(-i\Omega\xi)$$

$$\cdot \, T_{\omega_{k,n}}(\xi) \exp(-ik \cdot v\xi) \exp(-k \cdot \bar{L}_\xi \cdot k) \tag{3.5.2}$$

and $\bar{L}_\xi$ is given by (3.4.14) and (3.4.8). Thus, given $G(\Omega)$, (3.5.1) gives Z_ν, (3.4.8) then gives L_ξ, and (3.5.2) gives us back $G(\Omega)$. This is the fundamental self-consistency. For $E = 0$, $B = 0$ it reduces to minimizing the free energy; for expansions of one distribution (G_2) in terms of another (G_1), equation (3.5.2) gives the correct G (i.e., G_2) at once so that no iteration is necessary. Clearly in the absence of the static medium and confining potentials, V_m and V_c, $m^* = m$ and $K = 0$.

Applying this self-consistency to the full problem expressed in (3.3.8)–(3.3.11) requires some extensions. The gradients in both V_m and V_c involve the full influence functional S_{mc} as well as one of S_c or S_m in which the potential present in the gradient is absent in the action. Thus, oscillator distributions, effective masses, and effective confinement coupling for either V_c or V_m acting (as well as both) must be determined. Thus, to determine $G_c(\Omega)$, m_c^*, K_c, (3.3.8), (3.3.9), and (3.3.11) are used with $V_m = 0$. (Hence S_m and S_{mc} in (3.3.11) become S and S_c.) Clearly $m_c^* = m$ while $G_c(\Omega)$ and K_c will be found from equations similar to (3.5.1) and (3.5.2). To determine $G_m(\Omega)$, m_m^*, K_m, (3.3.8)–(3.3.10) are used in a similar manner: $K_m = 0$. Finally the full set of (3.3.8)–(3.3.11) can be applied to yield $G_{mc}(\Omega)$, m_{mc}^*, K_{mc} from which the complete approximate influence functional can be determined and the desired transport properties calculated.

3.6. CARRIER-ENERGY DISTRIBUTIONS

With the renewed interest in high-field, carrier transport in polar materials,[17,18] it is important to be able to calculate the energy distribution of such electrons when emitted into a vacuum. These calculations are complicated by so-called final-state interactions, which account for the conversion of the state of the interacting electron and its associated polarization into a state of the free electron and the relaxing (not necessarily adiabatically) polarization. This is _not_ a "quasiparticle" energy, since the emission process has annihilated the "quasiparticle". (In addition, in the presence of strong applied fields and the corresponding induced dissipation, a quasiparticle representation is usually unworkable, its lifetime being too short.) Nonetheless, if the medium-vacuum interface does not alter the electron's energy (elastic scattering), then it is relatively straight-forward in second-quantized representation to write an exact expression for the distribution including all final-state interactions.[9] (If interface scattering is important, such effects can also be included,[9] but such considerations would take us somewhat afield.)

The expression for the energy distribution $N(E)$ of carriers dissociated from an interactive media is given by

$$N(E) = 2 \operatorname{Re} \sum_j \int_0^\infty d\tau \, \langle c_j^\dagger(\tau) c_j(0) \rangle \frac{\exp(-iE\tau/\hbar)}{2\pi\hbar} \qquad (3.6.1)$$

Here $c_j^\dagger$ and c_j create and destroy an electron in the single-particle state $\phi_j(x)$, the $\{\phi_j(x)\}$ serving as a basis for the second quantization. Expression (3.6.1), which is quite similar to relations for density of states under thermal equilibrium conditions, is derived by weakly coupling the system of interest to a measuring system. In the latter, each state is independent and at energy E. If the coupling is independent of the state, then the expected rate of transfer of electrons from the system of interest to the measuring system is proportional to the number of electrons which can decouple from the interacting system at energy E. Similar reasoning leads to a relation between expected photoemission current densities and the rate of change of the rate of transfer, etc.[9] Unfortunately, it is not clear at this time what expectation value should be calculated in the path integral representation to obtain $N(E)$.

The energy-momentum structure of the electron interacting with the driven dissipative system discussed in this chapter can be characterized by letting the electron couple weakly to (additional) phonons of energy E and momentum k according to

$$H_T = H + H_p \qquad (3.6.2)$$

where H is given by (3.2.1) and

$$H_p = Ea^\dagger a + Ca \exp(ik \cdot x) + C^*a^\dagger \exp(-ik \cdot x) \qquad (3.6.3)$$

Eliminating all phonons (exactly) [recall (3.2.1)–(3.2.5)], one finds for the rate of change of the number of probe phonons

$$\dot{n}_p(t) \equiv \frac{i}{\hbar}[H_T, a^\dagger a]_t = \frac{i}{\hbar}(C \exp(ik \cdot x)a - C^* \exp(-ik \cdot x)a^\dagger)_t$$

$$= (N_p(k, E) + 1)\frac{|C|^2}{\hbar^2}\int_{-\infty}^{t} dt' \exp(-ik \cdot (x_t - x'_{t'}))\exp\left(\frac{iE(t - t')}{\hbar}\right)$$

$$-N_p(k, E)\frac{|C|^2}{\hbar^2}\int_{-\infty}^{t} dt' \exp(ik \cdot (x_t - x'_{t'}))\exp\left(\frac{-iE(t - t')}{\hbar}\right) \qquad (3.6.4)$$

where $N_p(k, E) \geq 0$ is the number of probe phonons with energy E and momentum $\hbar k$. [This result is somewhat more general than the response of the system to spatially periodic, oscillatory probe fields (ω, t).] Dividing (3.6.4) by $2\pi|C|^2/\hbar$, the appropriate rate-per-unit-energy density, results in a measure of the system transfer of energy E at momentum $\hbar k$. This is analogous, but, of course, not equivalent, to the transfer of energy E at momentum $\hbar k$ exterior to the dissipative system in the form of a free electron. The term $\ddot{n}_p$, suitably normalized[9] provides a measure of the transfer of $(E, \hbar k)$ as modulated by an applied field (ω, t), a probe akin to photoemission, but calculable using the path integral method.

3.7. CONCLUDING REMARKS

In the above we have outlined an approach for studying quantum transport phenomena using Feynman path integrals. Older results were formulated in order to indicate how newer problems such as transport under conditions of carrier confinement could be handled. In addition, integral invariants were derived which can serve as a basis for expansions of the Liouville equation for the density matrix in a representation corresponding to the approximate influence functionals used to analyze the problems of interest. These invariants were also written out for any, including the exact, influence functional. For approximate functionals derived from exact ones according to the self-consistency principle described, a number of rather demanding relations are satisfied analytically. This is remarkable, since no other many-body problem is known to possess solutions with these properties.

REFERENCES

1. R. P. FEYNMAN, *Rev. Mod. Phys.* **20**, 367 (1948).
2. R. P. FEYNMAN, *Phys. Rev.* **84**, 108 (1951).
3. R. P. FEYNMAN, *Phys. Rev.* **97**, 660 (1955).
4. R. P. FEYNMAN, *Phys. Rev.* **127**, 1004 (1962).
5. R. P. FEYNMAN and A. R. HIBBS, *Quantum Mechanics and Path Integrals*, McGraw-Hill, New York (1965).
6. R. P. FEYNMAN, *Statistical Mechanics, A Set of Lectures*, Benjamin (1972).
7. K. K. THORNBER and R. P. FEYNMAN, *Phys. Rev. B* **1**, 4099 (1970); *Phys. Rev. B* **4**, 674 (1971).
8. K. K. THORNBER, *Phys. Rev. B* **3**, 1929 (1971); *Phys. Rev. B* **4**, 675 (1971).
9. K. K. THORNBER, *Phys. Lett.* **34 A**, 205 (1971).
10. K. K. THORNBER, *Phys. Rev. B* **9**, 3489 (1974).
11. K. K. THORNBER, *Solid-State Electronics* **21**, 259 (1978).
12. K. K. THORNBER, in: *Polarons in Ionic Crystals and Polar Semiconductors* (J. T. Devreese, ed.), North-Holland, Amsterdam (1972).
13. K. K. THORNBER, in: *Linear and Nonlinear Electron Transport in Solids* (J. T. Devreese and V. E. van Doren, eds.), Plenum, New York (1976).
14. K. K. THORNBER, in: *Path Integrals and Their Applications in Quantum, Statistical, and Solid-State Physics* (G. J. Papadopoulos and J. T. Devreese, eds.), Plenum, New York (1978).
15. L. F. LEMMENS and J. T. DEVREESE, *Solid-State Communications* **12**, 1067 (1973).
16. L. F. LEMMENS, J. DE SITTER, and J. T. DEVREESE, *Phys. Rev. B* **8**, 2717 (1973).
17. M. V. FISCHETTI, *Phys. Rev. Lett.* **53**, 1755 (1984).
18. M. V. FISCHETTI and D. J. MARIA, *Phys. Rev. Lett.* **55**, 2475 (1985).

Chapter 4

Quantum Transport in Solids: The Density Matrix

Gerald J. Iafrate

4.1. INTRODUCTION

During the past decade, as microelectronics technology has continued to pursue the scaling down of IC device dimensions into the submicron and ultrasubmicron regions, many new and interesting questions have emerged[1] concerning the solid state dynamics and quantum transport of carriers in semiconductors subjected to rapidly varying, spatially inhomogeneous electric fields, and non-steady-state temporal conditions.

Semiconductor transport in the ultrasubmicron regime approaches the category of quantum transport. This is suggested by the fact that, within the effective-mass approximation, the thermal de Broglie wavelength for electrons in semiconductors is of the order of ultrasubmicron dimensions. Whereas classical transport physics is based on the concept of a probability distribution function that is defined over the phase space of the system, the concept of a phase-space distribution function in the quantum formulation of transport physics is not possible, inasmuch as the noncommutation of the position and momentum operators, the Heisenberg uncertainty principle, precludes the precise specification of a point in phase space. However, within the matrix formulation of quantum mechanics, it is possible to construct a "probability" density matrix, which is often interpreted as the analog of the classical function.

Ever since the early development of quantum theory, there has been considerable interest in deriving the quantum mechanical analog of the

Gerald J. Iafrate • U.S. Army Research Office, Research Triangle Park, North Carolina 27709, USA.

Quantum Transport in Semiconductors, edited by David K. Ferry and Carlo Jacoboni. Plenum Press, New York, 1991.

Boltzmann transport equation (BTE) for electric current flow in solids. Significant progress in this direction was made by Kohn and Luttinger[2] (KL), who considered a system of electrons in the presence of a weak homogeneous electric field interacting with randomly distributed impurities. By linearizing the Liouville equation of motion in the field, they were able to show that, for electrons in a periodic potential, the diagonal elements of the density matrix (i.e., the semiclassical distribution function in the Bloch representation) were much larger than the off-diagonal matrix elements in the weak scattering limit. Using this result, they obtain a closed-form equation for the diagonal elements alone that closely resembles the linearized semiclassical BTE employed in steady-state electric current calculation in solids. The KL theory was subsequently extended to include the possibility of electron-phonon scattering by Argyres.[3] In addition, KL also derive for free electrons the time-dependent (linearized in the field) BTE for both the case when a constant electric field is abruptly turned on, and the case in which the field varies sinusoidally in time. Price[4] has employed the KL method to the quantum mechanical extension of the BTE for electrons moving in a single band interacting with static impurities and electric fields which depended on both wave vector and frequency. In the limit of zero wave vector, he derives a time-dependent BTE which is slightly different from that of KL. In either case, these equations include a memory effect in the scattering term—that is, a collision integral depending on the distribution function at all previous times.

In order to understand nonlinear hot-electron phenomena, it is necessary to obtain a BTE that is valid for electric fields of arbitrary strength. Hasegawa and Yamashita[5] have shown that, for free electrons, if the density matrix is expanded in a power series in the field, the higher-order terms obey equations similar to that obeyed by the linear term in the weak scattering limit, and that if these terms are summed, the resulting equations for the steady-state distribution function is the usual BTE.

Further progress in obtaining a BTE for the time-dependent distribution function for homogeneous time-dependent electric and magnetic fields of arbitrary strength has been made by Levinson,[6] who showed that, for free electrons interacting with long-wavelength phonons, the collision integral included the effect of the fields on the **k** vector of electrons, the so-called intracollisional field (ICFE). Similar results have been derived by Barker[7] for electrons with constant effective mass moving in time-dependent homogeneous electric fields in the presence of arbitrary phonon and impurity scattering by employing resolvent superoperator techniques on the Liouville equation. The extension of Barker's results to the case of electrons moving in a periodic potential and interacting with phonons has been derived by Calecki and Pottier[8] in a one-band approximation employing more elementary considerations. With Levinson, they represent the contribution of the electric field to the Hamiltonian by a vector potential, so that, unlike the case when the electric field is described by a scalar potential, the resulting Hamiltonian (excluding the scattering), including the crystal periodic potential, maintains its periodicity. The instantaneous eigenfunctions of this time-dependent Hamiltonian are then easily shown to be the usual Houston[9] states multiplied by a field-dependent phase factor. Then, using these accelerated Bloch states as a basis, they show that in a one-band approximation, the diagonal matrix elements of the equation for the electron density matrix derived by Levinson to lowest order in the electron-phonon interaction yields a closed-form equation for the one-band distribution function. When this equation is transformed into the Bloch representation, the resulting equation is the one-band generalization of Barker's result for arbitrary energy-band structure and includes electron-phonon matrix elements between Bloch states instead of between plane waves.

In this chapter, quantum transport equations are derived for independent electrons moving in homogeneous electric fields of arbitrary time dependence and arbitrary strength, which include all possible quantum effects in lowest order in the scattering strength—i.e., intraband and interband scattering, interband Zener tunneling,[10] and nonlinear transient transport. In order to obtain a transport equation which correctly includes the possibility of interband tunneling, it is essential that the equations of motion for the density matrix are not linearized in the field. This follows from the fact that the tunneling probability goes as $\sim e^{-b/F}$, where F is the force on the electron due to the field and b depends on energy-band parameters.[9,10] Thus, the tunneling probability goes to zero nonanalytically as F goes to zero and has no power series expansion in the small-F limit. This is the reason why tunneling effects do not appear in the KL derivation, i.e., once the equations of motion are linearized in the field and a power series solution for the density matrix is assumed, the effects of tunneling are automatically eliminated because the tunneling probability is zero to all orders in F if it is expanded about $F = 0$. Thus, even if the series were summed to get the full distribution function, as suggested by Hasegawa and Yamashita,[5] these terms would be absent.

However, the use of a vector potential to describe the effects of the electric field in the Hamiltonian gives rise to instantaneous eigenfunctions, i.e., the accelerated Bloch states, which include the effect of the field to all order in the electron motion if interband tunneling is neglected.[11] Furthermore, if an accelerated Bloch state representation (ABR) is employed, we find[11] that both the tunneling probability to other bands and the ladderlike structure in the optical absorption are obtained without any assumption concerning the existence of the controversial Wannier–Stark energy levels.[12] It thus appears that the use of the ABR not only leads to a simple derivation of the transport equation in the one-band approximation, but also leads to a significant simplification of the theory of interband tunneling and optical absorption, which suggests its application to the derivation of a quantum transport equation (QTE) including interband scattering and tunneling effects.

4.2. ACCELERATED BLOCH REPRESENTATION: QUANTUM TRANSPORT

In this chapter, a novel formalism[13] is discussed for treating Bloch electron dynamics and quantum transport in inhomogeneous electric fields of arbitrary strength and time dependence. In this formalism, the electric field is described through the use of the vector potential. In this regard, this work is an expansion of methodology previously developed by the authors[14-16] to describe solid-state dynamics and quantum transport for Bloch electrons in a homogeneous electric field of arbitrary strength and time dependence, including weak scattering from randomly distributed impurities and phonons.

The Hamiltonian for a single electron in a periodic crystal potential subject to a general inhomogeneous electric field of arbitrary time dependence and strength is

$$H = \frac{1}{2m}\left(\mathbf{P} - \frac{e}{c}\mathbf{A}(\mathbf{x}, t)\right)^2 + V_c(\mathbf{x}) \tag{4.2.1}$$

Here $V_c(\mathbf{x})$ is the periodic crystal potential, and $\mathbf{A}(\mathbf{x}, t)$ is the vector potential for the inhomogeneous field $\mathbf{E}(\mathbf{x}, t)$ where

$$\mathbf{A}(\mathbf{x}, t) = -c\int_{t_0}^{t} \mathbf{E}(\mathbf{x}, t')\, dt'$$

In developing a quantum transport formalism, we start with the Liouville equation for the density matrix

$$i\hbar \frac{\partial \hat{\rho}}{\partial t} = [H, \hat{\rho}] \tag{4.2.2}$$

from which we seek to derive an equation for the distribution function, i.e., the matrix elements of $\hat{\rho}$, $\rho_{m'm}$, in a convenient representation so that we can establish expectation values of any operator B through

$$\bar{B} = \text{trace}(\hat{\rho}\hat{B}) \equiv \sum_{m'}\sum_{m} \rho_{m'm}B_{mm'} \tag{4.2.3}$$

Specifically, we choose $\mathbf{B} \equiv \mathbf{v} = (1/i\hbar)[\mathbf{x}, H]$, the velocity operator, which, for the Hamiltonian of equation (4.2.1) is

$$\mathbf{v} = \frac{1}{m}\left(\mathbf{P} - \frac{e}{c}\mathbf{A}\right) \tag{4.2.4}$$

In previous work,[14-16] when considering Bloch dynamics in a spatially homogeneous electric field, the authors exploited, as a basis set, the instantaneous eigenstates of the Hamiltonian of equation (4.2.1) with $\mathbf{A} \equiv \mathbf{A}_0(t) = -c \int_{t_0}^{t} \mathbf{E}_0(t')\, dt'$; explicitly, these basis states are

$$\phi_{n\mathbf{K}}(\mathbf{r}, t) = \frac{e^{i\mathbf{K}\cdot\mathbf{r}}}{\Omega^{1/2}} U_{n\mathbf{k}(t)}(\mathbf{r}) \tag{4.2.5}$$

Here, $U_{n\mathbf{k}(t)}$ is the periodic part of the usual Bloch function with band index n and wavevector $\mathbf{k}$; $\mathbf{k}(t)$ is determined by

$$\mathbf{k}(t) = \mathbf{K} - \frac{e}{\hbar c} \mathbf{A}_0(t) = \mathbf{K} + \frac{e}{\hbar} \int_{t_0}^{t} \mathbf{E}_0(t')\, dt' \tag{4.2.6}$$

where $\mathbf{K}$ is a constant determined by periodic boundary conditions in a box of volume Ω. (In subsequent equations, the vector nature of $\mathbf{K}$ and $\mathbf{k}$ will be understood but not always indicated.)

In this work, we note that a set of basis functions for Bloch electron dynamics in an inhomogeneous electric field can also be established based on the vector potential choice of gauge. Letting $\mathbf{A}(\mathbf{x}, t) = \mathbf{A}_0(t) + \mathbf{A}_1(\mathbf{x}, t)$ so that

$$\mathbf{E}(\mathbf{x}, t) = -\frac{1}{c}\frac{\partial \mathbf{A}}{\partial t} = \mathbf{E}_0(t) + \mathbf{E}_1(\mathbf{x}, t)$$

a convenient basis for describing transport in this field having a spatially homogeneous and inhomogeneous part is

$$\psi_{n\mathbf{K}} = \exp\left(i\frac{e}{\hbar c} \int_0^{\mathbf{x}} \mathbf{A}_1 \cdot d\mathbf{l} \right) \phi_{n\mathbf{K}} \tag{4.2.7}$$

where $\phi_{n\mathbf{K}}$ is given by equation (4.2.5) and $\mathbf{K}$ is defined through equation (4.2.6). Here, the external magnetic field, defined through $\mathbf{B} = \nabla \times \mathbf{A}$, is zero so that the line integral in equation (4.2.7) is independent of the path. This basis set, $\{\psi_{n\mathbf{K}}\}$, is complete and orthonormal since each element is obtained from the corresponding element of the complete orthonormal set $\{\phi_{n\mathbf{K}}\}$ by the multiplication of a common temporal and spatially dependent phase factor which is independent of n and $\mathbf{K}$.

Taking matrix elements of equation (4.2.2), the Liouville equation in this representation is found to be

$$i\hbar \frac{\partial \rho_{m'm}}{\partial t} = (\varepsilon_{m'} - \varepsilon_m)\rho_{m'm} + \sum_{m''} (S_{m''m}\rho_{m'm''} - S_{m'm''}\rho_{m''m}) \tag{4.2.8}$$

where $m \equiv (n\mathbf{K})$; also $\varepsilon_m \equiv \varepsilon_n(\mathbf{K} - (e/\hbar c)\mathbf{A}_0)$ is the energy band function of the nth band, and $S_{m'm}$ is given by

$$S_{m'm} \equiv S_{n'\mathbf{K}'n\mathbf{K}} = e\mathbf{E}_0(t) \cdot \mathbf{R}_{n'n}(\mathbf{k})\delta_{\mathbf{K}'\mathbf{K}} - eV_{n'\mathbf{K}'n\mathbf{K}} \tag{4.2.9}$$

In equation (4.2.9),

$$\mathbf{R}_{n'n}(\mathbf{k}) = \frac{i}{\Omega} \int d\mathbf{x} \; U_{n'\mathbf{k}}^{*} \nabla_k U_{nk} \tag{4.2.10}$$

and

$$V_{n'\mathbf{K}'n\mathbf{K}} = \int \psi_{n'\mathbf{K}'}^{*} V \psi_{n\mathbf{k}} \, d\mathbf{x} \equiv \int \phi_{n'\mathbf{K}'}^{*} V \phi_{n\mathbf{K}} \, d\mathbf{x} \tag{4.2.11}$$

where $V(\mathbf{x}, t)$ is the potential derived from the inhomogeneous electric field through the relation

$$\mathbf{E}_1(\mathbf{x}, t) = -\nabla V(\mathbf{x}, t) \tag{4.2.12}$$

The Liouville equation given by equation (4.2.8) includes an exact description of the time evolution of the density matrix elements for a Bloch electron in a spatially homogeneous electric field; included explicitly are the effects of Zener tunneling, effective-mass dressing, and scattering from the inhomogeneous field. The authors have solved this equation analytically under the conditions of short times (after turning on the electric field) and weak inhomogeneity.[15–16]

Although the basis states of equation (4.2.7), along with the resultant Liouville equation, equation (4.2.8), provide general applicability for problems involving Bloch dynamics in inhomogeneous electric fields, it is particularly difficult to use in situations where the inhomogeniety is localized and strong such as in the case of impurities, heterojunctions, or band-engineered quantum wells and barriers; this difficulty arises mainly due to the extended nature of the basis states. Therefore, we use the basis defined in equation (4.2.7), along with the equivalent Wannier representation, to develop quantum transport equations for such local inhomogeneities.[17–18]

4.3. DYNAMICAL WANNIER REPRESENTATION: QUANTUM TRANSPORT

In the Wannier representation,[19–20] the Bloch functions are equivalently expressed as a Fourier decomposition of localized functions, the Wannier functions, defined on each periodic lattice site l of the crystal. Wannier functions can also be defined for Bloch electrons in a spatially homogeneous electric field; as such, it can be shown[21] that

$$\phi_{n\mathbf{K}} = \frac{1}{\sqrt{N}} \sum_{\mathbf{l}} e^{i\mathbf{K} \cdot \mathbf{l}} A_n(\mathbf{x} - \mathbf{l}, t) \tag{4.3.1}$$

where the $\mathbf{K}$'s are defined in equation (4.2.6) and N is the number of lattice sites. With the use of the completeness relations

$$\sum_{\mathbf{K}} e^{i\mathbf{K}\cdot(\mathbf{l}-\mathbf{l}')} = N\delta_{\mathbf{l},\mathbf{l}'} \tag{4.3.2a}$$

and

$$\sum_{\mathbf{l}} e^{-i\mathbf{l}\cdot(\mathbf{K}-\mathbf{K}')} = N\delta_{\mathbf{K},\mathbf{K}'} \tag{4.3.2b}$$

equation (4.3.1) can be inverted so that

$$A_n(\mathbf{x} - \mathbf{l}, t) = \frac{1}{\sqrt{N}} \sum_{\mathbf{K}} e^{-i\mathbf{K}\cdot\mathbf{l}} \phi_{n\mathbf{K}} \tag{4.3.3}$$

Since $(\phi_{n'\mathbf{K}'}, \phi_{n\mathbf{K}}) = \delta_{n'n}\delta_{\mathbf{K}'\mathbf{K}}$, it immediately follows from equation (4.3.3) that

$$\int d\mathbf{x}\, A_{n'}^{*}(\mathbf{x} - \mathbf{l}', t)A_n(\mathbf{x} - \mathbf{l}, t) = \delta_{n'n}\delta_{\mathbf{l}'\mathbf{l}}$$

It is useful to note that

$$A_n(\mathbf{x} - \mathbf{l}, t) = \exp\left[i\frac{e}{\hbar c}\mathbf{A}_0 \cdot (\mathbf{x} - \mathbf{l})\right] A_n^0(\mathbf{x} - \mathbf{l})$$

where $A_n^0(\mathbf{x} - \mathbf{l})$ is the usual time-independent Wannier function.

The localized functions in equation (4.3.3) are the instantaneous Wannier functions for Bloch electrons in a spatially homogeneous electric field. As such, it can be shown that

$$H_0 A_n(\mathbf{x} - \mathbf{l}, t) = \sum_{\mathbf{l}'} \varepsilon_n(\mathbf{l} - \mathbf{l}', t)A_n(\mathbf{x} - \mathbf{l}', t) \tag{4.3.4}$$

where

$$\varepsilon_n(\mathbf{l} - \mathbf{l}', t) = \frac{1}{N}\sum_{\mathbf{K}} e^{-i\mathbf{K}\cdot(\mathbf{l}-\mathbf{l}')}\varepsilon_n\left(\mathbf{K} - \frac{e}{\hbar c}\mathbf{A}_0\right) \tag{4.3.5}$$

Here H_0 is the Hamiltonian of equation (4.2.1) with $\mathbf{A} = \mathbf{A}_0(t)$ and $\varepsilon_n(\mathbf{K})$ is the nth Bloch energy band with crystal momentum $\mathbf{K}$. (In subsequent equations, the explicit time dependence will be assumed but not indicated).

A set of instantaneous Wannier functions for the inhomogeneous electric field can also be established. Using the basis set of equation (4.2.7) for the inhomogeneous field and the Wannier expansion of equation (4.3.1), we can identify these localized functions for the inhomogeneous electric field as

$$W_n(\mathbf{x}, \mathbf{l}, t) = \frac{1}{\sqrt{N}}\sum_{\mathbf{K}} e^{-i\mathbf{K}\cdot\mathbf{l}}\psi_{n\mathbf{k}} \equiv \exp\left(i\frac{e}{\hbar c}\int_0^{\mathbf{x}} \mathbf{A}_1 \cdot d\mathbf{l}\right)A_n(\mathbf{x} - \mathbf{l}, t) \tag{4.3.6}$$

where

$$\int d\mathbf{x}\, W_n^{*\prime}(\mathbf{x}, \mathbf{l}')\, W_n(\mathbf{x}, \mathbf{l}) = \delta_{n'n}\delta_{\mathbf{l}',\mathbf{l}} \tag{4.3.7}$$

Taking the matrix elements of equation (4.2.2), the Liouville equation in the localized Wannier representation is

$$i\hbar\frac{\partial \rho_{n'n}(l'l, t)}{\partial t} = \sum_{n''l''} \{[\varepsilon_{n''}(l' - l'')\delta_{n'n''} - e\mathbf{E}_0\cdot\mathbf{\Delta}_{n'n''}(l' - l'')$$

$$+ eV_{n'n''}(l', l'')]\rho_{nn''}(l'', l, t)$$

$$- [\varepsilon_{n''}(l'' - l)\delta_{nn''} - e\mathbf{E}_0\cdot\mathbf{\Delta}_{n''n}(l'' - l)$$

$$+ eV_{n''n}(l'', l)]\rho_{n'n''}(l', l'', t)\} \tag{4.3.8}$$

where $\varepsilon_n(l - l')$ is defined in equation (4.3.5); also

$$\mathbf{\Delta}_{nn'}(\mathbf{l} - \mathbf{l}') = \frac{1}{N}\sum_\mathbf{K} e^{-i\mathbf{K}\cdot(\mathbf{l}-\mathbf{l}')}\mathbf{R}_{nn'}\left(\mathbf{K} - \frac{e}{\hbar c}\mathbf{A}_0\right) \tag{4.3.9}$$

where $\mathbf{R}_{nn'}$ is defined in equation (4.2.10) and

$$V_{n'n}(l'l) = \int d\mathbf{x}\, W_{n'}^*(\mathbf{x}, \mathbf{l}')\, V(\mathbf{x}, t)\, W_n(\mathbf{x}, \mathbf{l})$$

$$\equiv \int d\mathbf{x}\, A_{n'}^*(\mathbf{x} - \mathbf{l}')\, V(\mathbf{x}, t)\, A_n(\mathbf{x} - \mathbf{l}) \tag{4.3.10}$$

the matrix elements of the inhomogeneous potential with respect to the localized basis.

The solution to equation (4.3.8) is amenable to separation of variables[†] in the pure-state situation; seeking solutions to equation (4.3.8) of the form

$$\rho_{n'n}(l', l, t) \equiv f_n^*(l, t)f_{n'}(l', t) \tag{4.3.11}$$

we find that $f_n(l, t)$ obeys the equation

$$i\hbar\frac{\partial f_n(l, t)}{\partial t} = \sum_{n''}\sum_{l''}[\varepsilon n''(l'' - l, t)\delta_{nn''} - e\mathbf{E}_0\cdot\mathbf{\Delta}_{nn''}(l'' - l, t)$$

$$+ eV_{nn''}(l, l'', t)]f_{n''}(l'', t) \tag{4.3.12}$$

Using the well-known Wannier theorem,[22]

$$\sum_{l''} G_{n''}(l'' - l)f_{n''}(l'', t) \equiv G_{n''}\left(-i\mathbf{\nabla} - \frac{e}{\hbar c}\mathbf{A}_0\right)f_{n''}(\mathbf{r})|_{\mathbf{r}=\mathbf{l}} \tag{4.3.13}$$

[†]In the case where the initial state of the density matrix is not separable, as in mixed state analyses, such a separation may not be very useful.

where G is assumed to be an analytic function of $\mathbf{K}$ [$\mathbf{K}$ defined by equation (4.2.6)], equation (4.3.12) reduces to the differential equation

$$i\hbar \frac{\partial f_n(\mathbf{r}, t)}{\partial t} = \varepsilon_n\left(-i\nabla - \frac{e}{\hbar c}\mathbf{A}_0\right)f_n(\mathbf{r}, t)$$

$$- e\mathbf{E}_0 \cdot \sum_{n''} \mathbf{R}_{nn''}\left(-i\nabla - \frac{e}{\hbar c}\mathbf{A}_0\right)f_{n''}(\mathbf{r}, t)$$

$$+ \sum_{n''}\sum_{l''} eV_{nn''}(\mathbf{l''}, \mathbf{r})f_{n''}(\mathbf{l''}, t) \qquad (4.3.14)$$

We point out that equations (4.3.8), (4.3.12) are identical to the result one would obtain by describing the *inhomogeneous* field by the scalar potential from the outset, and using, as a basis, the dynamical Wannier functions of equation (4.3.3) corresponding to the homogeneous field alone.

4.4. DISCUSSION AND SUMMARY

For a general nonparabolic band structure and an inhomogeneous electric field, equation (4.3.14) represents a set of coupled differential equations for the $\{f_n\}$. We have studied the solutions to equation (4.3.14) extensively and will report[21] the detailed analysis elsewhere. However, to show the utility of the method in this Chapter, we look at specific solutions to equation (4.3.14) under assumption that interband mixing from both the homogeneous and inhomogeneous electric fields are negligible; then equation (4.3.14) reduces to the single-band equation

$$i\hbar \frac{\partial f_n(\mathbf{r}, t)}{\partial t} = \varepsilon_n\left(-i\nabla - \frac{e}{\hbar c}\mathbf{A}_0\right)f_n(\mathbf{r}, t) + \sum_{l''} eV_{nn}(\mathbf{l''}, \mathbf{r})f_n(\mathbf{l''}, t) \quad (4.4.1)$$

Taking a trivial example to show the nature of the solution to equation (4.4.1), consider $V_{nn}(\mathbf{l''}, \mathbf{r}) = 0$ for all $\mathbf{l}$; then

$$f_n(\mathbf{r}, t) = \exp\left[-\frac{i}{\hbar}\int_0^t \varepsilon_n\left(\mathbf{K} - \frac{e}{\hbar c}\mathbf{A}_0\right) dt'\right]f_n(\mathbf{r}, 0) \qquad (4.4.2a)$$

where

$$f_n(\mathbf{r}, 0) = \frac{1}{\sqrt{N}} e^{i\mathbf{K}\cdot\mathbf{r}} \qquad (4.4.2b)$$

In a more significant application, if we consider the potential assumed by Slater and Koster[17] to describe a single impurity at the lattice site $\mathbf{l}_0$, namely,

$$V_{n''n}(\mathbf{l''}, \mathbf{r}) = V_0\delta_{\mathbf{l''},\mathbf{l}_0}\delta_{\mathbf{r},\mathbf{l}_0}\delta_{n''n}$$

then

$$f_n(\mathbf{r}, t) = \sum_{\mathbf{K}} \exp\left[-\frac{i}{\hbar} \int_0^t \varepsilon_n\left(\mathbf{K} - \frac{e}{\hbar c}\mathbf{A}_0\right) dt' \right] e^{i\mathbf{K}\cdot(\mathbf{r}-\mathbf{l}_0)} A_n(\mathbf{K}, t) \quad (4.4.3a)$$

where $A_n(\mathbf{K}, t)$ satisfies

$$i\hbar \frac{\partial A_n(\mathbf{K}, t)}{\partial t} = \frac{eV_0}{N} \sum_{\mathbf{K}'} A_n(\mathbf{K}', t) \exp\left\{ \frac{i}{\hbar} \int_0^t \left[\varepsilon_n\left(\mathbf{K} - \frac{e}{\hbar c}\mathbf{A}_0\right) \right.\right.$$
$$\left.\left. - \varepsilon_n\left(\mathbf{K}' - \frac{e}{\hbar c}\mathbf{A}_0\right) \right] dt' \right\} \quad (4.4.3b)$$

In situations where the inhomogeneity is a positive or negative constant value over several contiguous lattice sites, we have the possibility of tunneling or scattering, depending upon the magnitude of the inhomogeneity relative to the incident energy of the particle; in this case[21] $f_n(\mathbf{r}, t)$ will consist of the appropriate linear combination of plane waves or exponential functions.

Once $f_n(\mathbf{r}, t)$ has been determined, the density matrix elements are then calculated from equation (4.3.8) and the current density can be calculated with the use of equations (4.2.3), (4.2.4), and (4.3.11). Explicitly, using $W_n(\mathbf{x}, \mathbf{l})$ from equation (4.3.6), it can be shown that the matrix elements of the velocity defined in equation (4.2.4) are

$$(\mathbf{v})_{m'm} \equiv v_{n'n}(\mathbf{l}', \mathbf{l}) = \frac{1}{m_0}\left\{ \int A_{n'}^*(\mathbf{x} - \mathbf{l}')\mathbf{P}A_n(\mathbf{x} - \mathbf{l}) \, d\mathbf{x} - \frac{e}{c}\mathbf{A}_0\delta_{n'n}\delta_{\mathbf{l}'\mathbf{l}} \right\} \quad (4.4.5)$$

where $A_n(\mathbf{x}, \mathbf{l})$ is defined in equation (4.3.3) [the instantaneous Wannier functions for the homogeneous field $E_0(t)$], and

$$\int A_{n'}^*(\mathbf{x} - \mathbf{l}) \frac{\mathbf{P}}{m_0} A_n(\mathbf{x} - \mathbf{l}) \, dx = \frac{1}{N} \sum_{\mathbf{K}} e^{-i\mathbf{K}\cdot(\mathbf{l}-\mathbf{l}')}\{\mathbf{V}_n(\mathbf{K}, t)\delta_{n'n} + \mathbf{g}_{n'n}(\mathbf{K}, t)\}$$

with

$$\mathbf{V}_n(\mathbf{K}) = \frac{1}{\hbar}\mathbf{\nabla}_{\mathbf{k}}\varepsilon_n\left(\mathbf{K} - \frac{e}{\hbar c}\mathbf{A}_0\right) \quad (4.4.6)$$

and

$$\mathbf{g}_{n'n}(\mathbf{K}) = \frac{i}{\hbar}\left[\varepsilon_{n'}\left(\mathbf{K} - \frac{e}{\hbar c}\mathbf{A}_0\right) - \varepsilon_n\left(\mathbf{K} - \frac{e}{\hbar c}\mathbf{A}_0\right) \right]\mathbf{R}_{n'n}\left(\mathbf{K} - \frac{e}{\hbar c}\mathbf{A}_0\right) \quad (4.4.7)$$

Here, $\mathbf{R}_{n'n}$ is defined in equation (4.2.10). It then follows from equations (4.2.3), (4.3.11), and (4.4.5) that

$$\bar{\mathbf{v}} = \frac{1}{m_0} \sum_{\mathbf{K}}\left[\sum_n |\phi_n(\mathbf{K})|^2 \mathbf{V}_n(\mathbf{K}) + \sum_{n'}\sum_n \phi_{n'}^*(\mathbf{K})\phi_n(\mathbf{K})\mathbf{g}_{n'n}(\mathbf{K}, t) \right] \quad (4.4.8)$$

where

$$\phi_n(\mathbf{K}) = \frac{1}{\sqrt{N}} \sum_{\mathbf{l}} f_n(\mathbf{l}) e^{-i\mathbf{K}\cdot\mathbf{l}} \tag{4.4.9}$$

with $\mathbf{V}$ and $\mathbf{g}_{n'n}$ defined in equations (4.4.6) and (4.4.7), respectively.

In this formulation of quantum transport based on dynamical Wannier functions, the terms $\varepsilon_n(\mathbf{l} - \mathbf{l}', t)$ and $\mathbf{\Delta}_{nn'}(\mathbf{l} - \mathbf{l}', t)$ defined in equations (4.3.5) and (4.3.9), respectively, appear in the fundamental density matrix equation of equation (4.3.8). Upon substitution of variables $\boldsymbol{\xi} = \mathbf{K} - (e/\hbar c)\mathbf{A}_0$ and subsequent sum over $\boldsymbol{\xi}$, it follows that these terms can be reexpressed as

$$\varepsilon_n(\mathbf{l} - \mathbf{l}', t) = \exp\left[i\frac{e\mathbf{E}_0}{\hbar} \cdot (\mathbf{l} - \mathbf{l}')t \right] \varepsilon_n(\mathbf{l} - \mathbf{l}', 0) \tag{4.4.10}$$

with

$$\varepsilon_n(\mathbf{l} - \mathbf{l}', 0) = \frac{1}{N} \sum_{\mathbf{K}} e^{-i\mathbf{K}\cdot(\mathbf{l}-\mathbf{l}')} \varepsilon_n(\mathbf{K}) \tag{4.4.11}$$

and

$$\mathbf{\Delta}_{nn'}(\mathbf{l} - \mathbf{l}', t) = \exp\left[i\frac{e\mathbf{E}_0}{\hbar} \cdot (\mathbf{l} - \mathbf{l}')t \right] \mathbf{\Delta}_{nn'}(\mathbf{l} - \mathbf{l}', 0) \tag{4.4.12}$$

with

$$\mathbf{\Delta}_{nn'}(\mathbf{l} - \mathbf{l}', 0) = \frac{1}{N} \sum_{\mathbf{K}} e^{-i\mathbf{K}\cdot(\mathbf{l}-\mathbf{l}')} \mathbf{R}_{nn'}(\mathbf{K}) \tag{4.4.13}$$

As seen from equations (4.4.10) and (4.4.12) for strong electric fields, $\varepsilon_n(\mathbf{l} - \mathbf{l}', t)$ and $\mathbf{\Delta}_{nn'}(\mathbf{l} - \mathbf{l}', t)$ oscillate rapidly with time; thus, only terms with $\mathbf{l} = \mathbf{l}'$ contribute to the sum over $\mathbf{l}'$ in equations (4.3.8) and (4.3.12) thereby leading to cell localization and hopping from cell to cell through Stark ladder transitions.

The manifestation of lattice site localization and Stark ladder hopping is evident by considering the time evolution of an electron initially in the energy-band state n_0 at lattice position $\mathbf{l}_0$, and with no inhomogeneous field present; then the solution to equation (4.3.12) (the appropriate equation for pure-state initial conditions), in the early-time approximation, is

$$f_n(\mathbf{l}, t) = \delta_{n,n_0}\delta_{\mathbf{l},\mathbf{l}_0} + \frac{F_{nn_0}(\mathbf{l} - \mathbf{l}_0)}{e\mathbf{E}_0 \cdot (\mathbf{l} - \mathbf{l}_0)}\left[1 - \exp\left\{ i\frac{e}{\hbar}\mathbf{E}_0 \cdot (\mathbf{l}_0 - \mathbf{l})t \right\} \right]$$

where

$$F_{nn_0}(\mathbf{l} - \mathbf{l}_0) = \varepsilon_{n_0}(\mathbf{l}_0 - \mathbf{l}, 0)\delta_{nn_0} - e\mathbf{E}_0 \cdot \mathbf{\Delta}_{nn_0}(\mathbf{l}_0 - \mathbf{l}, 0)$$

with $\varepsilon_{n_0}(l_0 - l, 0)$ and $\Delta_{nn_0}(l_0 - l, 0)$ given by equations (4.4.11) and (4.4.13) respectively. In this case, the density matrix in equation (4.3.11) can be derived; for the diagonal elements of the density matrix, we find that

$$\rho_{nn}(l, l) = \delta_{n,n_0}[\delta_{l,l_0} + |\varepsilon_{n_0}(l_0 - l, 0)|^2 \alpha^2(t)] + e^2 E_0^2 |\Delta_{nn_0}(l_0 - l, 0)|^2 \alpha^2(t)$$

where

$$\alpha(t) = \frac{\sin\{e\mathbf{E}_0 \cdot (l_0 - l)t/2\}}{e\mathbf{E}_0 \cdot (l_0 - l)}$$

The first term describes hopping from the initial site l_0 to a subsequent site "l" within the initial energy band n_0, whereas the second term describes hopping from site l_0 to site l accompanied by Zener tunneling from the initial energy band n_0 to the energy band n. As well, for large electric fields, the oscillatory term $\alpha(t)$ will lead to vanishing contributions to the density matrix elements unless $l = l_0$, which results in localization at the initial site.

A similar analysis can be performed with the inhomogeneous field included; a detailed discussion of this case will be considered elsewhere.[21]

The time-independent contributions to $\varepsilon_n(l - l', t)$ and $\Delta_{nn'}(l - l', t)$, as defined in equations (4.4.11) and (4.4.13) are band-structure-dependent quantities dependent upon the specific energy bands of the periodic potential in question as well as the related interband coupling, respectively. For a given energy band $\varepsilon_n(\mathbf{K})$, $\varepsilon_n(l - l', 0)$, defined in equation (4.4.11), is the Fourier transform of $\varepsilon_n(\mathbf{K})$ with respect to the lattice sites of the crystal potential. As an example, using the familiar nearest-neighbor tight-binding approximation

$$\varepsilon_n(\mathbf{K}) = \varepsilon_n(0) + W \sin^2 \frac{\mathbf{K} \cdot \mathbf{a}}{2} \tag{4.4.14a}$$

with a denoting as the basic lattice vector in a specific crystallographic direction and with $\varepsilon_n(0)$ and W denoted as the energy-band minimum and width, respectively, $\varepsilon_n(l - l', 0)$ in equation (4.4.11) becomes

$$\varepsilon_n(l - l', 0) = [\varepsilon_n(0) + W]\delta_{ll'} - \tfrac{1}{4} W[\delta_{l',l+a} + \delta_{l',l-a}] \tag{4.4.14b}$$

which includes the salient features of the specific band description.

As well, an explicit form for $\Delta_{nn'}(l - l', 0)$ can be obtained in terms of band-structure parameters by employing the two-band model first developed by Kane[23] and later extended by Krieger.[24] Specifically, Kane shows that an explicit form for $R_{nn'}(k)$, defined in equation (4.2.10), can be written, for a specific direction, as

$$R_{nn'}(\mathbf{K}) = \tfrac{1}{2} q_{nn'}/(K^2 + q_{nn'}^2) \tag{4.4.15}$$

with $q_{nn'} = \frac{1}{2}E_g / P_{nn'}(0)$; here E_g is the energy gap, and $P_{nn'}(0)$ is the momentum matrix element, between the nth and n'th bands at the assumed band extrema, $\mathbf{K} = 0$. Utilizing $\mathbf{R}_{nn'}(\mathbf{K})$ in equation (4.4.13) results in

$$\Delta_{nn'}(\mathbf{l} - \mathbf{l'}, 0) = \frac{2\Pi^2}{N} q_{nn'} e^{-q_{nn'}|\mathbf{l}-\mathbf{l'}|} \tag{4.4.16}$$

in the direction of the electric field; $\Delta_{nn'}(\mathbf{l} - \mathbf{l'}, 0)$ depends explicitly on the interband parameter $q_{nn'}$ and decreases exponentially with energy gap as $\mathbf{l} - \mathbf{l'}$ increases, a result which leads to the one-band approximation in the large energy-gap limit.

Although we have reported a method for describing quantum transport in the presence of a local but strong inhomogeneous electric field, a more detailed illustration of the power of this method, which not only incorporates the ability to handle tunneling in the presence of a homogeneous electric field, but, along with tunneling, also allows for the simultaneous possibility of including weak scattering from a random distribution of defects or phonons during the tunneling process, will be published shortly.[21]

ACKNOWLEDGMENT

The author acknowledges the close collaboration with J. B. Krieger; he also acknowledges the invaluable assistance from C. S. Kavina for the preparation of the manuscript.

REFERENCES

1. G. J. IAFRATE, in: *Gallium Arsenide Technology* (D. K. Ferry, ed.), Ch. 12, Sams, Indianapolis (1985).
2. W. KOHN and J. M. LUTTINGER, *Phys. Rev.* **108**, 590 (1957).
3. P. N. ARGYRES, *J. Phys. Chem. Solids* **19**, 66 (1961).
4. P. J. PRICE, *IBM J. Res. Dev.* **10**, 395 (1966).
5. A. HASEGAWA and J. YAMASHITA, *J. Phys. Chem. Solids* **23**, 875 (1962).
6. I. B. LEVINSON, *Zh. Eksp. Teor. Fiz.* **30**, 660 (1969) [*Sov. Phys. JETP* **30**, 362 (1970)].
7. J. R. BARKER, *J. Phys. C* **6**, 2663 (1973); in: *Physics of Non-Linear Transport in Semiconductors*, Plenum, New York (1979).
8. D. CALECKI and N. POTTIER, *J. Phys. (Paris) Colloq.* **42**, C7-271 (1981). N. POTTIER and D. CALECKI, *Physica* **110A**, 471 (1982).
9. W. V. HOUSTON, *Phys. Rev.* **57**, 184 (1940).
10. C. ZENER, *Proc. R. Soc. London* **145**, 523 (1934).
11. J. B. KRIEGER and G. J. IAFRATE, *Physica* **134B**, 228 (1985); *Phys. Rev. B* **33**, 5494 (1986).
12. G. H. WANNIER, *Rev. Mod. Phys.* **34**, 645 (1962); J. ZAK, *Phys. Rev. Lett.* **20**, 1477 (1968); G. H. WANNIER, *Phys. Rev.* **181**, 1364 (1969); J. ZAK, *Phys. Rev.* **181**, 1366 (1969).
13. G. J. IAFRATE and J. B. KRIEGER, *Bull. Am. Phys. Soc.* **S13-3**, 814 (1988).
14. J. B. KRIEGER and G. J. IAFRATE, *Phys. Rev. B* **33**, 5494 (1986).

15. J. B. KRIEGER and G. J. IAFRATE, *Phys. Rev. B* **35**, 9644 (1987).
16. G. J. IAFRATE and J. B. KRIEGER, *Solid-State Electronics* **31**, 517 (1988).
17. G. F. KOSTER and J. C. SLATER, *Phys. Rev.* **95**, 1167 (1954).
18. J. G. GAY and J. R. SMITH, *Phys. Rev. B* **11**, 4906 (1975).
19. G. H. WANNIER, *Phys. Rev.* **52**, 191 (1937).
20. J. M. ZIMAN, *Principles of the Theory of Solids*, Cambridge University Press, Cambridge (1964).
21. G. J. IAFRATE and J. B. KRIEGER, to be published.
22. P. T. LANDSBERG, *Solid State Theory—Methods and Applications*, Wiley-Interscience, New York (1969).
23. E. O. KANE, *J. Phys. Chem. Solids* **12**, 181 (1959).
24. J. B. KRIEGER, *Phys. Rev.* **156**, 776 (1967).

Chapter 5

The Quantum Hall and Fractional Quantum Hall Effects

N. d'Ambrumenil

5.1. INTRODUCTION

The range of phenomena observed in electronic systems in magnetic fields is large and spectacular. Perhaps most spectacular of all are the quantum Hall effect (QHE) and fractional quantum Hall effect. This chapter is concerned with the theoretical description of these phenomena.

The plateaux observed at low temperatures in the Hall resistance of semiconductor heterostructures at multiples $1/i$ of h/e^2, with i integer,[1] and the related drastic reduction in the longitudinal resistance are known as the quantum Hall effect. The QHE is a macroscopic quantum phenomenon. The QHE can be understood in terms of the energy levels for a single electron or quasiparticle in the presence of a magnetic field, host lattice and electric field, and the interplay between them.

Plateaux are also observed in the Hall resistance of some semiconductor heterostructures at rational multiples m of h/e^2.[2] These and the related anomalies in the longitudinal resistance are known as the fractional quantum Hall effect. Again this is a macroscopic quantum phenomenum, but, unlike the QHE, it cannot be understood in terms of a modified one-electron picture.

The observation of the fractional quantum Hall effect implied an unsuspected role for many-body effects and at first appeared mysterious. The initial mystery has now been more or less resolved, although the theory

N. d'Ambrumenil • Department of Physics, University of Warwick, Coventry CV4 7AL, United Kingdom.

Quantum Transport in Semiconductors, edited by David K. Ferry and Carlo Jacoboni. Plenum Press, New York, 1991.

of the effect is certainly not complete. The most widely accepted explanation invokes the existence of a correlated ground state, probably a fluid, which is preferentially stabilized close to certain magnetic field strengths.

The quantum Hall effect and the fractional quantum Hall effect are presented in the next two sections. Standard results for a single electron in a magnetic field are included in Section 5.4.

5.2. THE QUANTUM HALL EFFECT

The quantum Hall effect, discovered by von Klitzing *et al.*,[1] is now reasonably well understood as a macroscopic phenomenon. This section reviews the basis for our understanding of the QHE.

5.2.1. The Measurement

Plateaux in the Hall resistance of suitably doped inversion layers and heterostructures are observed at low temperatures when plotted as a function either of applied magnetic field or of particle density (Figure 5.1). At the same time the longitudinal resistivity appears to vanish, dropping to values seven orders of magnitude or more smaller than at magnetic fields at which there are no plateaux.

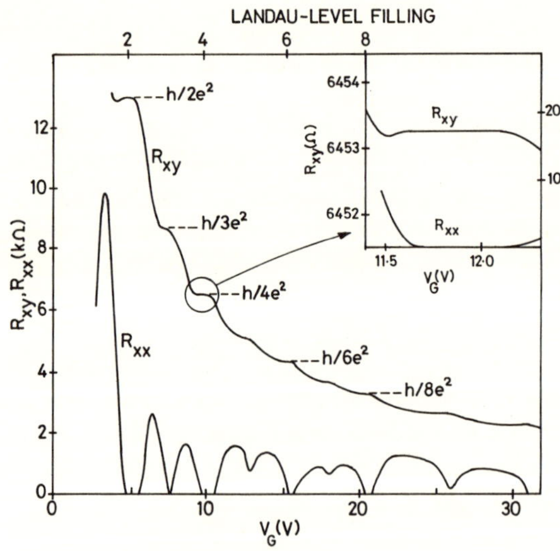

FIGURE 5.1. The Hall resistivity ρ_{xy} and longitudinal resistivity ρ_{xx} of a Si-MOSFET as a function of applied gate voltage V_g. V_g controls the number density of electrons in the quasi-two-dimensional electron gas in the channel region of the transistor (after von Klitzing[5]).

The experiments are usually performed in a geometry similar to the one shown in Figure 5.2. I will use the orientation of the x, y, and z axes shown in Figure 5.2.

5.2.2. Interpretation of the Measurement

Not surprisingly the observation of plateaux in the Hall resistance of a system in a magnetic field says something fundamental about that system. Quite what was first stated by Laughlin.[3]

Laughlin deduced from the implications of a Gedankenexperiment, outlined below, that the Hall response σ_{xy} of a system of charged particles must be quantized in units of e^2/h if the Fermi energy lies in a mobility gap. Alternatively one may state that the observation of a quantized Hall resistance implies the existence of a mobility gap. By itself the original argument of Laughlin does not give the level quantization; i.e., it does not give the value of n where

$$\sigma_{xy} = ne^2/h \qquad (5.2.1)$$

This quite general result coupled with hindsight explains why some approximate treatments like that of Ando *et al.*[4] appeared to anticipate the discovery of the QHE. Irrespective of the details of any calculation or approximation scheme, the response to crossed electric and magnetic fields, the Hall conductivity σ_{xy}, can only take values consistent with equation (5.2.1) when the Fermi energy is in a mobility gap.

5.2.3. Laughlin's Gedankenexperiment

Suppose a system of electrons is confined on the surface of a cylinder with a constant magnetic field B perpendicular to the cylinder surface at all points (Figure 5.3). Assume also that the total flux threading the circuit

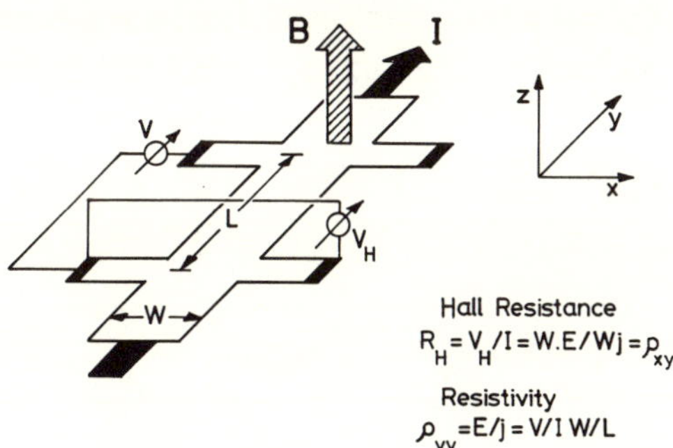

FIGURE 5.2. A standard sample configuration for the Hall measurement (after Störmer and Tsui[6]).

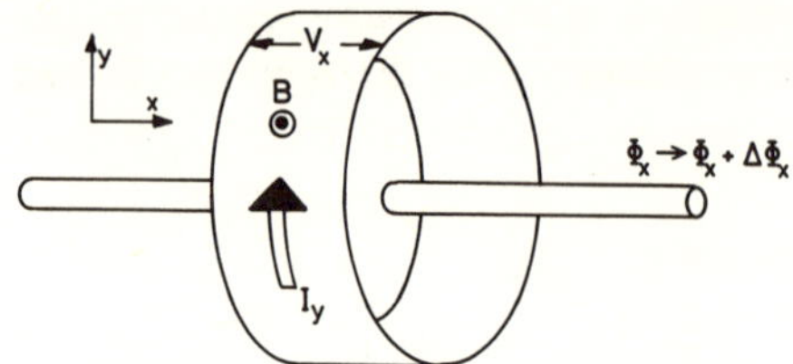

FIGURE 5.3. Laughlin's[3] Gedankenexperiment. With the Fermi energy in a mobility gap, the voltage V_x induces a current I_y to flow around the ring. By considering the gauge change $\Phi_x \to \Phi_x + \Delta\Phi_x$, one can deduce that V_x/I_y takes only quantized values.

of the cylinder is Φ_x, but that none of the flux lines of Φ_x come close to the surface of the cylinder. Let the application of the transverse voltage V_x induce a current to flow around the ring as shown.

Imagine causing Φ_x to be changed adiabatically by an amount $\Delta\Phi_x = h/e$. The change in Φ_x implies a change in the vector potential for the electrons on the cylinder. The change in vector potential,

$$\mathbf{A} \to \mathbf{A} - \frac{h}{e}\frac{\hat{\mathbf{y}}}{L}\ \left(=\mathbf{A}+\frac{\hbar}{e}\nabla\theta\right) \tag{5.2.2}$$

corresponds to the gauge change [see Section 5.4.2 $(q = -e)$]

$$\psi \to \psi \exp(-i\theta(x,y)) = \exp(-2\pi i y/L)\psi \tag{5.2.3}$$

But the gauge change accompanying the change, $\Phi_x \to \Phi_x + \Delta\Phi_x$, does not change the energy eigenvalues of H_0, so the physical properties of the system must be the same after the gauge change as they were before. When the Fermi energy of the system lies within a band of localized states, the occupied bands of delocalized states before the gauge change must be occupied afterwards. The occupation numbers of the localized states certainly do not change. The only possible change is in the occupation numbers of the edge states; i.e., particles can be transferred from one edge to the other.

Suppose n particles with charge q are transferred from one edge to the other. Then the potential energy of the system is increased by ΔU,

$$\Delta U = nqV_x \tag{5.2.4}$$

as a result of the gauge change. As the change was effected adiabatically, this energy must compensate exactly the work done by the system, which by Faraday's law is ΔW, with

$$\Delta W = -\int dt\, I_y \frac{d\Phi}{dt} = -I_y\frac{h}{e} \tag{5.2.5}$$

So

$$0 = \Delta U + \Delta W = nqV_x - I_y\frac{h}{e}$$

or

$$\frac{I_y}{V_x} = -\sigma_{xy} = n\frac{qe}{h} \tag{5.2.6}$$

Equation (5.2.6) is Laughlin's result. For electrons or holes, $|q| = e$, the Hall response $|\sigma_{xy}| = ne^2/h$, with n integer. This quantization of the Hall resistance followed directly from the assumption that the Fermi energy was within some mobility gap.

5.2.4. Aspects of a Microscopic Theory of the Quantum Hall Effect

The motion of electrons or holes in the inversion layers or at the interface in a semiconductor heterostructure is essentially two-dimensional because the motion perpendicular to the interface is quantized into discrete levels. The behavior of the system is then determined by the motion in the plane of the interface. One may write an effective Hamiltonian H_0 describing the motion of the particles in the plane of the interface with coordinates x and y; see Section 5.4.1.

The energy spectrum for these two-dimensional particles in a perpendicular magnetic field B, in the absence of impurities or an electric field, is just that of the Landau levels:

$$\varepsilon_n = \hbar\omega_c(n + \tfrac{1}{2}) \tag{5.2.7}$$

In each Landau level there are ρ_0 states per unit area with (see Section 5.4.1

$$\rho_0 = Be/h \tag{5.2.8}$$

If one were to plot the density of states in energy, one would find a series of delta functions at energies ε_n.

In the presence of an electric field E parallel, say, to the x-direction the energies form bands centered on the energies ε_n (Section 5.4.4) with energies

$$\varepsilon_{nk} = \varepsilon_n - kv \tag{5.2.9}$$

where $v = eEl_0/\hbar\omega_c$. The delta functions in the density of states are replaced by these bands.

If there are also impurities in the system, these may give rise to energy shifts for those cyclotron orbits which remain delocalized. They may also give rise to some localized states centered on the individual scattering centers. These localized states will have been split off from the continua of delocalized states. The density of states could then be expected to look like Figure 5.4.

The localized states will not contribute to any net current in the system, so that between the lower end of one continuum and the next there will be a mobility gap. If the Fermi energy is in one of these mobility gaps, then Laughlin's argument tells us that σ_{xy} $(= I_y/V_x)$ is ne^2/h. This is independent of the number of localized states.

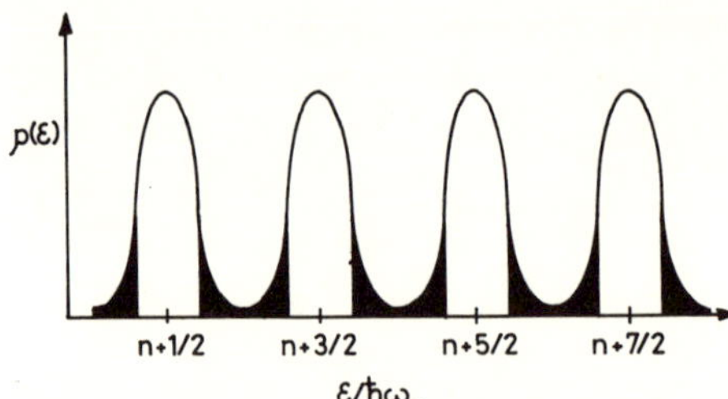

FIGURE 5.4. Possible density of electronic states in a magnetic field of a quasi-two-dimensional electron gas containing impurities. The shaded regions indicate bands of localized states.

5.2.4.1. Contribution of the Delocalized States

The contributions to the current from the individual delocalized states can be calculated as follows. The argument is adapted from Prange's original argument.[7,8,9]

For a system of electrons write the Hamiltonian H_0 in the Landau gauge including an additional gauge parameter a and an impurity scattering potential V_{sc} as

$$H_0(a) = \frac{p_x^2 + [p_y + eB(x - a)]^2}{2m^*} + eEx + V_{sc}(x, y) \qquad (5.2.10)$$

Equation (5.2.10) is as equation (5.4.41) except for the parameter a. A change in the parameter a corresponds to a simple gauge transformation (see section 5.4.2 and remember $q = -e$),

$$a \to a + \Delta a$$

$$\mathbf{A} \to \mathbf{A} - B\,\Delta a\,\hat{\mathbf{y}}$$

$$\psi \to \exp\left(+i\frac{a}{l_0}y'\right)\psi \qquad (5.2.11)$$

with $y' = y/l_0$.

The eigenstates of $H_0(a)$ in the absence of V_{sc} may be written as

$$\psi_{nk} \sim e^{i(k-v)y'}H_n(x' + k - \eta)e^{-(x'+k-\eta)^2/2} \qquad (5.2.12)$$

where $x' = x/l_0$, $y' = y/l_0$, $v = eEl_0/\hbar\omega_c$, $\eta = a/l_0$, and H_n are the Hermite polynomials. The energies are

$$\varepsilon_{nk} = \hbar\omega_c(n + \tfrac{1}{2}) - eEl_0k + eEa \qquad (5.2.13)$$

These energies form quasicontinuous bands centered on the cyclotron energies $\hbar\omega_c(n + \tfrac{1}{2})$ and bounded by

$$\varepsilon^l < \varepsilon_{nk} < \varepsilon^u \qquad (5.2.14)$$

where, if the system is assumed bounded at $x = \pm x_0$,

$$\varepsilon^l = \hbar\omega_c(n + \tfrac{1}{2}) - eEx_0$$
$$\varepsilon^u = \hbar\omega_c(n + \tfrac{1}{2}) + eEx_0 \qquad (5.2.15)$$

Direct differentiation of equation (5.2.9) gives

$$\frac{\partial H_0}{\partial a} = -\frac{p_y + eB(x-a)}{m^*}\, eB = Bj_y \tag{5.2.16}$$

an operator equation for the current j_y.

Now the Feynman–Hellmann theorem states that for some eigenstate of $H_0(a)$ characterized by quantum numbers α and with energy ε_α,

$$\langle \psi_\alpha(a)|\frac{\partial H_0}{\partial a}|\psi_\alpha(a)\rangle = \frac{\partial \varepsilon_\alpha}{\partial a} \tag{5.2.17}$$

This gives

$$j_y^\alpha = \frac{1}{B}\frac{\partial \varepsilon_\alpha}{\partial a} \tag{5.2.18}$$

for the current carried by the state characterized by α.

For the eigenstates of the Hamiltonian H_0 with $V_{sc} = 0$, we have that the states $\psi_\alpha = \psi_{nk}$. Inserting equation (5.2.13) into equation (5.2.17) gives

$$j_y^\alpha = j_y^0 = eE/B \tag{5.2.19}$$

Each state makes the same contribution to the current.

The total current j_y is the sum of the contributions from all occupied states. In each Landau level there are $N = B/(h/e)$ states per unit area, so that with exactly n Landau levels full,

$$j_y = nNj_y^0 = n\frac{e^2}{h}E \tag{5.2.20}$$

in agreement with equation (5.2.6).

For $V_{sc} \neq 0$ one can treat the problem as one in scattering theory. The effect of impurity scattering on the electronic states ψ_{nk} is severely restricted in the presence of the magnetic and electric fields because the states ψ_{nk} are nondegenerate. The usual degeneracies associated with time reversal and rotational symmetries have been lifted. Reflection of a particle in state ψ_{nk} is therefore impossible.

The only possibility for an electron in a delocalized state, with energy in one of the continua of equation (5.2.14) is to be scattered into itself. Its asymptotic form can therefore differ from its form in the absence of the scattering potential by at most a phase factor. At energies outside of these continua there may be localized states, cf. Figure 5.4.

Suppose the system is bounded somewhere in the y-direction far from the region in which V_{sc} is nonzero, say at $y = \pm L$. We may then write the asymptotic form for an eigenstate ψ_α, including the effect of a possible gauge change $a \to a + \Delta a$ [see equation (5.2.11)], as

$$y \to +L$$

$$\psi_\alpha(\mathbf{r}, \varepsilon_\alpha) \sim \psi_{nk}(\mathbf{r}, \varepsilon_{nk}) \sim C_\alpha \exp\left[i\left(k_\alpha - v + \frac{\Delta a}{l_0} y' \right) \right]$$

$$y \to -L$$

$$\psi_\alpha(\mathbf{r}, \varepsilon_\alpha) \sim e^{-i\delta(\varepsilon_\alpha)} \psi_{nk}(\mathbf{r}, \varepsilon_{nk}) \sim C_\alpha \exp\left[i\left(k_\alpha - v + \frac{\Delta a}{l_0} y' - \delta(\varepsilon_\alpha) \right) \right] \tag{5.2.21}$$

for an energy $\varepsilon_\alpha = \varepsilon_{nk}$ for some nk in the continua of equation (5.2.14). The phase shifts $\delta(\varepsilon_\alpha)$ can be found by solving first for the Green's function of the system without impurities and then for $\psi_\alpha(\mathbf{r}, \varepsilon_\alpha)$ from Dyson's equation, as outlined in section 5.4.4.

The phase shifts are related by Levinson's theorem[10] to the number of bound states N_b split off from the quasicontinuum by the impurity potential. They also determine the contribution to the current from the state α as shown below.

Imposing some boundary condition at $y = \pm L$, allows one to label the states α by the integer variable m_α. For example, one might choose periodic boundary conditions:

$$\psi_\alpha(+L) = \psi_\alpha(-L) \tag{5.2.22}$$

in which case

$$(k_\alpha - v)\frac{2L}{l_0} + \delta(\varepsilon_\alpha) = -2\pi m_\alpha - \Delta a\frac{2L}{l_0^2} \tag{5.2.23}$$

In the absence of impurities, m_α takes N successive values, with N the number of states in the Landau level. In the presence of impurities m_α will only take $(N - N_b)$ successive values where N_b is the number of localized states split off from the continuum.

It is also evident from equation (5.2.23) that a gauge change

$$\Delta a = l_0^2/2L \tag{5.2.24}$$

takes the state labeled by m_α to the state labeled by $m_\alpha + 1$.

Treating the differentials in equation (5.2.18) as finite differences, we may write

$$\begin{aligned}
j_y^\alpha &= \frac{\Delta\varepsilon}{B\,\Delta a} = \frac{\varepsilon_{\alpha+1} - \varepsilon_\alpha}{B\,\Delta a} \\[2mm]
&= -eEl_0 \frac{k_{\alpha+1} - k_\alpha}{B\,\Delta a} \\[2mm]
&= -eEl_0 \frac{l_0}{2L} \frac{-2\pi - \delta(\varepsilon_{\alpha+1}) + \delta(\varepsilon_\alpha)}{B\,\Delta a} \\[2mm]
&= j_y^0 \left(1 + \frac{\delta(\varepsilon_{\alpha+1}) - \delta(\varepsilon_\alpha)}{2\pi} \right)
\end{aligned} \tag{5.2.25}$$

showing that the contribution to the current from state α has an added contribution determined by the phase shifts $\delta(\varepsilon_\alpha)$ when compared to its value in the absence of impurities [equation (5.2.19)].

The total current if all the delocalized states in the Landau level are occupied is then given by

$$j_y = (N - N_b)j_y^0 + \frac{\delta^u - \delta^l}{2\pi}j_y^0 \qquad (5.2.26)$$

where $\delta^{u,l}$ are the values of δ at the upper and lower edges of the band. The first term is as in equation (5.2.20) but with the contributions from the N_b localized states subtracted. The second term is the extra contribution of the remaining delocalized states in the presence of the impurities.

Using equation (5.2.23) and writing k^u and k^l for the momenta at the upper and lower edges of the band gives

$$(k^u - k^l)\frac{L}{l_0} + \delta^u - \delta^l = -2\pi(N - N_b) \qquad (5.2.27)$$

When compared to the equivalent equation for the system without impurities, this gives

$$\delta^u - \delta^l = 2\pi N_b \qquad (5.2.28)$$

Equations (5.2.26) and (5.2.28) give as expected the result that the total current is independent of N_b. The loss of contributions to the current from the states which become localized in the presence of impurities is exactly compensated by the extra contributions from the delocalized states.

Since this result was first derived by Prange, it has been extended and made more rigorous. In particular, Thouless *et al.*[11] have shown how to relate the response σ_{xy} to results in differential geometry. By placing the system on a torus (imposing periodic boundary conditions in both the x and y directions), they were able to eliminate the rather vague account of the boundaries, which was used here. On the torus the quantity corresponding to σ_{xy} is a topological invariant independent of the number of localized states in the Landau level. This connection has, not surprisingly, attracted interest from other branches of physics.

Perhaps more interesting from the viewpoint of quantum transport in general is to look more closely at the role played by boundary states. This question has been addressed by Halperin[12] and is discussed in the next section.

5.2.5. Edge States

For the sake of simplicity assume the simple boundary conditions at the edges $x = \pm x_0$:[12]

$$\psi(\pm x_0) = 0 \qquad (5.2.29)$$

For x close to x_0 the eigenstates of $H_0(a)$, equation (5.2.10), with no applied electric field and $V_{sc} = 0$ will now be of the form

$$\psi_{nk} \sim e^{iky'} g_n((x' + k - \eta), x_k - x_0)e^{-(x'+k-\eta)^2/2} \tag{5.2.30}$$

where

$$x_k = l_0(\eta - k) = a - kl_0 \tag{5.2.31}$$

is the center of the cyclotron orbit with "wave vector" k. The other symbols are as in equation (5.2.12). The function g_n satisfies the same equation as the H_n in equation (5.2.12) but subject to the added boundary condition equation (5.2.29). The solutions for g_n depend on how far the center of the cyclotron orbit, x_k, is from the boundary as well as on $x - x_k$.

For $x_k - x_0 \ll l_0$, g_n will take its bulk form $H_n(x' - x'_k)$ with eigenvalue $\varepsilon_{nk} = (n + \frac{1}{2})\hbar\omega_c$. As x_k approaches x_0, ε_{nk} will change as the boundary causes g to involve admixtures of the various $H_{n'}$. Precisely at $x_k = x_0$, only those solutions of $H_0(a)\psi = \varepsilon\psi$ which have $g(0, 0) = 0$ are possible. These are the $H_{n'}$ with $n' = 2n + 1$, so that

$$\varepsilon_n(x_k = x_0) = (2n + \tfrac{3}{2})\hbar\omega_c \tag{5.2.32}$$

For $x_k \gg x_0$, ε_{nk} will vary as $(x_k - x_0)^2$. Figure 5.5 shows how ε_{nk} can be expected to vary as a function of x_k.

The variation of ε_{nk} with x_k and therefore with the gauge parameter a implies that these states carry nonzero current even in the absence of an applied electric field [see equation (5.2.18)]:

$$j_y^k = \frac{1}{B}\frac{\partial\varepsilon_{nk}}{\partial a} = -\frac{1}{B}\frac{\partial\varepsilon_{nk}}{\partial x_k} \tag{5.2.33}$$

Multiplying by the density of states per unit area $B/(h/e)$ and integrating equation (5.2.33) from the edges at $\pm x_0$ into the bulk gives that for each

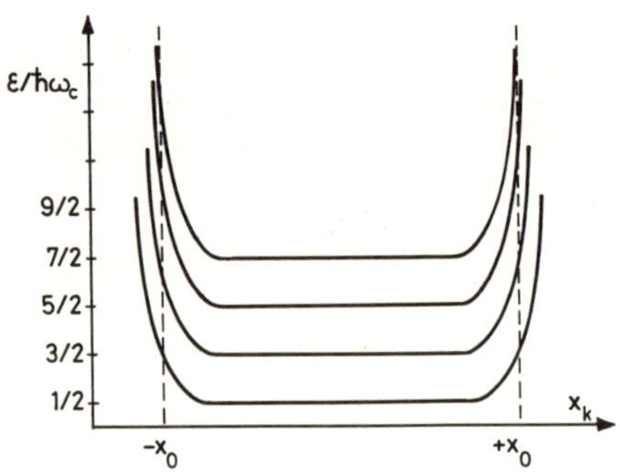

FIGURE 5.5. Energy levels of a two-dimensional electron gas confined to a ribbon bounded at $\pm x_0$, as a function of the "cyclotron orbit center" coordinate, $x_k = a - kl_0$ (after Halperin[12]).

Landau level m there is a contribution to the current from the occupied edge states close to $-x_0$:

$$j_y = +(e/h)\delta E_n^- \qquad (5.2.34)$$

and a contribution $-(e/h)\delta E_n^+$ from the states close to x_0. $\delta E_n^{\pm}$ is the difference in energy between the highest occupied state in the nth level and the bulk energy for the level at the two edges. If $\delta E_n^+ \neq \delta E_n^-$, there will be a net current.

When a voltage is applied between the edges, one expects that $\delta E_n^+ \neq \delta E_n^-$ so that there is a contribution from the edge states to the total current. The voltage dropped across the edge states, ΔV, as a proportion of the total applied voltage V can be written approximately in terms of the effective capacitance per unit length of the edge states, C. With the Fermi energy in the bulk of the sample between the nth and $(n-1)$th Landau level, the net charge accumulated per unit length at the edges, σ, is

$$\sigma \approx \frac{e}{l_0} \frac{e\,\Delta V}{\hbar\omega_c(n+1)} \qquad (5.2.35)$$

The second term in equation (5.2.35) measures the proportion of states of the nth level whose occupation is changed by the voltage drop ΔV. comparing this with $\sigma = CV$ gives

$$\frac{\Delta V}{V} \sim \frac{l_0}{e^2}\,\hbar\omega_c(n+1)C \qquad (5.2.36)$$

For a disordered sample ($V_{\mathrm{sc}} \neq 0$) in equation (5.2.10), the effect of any scattering on edge states will be much as for bulk states. If localized states are formed close to $x = \pm x_0$, then the remaining delocalized edge states compensate exactly for the lost contribution of the localized states, as shown explicitly for bulk states in section 5.2.4.1. This case has been argued by Halperin.[12]

5.3. THE FRACTIONAL QUANTUM HALL EFFECT

At low temperatures plateaux can be observed in the Hall resistivity ρ_{xy} of some GaAs–GaAl$_x$As$_{1-x}$ heterostructures.[2] These tend to be those suitably doped structures with very high electron mobilities $> \sim 10^5$ cm^2/V-sec. The plateaux are observed to occur at rational multiples m of h/e^2, with m^{-1} noninteger. This phenomenon and related effects are known as the fractional quantum Hall effect. Figure 5.6 shows some of the experimental results of Chang *et al.*[14]

If the quantum Hall effect can be thought of as the quantized response of filled bands of (delocalized) states of quasi two-dimensional electrons,

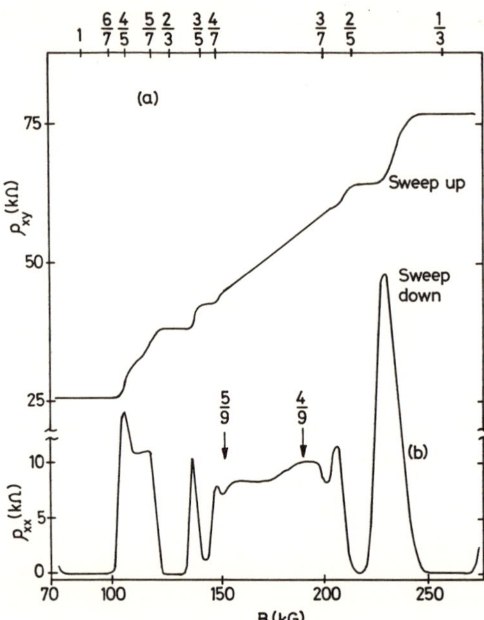

FIGURE 5.6. The Hall resistivity ρ_{xy} and longitudinal resistivity ρ_{xx} of a GaAs–Al$_{0.3}$Ga$_{0.7}$As heterojunction as a function of applied magnetic field or filling fraction of the lowest Landau level ν. Plateaux observed in ρ_{xy} and minima in ρ_{xx} indicate fractional quantized Hall states (after Chang *et al.*[14]).

the Landau levels, the fractional quantum Hall effect appears to be the quantization of the response of an apparently partially filled lowest Landau level. The plateaux are found for magnetic fields close to those at which one would expect a fraction ν of the states in the lowest Landau level to be occupied. For these magnetic fields the resistivity ρ_{xy} is found to be

$$\rho_{xy} = h/\nu e^2 \tag{5.2.37}$$

Plateaux have been observed for many rational ν, all with odd denominators. The most pronounced and readily observed is at $\nu = \frac{1}{3}$, followed in approximate order by $\frac{2}{3}, \frac{2}{5}, \frac{2}{7}, \frac{3}{5}, \ldots$[14] Plateaux have also been observed at filling fractions $\frac{5}{3}, \frac{4}{3}$ and a few others with $1 < \nu < 2$, and recently "forerunners" of plateaux at $2 < \nu < 3$.[15]

At the high fields involved it is generally assumed that the electrons in these systems are completely spin-polarized. One expects that the results for a spin-up ν-partially-filled zeroth Landau level, $0 < \nu < 1$, to be mirrored approximately by those for the corresponding spin-down $(2 - \nu)$-partially-filled zeroth Landau level $(1 < \nu < 2)$. The results for $2 < \nu < 3$ would then correspond to a partially filled first Landau level.

Note: In the following all quantities will be assumed to be in magnetic units; see section 5.4.1.

5.3.1. Interpretation of the Measurement: Many-Body Gap and Fractional Charge

Faced with the observed quantization, one's first reaction might be to invoke Laughlin's Gedankenexperiment. The observation of a quantization of the Hall resistance would then imply the existence of a mobility gap in the excitation spectrum as in the case of the quantum Hall effect. This gap would have to be caused by the particle–particle interaction because the single-particle states, the cyclotron orbits, are degenerate. With the Fermi energy in this gap ρ_{xy} is given (neglecting signs) by equation (5.2.6):

$$\rho_{xy} = |V_x/I_y| = |h/nqe| \tag{5.3.1}$$

Comparing with the observed value, at, say, $\nu = \frac{1}{3}$ gives

$$n|q/e| = \tfrac{1}{3} \tag{5.3.2}$$

n is the number of particles of charge q transferred in the Gedankenexperiment between edge states when the flux threading the ring is changed by h/e.

Simple application of this result at $\nu = \frac{1}{3}$ implies the existence of fractionally charged particles with charge $|q| = e/3$. These particles might be expected to behave like the electrons in the case of the normal quantum Hall effect. Impurities would be expected to localize these particles, giving rise to a mobility gap, and so on. This appealing interpretation was developed originally by Laughlin,[16] who identified possible forms for the ground state and excitation wave functions, which do indeed describe fractionally charged objects.

The Gedankenexperiment presupposes that the bulk occupation numbers cannot change when the gauge is changed. However, if there is more than one equivalent ground state, it could be that the gauge change causes the electrons in the bulk to be mapped not into the same ground state but into one of the equivalent ground states. Indeed, it turns out that at rational filling fraction, $\nu = p/t$, and with the application of certain types of boundary conditions there is a t-fold center-of-mass degeneracy of the ground state. Only changes of flux by multiples of th/e map these systems into themselves, again with a corresponding change of one electron/hole in the occupation of the edge states. Relating the measurement to this center-of-mass degeneracy was suggested later by Niu et $al.$[17]

The connection between ground-state degeneracy and topological defects with fractional quantum numbers have been discussed in this and other contexts by Schrieffer.[18]

Either way, both interpretations of the Gedankenexperiment imply the existence of an excitation gap. (This result can also be deduced on quite general thermodynamic grounds.[19]) We can also expect to find locally well-defined excitations with apparent fractional charge.

5.3.2. Zeros and Flux Quanta

All current theories associate the gaps in the excitation spectra implied by the experiments with the density of "zeros" permitted in the N-particle wave functions $\psi(\mathbf{r}_1, \ldots, \mathbf{r}_N)$ as a function of filling fraction of the lowest Landau level ν.

To identify the role of zeros in the wave function, suppose a system of electrons is confined in two dimensions to some area Ω, and a large magnetic field leads to complete spin polarization of the electrons. At filling fraction ν there are then ν electrons per cyclotron orbit. Equivalently, one may say that there are ν electrons per flux quantum threading the area Ω (see section 5.4.1).

Now imagine performing the following Gedankenexperiment: take one of the electrons, say the ith one, counterclockwise along the closed path $\sum_l \mathbf{a}_l$, which, assume, defines the perimeter of the system enclosing Ω. Then $\psi(\mathbf{r}_1, \ldots, \mathbf{r}_N)$ is transformed to $\psi'(\mathbf{r}_1, \ldots, \mathbf{r}_N)$:

$$\psi(\mathbf{r}_1, \ldots, \mathbf{r}_N) \to \psi'(\mathbf{r}_1, \ldots, \mathbf{r}_N) = \left(\prod_l T(\mathbf{a}_l) \right) \psi(\mathbf{r}_1, \ldots, \mathbf{r}_N) \quad (5.3.3)$$

where the $T(\mathbf{a}_l)$ are the magnetic translation operators defined in section 5.4.3. Successive application of the combination rule for the $T(\mathbf{a}_l)$, equation (5.4.27), yields, with $q = -e$,

$$\psi' = \exp\left(2\pi i \frac{q}{e} \oint \frac{\mathbf{A} \cdot d\mathbf{l}}{\phi_0} \right) \psi = \exp\left(-2\pi i \frac{\Phi}{\phi_0} \right) \psi \quad (5.3.4)$$

where Φ $(= B\Omega)$ is the total flux threading Ω and $\phi_0 = h/e$. But

$$\Phi/\phi_0 = N\nu^{-1} \equiv Nm \quad (5.3.5)$$

The phase of ψ therefore changes as a result of the Gedankenexperiment by $-2\pi m$ for each electron contained in Ω $(m = \nu^{-1})$.

The wave function of an electron in the lowest Landau level can be written in the symmetric gauge, $\mathbf{A} = (y, -x, 0)/2$ (see section 5.4.2), as

$$\psi(\mathbf{r}_i) \sim P(z_i) e^{-|z_i^2|/4} \quad (5.3.6)$$

with $P(z_i)$ a polynomial in the variable

$$z_i = x_i + i(q/e)y_i = x_i - iy_i \quad (5.3.7)$$

The wave function $\psi(\mathbf{r}_i)$ changes phase by -2π around a point at which $P(z_i)$ has a simple zero, and by $-2\pi n$ around a point at which $P(z_i)$ has a zero of order n. From equation (5.3.5) we then see that for each electron within Ω, the N-particle wave function $\psi(\mathbf{r}_1, \ldots, \mathbf{r}_N)$ must have Nm zeros as a function of the position coordinates z_i. This is equivalent to one zero per flux quantum or m zeros per particle.

The fractional quantum Hall effect is thought to result from the "binding" of the zeros in the wave function as a function of the coordinates z_i to the coordinates of each of the other electrons. This will only be possible at a certain odd integer m $(= \nu^{-1})$ if the wave function is to retain an analytic form and describe fermions. The binding of m zeros coincidentally to particle positions will tend to minimize the interaction energy of two particles, since the probability of particle i approaching particle j vanishes as

$$|\psi(\mathbf{r}_i, \mathbf{r}_j)|^2 \sim |\mathbf{r}_i - \mathbf{r}_j|^{2m} \tag{5.3.8}$$

as $|\mathbf{r}_i - \mathbf{r}_j| \to 0$. If the zeros were somehow bound close but not coincidentally, the probability of two particles approaching would also be small.

The system would then be expected to have a gap in the excitation spectrum for ν close to $1/m$, because it would not be possible to vary the number of electrons or flux quanta without adding or removing zeros to the wave function. Excitations involving the addition or removal of flux quanta from the system would lead to excitations which would appear at least to carry fractional charge. For, whereas in the ground-state wave function m zeros are associated with the electron positions, an added or missing zero would be associated with $\pm 1/m$ electronic charges.

5.3.3. Laughlin's Wave Function

In 1983 Laughlin constructed his trial wave function:

$$\psi_L(\mathbf{r}_1, \ldots, \mathbf{r}_N) = \prod_{i<j}^{N} (z_i - z_j)^m \exp\left(-\sum_l \frac{|z_l|^2}{4}\right) \tag{5.3.9}$$

in which all the zeros as a function of coordinates z_i are clearly coincidental with the position coordinates of the other electrons. The function ψ_L is of the form of equation (5.3.6) for each z_j and so describes electrons in the lowest Landau level; ψ_L describes a fluid, as can be seen by writing $|\psi_L|^2$ as a classical distribution function:

$$|\psi_L|^2 \sim e^{-\beta V} \tag{5.3.10}$$

with $\beta = 2/m$ and

$$V = m^2\left(\sum_{i<j}^{N} -\ln|z_i - z_j| + \frac{1}{4m}\sum_l^{N} |z_l|^2\right) \tag{5.3.11}$$

V is the potential energy of the two-dimensional one-component plasma (2DOCP). This has particles of charge m interacting via a logarithmic interaction in a neutralizing background with number density $1/2\pi m$. For $2m < {\sim}140$ it is known that equation (5.3.10) describes a fluid.[20] This fluid is incompressible because the length scale l_0 cannot change without introducing or removing flux quanta. As stated above, this would have to involve excitations across a gap.

Laughlin[16] also suggested explicit forms for excitations he called quasiparticles and quasiholes. Imagining removing and adding flux quanta from the system at some point z_0 $(=x_0 + i(q/e)y_0)$, he suggested the wave functions

$$\psi_L^{+z_0} = \prod_i (z_i - z_0)\psi_L$$

$$\psi_L^{-z_0} = \prod_i \left(2\frac{\partial}{\partial z_i} - z_0\right)\psi_L$$

(5.3.12)

Now $|\psi_L^{+z_0}|^2$ describes a classical system as before, but with an added particle with charge 1 at z_0. This will be screened by the charge m particles, so that in the electron system there will be an absence of $1/m$ of an electron in the neighborhood of z_0, and hence a local excitation with effective charge $+e/m$. Interpretation of $|\psi_L^{-z_0}|^2$ along these lines is also possible, though not so straightforward.

The wave functions $|\psi_L^{+z_0}|^2$ together with $|\psi_L|^2$ and $|\psi_L^{-z_0}|^2$ have been extensively studied by Monte Carlo simulation and hypernetted chain calculations for the corresponding classical systems. All these studies show that the Laughlin picture is at least consistent. There appears to be a well-defined thermodynamic limit for the quasiparticle, quasihole, and gap excitation energies, ε^+, ε^-, and ε_g. Perhaps the most extensive studies to date have been made by Morf and Halperin,[21] who found $\varepsilon^+ \approx 0.026e^2/\varepsilon l_0$ and $\varepsilon^- \approx 0.073e^2/\varepsilon l_0$.

Their Monte Carlo simulations also showed that a "pairing" ansatz for ψ^-,

$$\psi^- \sim A\frac{1}{(z_1 - z_2)^2}\prod_{j=3}^{N}\frac{z_j - (z_1 + z_2)/2}{(z_1 - z_j)(z_2 - z_j)}\psi_L$$

(5.3.13)

gives at least as low an estimate for the quasiparticle excitation as Laughlin's $\psi_L^{-z_0}$. Here A is the antisymmetrizing operator. Both wave functions describe an excitation with a local accumulation of charge $-e/m$ distributed on a ring about z_0 for $\psi_L^{-z_0}$ or about the origin for ψ^-, again consistent with Laughlin's original interpretation.[21]

The low-lying collective excitations at constant field and particle number might be thought of as quasiexcitons. The quasiexciton dispersion $\varepsilon(\mathbf{k})$ would then be the energy wave vector dependence of bound pairs of fractionally charged quasiparticles and quasiholes.

In the spirit of Laughlin's theory, Girvin, MacDonald, and Platzmann (GMP)[22] have suggested using a Feynman–Bijl ansatz to describe the quasiexcitons. They obtain an $\varepsilon(\mathbf{k})$, which is in agreement with results from the direct diagonalization of the Hamiltonian for small systems.

The Feynman–Bijl ansatz assumes that the density operator $\rho(\mathbf{k})$ only couples a single mode to the homogeneous ground state ψ_{gs}. So

$$\psi_{\mathbf{k}} = \rho_{\mathbf{k}}\psi_{gs} \qquad (5.3.14)$$

is an eigenstate of the system with energy $\varepsilon(\mathbf{k})$. The approximation is sometimes known as the single-mode approximation (SMA). One may also think of $\psi_{\mathbf{k}}$ as a trial wave function for the lowest-lying energy state with wave vector $\mathbf{k}$.

GMP make the further approximations of treating ψ_L as the exact ground state and replacing $\rho_{\mathbf{k}}$ by the projection onto the lowest Landau level $\tilde{\rho}_{\mathbf{k}}$. The variational estimate for $\varepsilon(\mathbf{k})$ $(=\langle\psi_{\mathbf{k}}|H - \varepsilon_{gs}|\psi_{\mathbf{k}}\rangle/\langle\psi_{\mathbf{k}}|\psi_{\mathbf{k}}\rangle)$ is then a function of the static structure factor $s(k)$. GMP took $s(k)$ from results obtained for the 2DOCP. Their results together with exact results for small systems obtained by Haldane and Rezayi are shown in Figure 5.7.

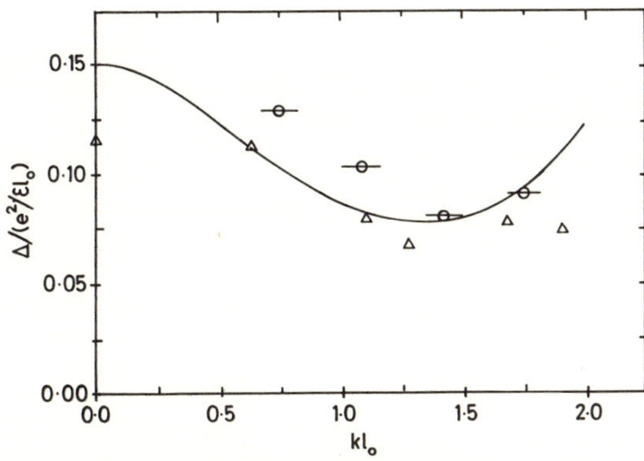

FIGURE 5.7. The dispersion relation $\varepsilon(k)$ for excitations (quasiexcitons) of the two-dimensional electron gas at constant magnetic field and particle number for $\nu = \frac{1}{3}$. The continuous line is the variational estimate of $\varepsilon(k)$ based on a single-mode approximation. The points are results from exact calculations on a system of six particles (after Girvin et al.[22] and Haldane and Rezayi[23]).

5.3.4. Haldane's Argument

A slightly different justification of the Laughlin picture has been given by Haldane and Rezayi[23] and others. They pointed out that ψ_L is the exact ground-state wave function at $\nu = 1/m$ for a system of particles interacting through a kind of hard-core interaction.

One might then imagine a Gedankenexperiment in which the range of the interaction was adiabatically changed from this hard-core interaction to the interaction of the physical system. Provided no crossover occurred between states of different symmetry, the physical system would be qualitatively the same as the system for which exact solutions are known. There would only be some renormalization of parameters such as excitation energies. This would be as in the Landau-Fermi liquid theory of interacting systems, where, of course, the system whose solutions are known exactly is the one made up of noninteracting particles.

Haldane and Rezayi simulated, by direct numerical diagonalization of the Hamiltonian for small systems, the adiabatic variation of the interaction. They defined the interaction

$$V(\lambda) = V^{\text{Coulomb}} + \lambda(V^{\text{h.c.}} - V^{\text{Coulomb}}) \qquad (5.3.15)$$

and adiabatically varied λ. Here, V^{Coulomb} describes a Coulomb interaction, and $V^{\text{h.c.}}$ describes the hard-core interaction defined in section 5.4.5. The ground state and excited states of six particles on the surface of a sphere, labeled by total angular momentum, are plotted as a function of λ in Figure 5.8.

Figure 5.8 shows that at least for a system of only six particles there is no crossover between states with different quantum numbers for the range of interactions likely in physical systems. It is also suggestive of an excitation gap between the Laughlin-like ground state ($L = 0$) and the first excited states.

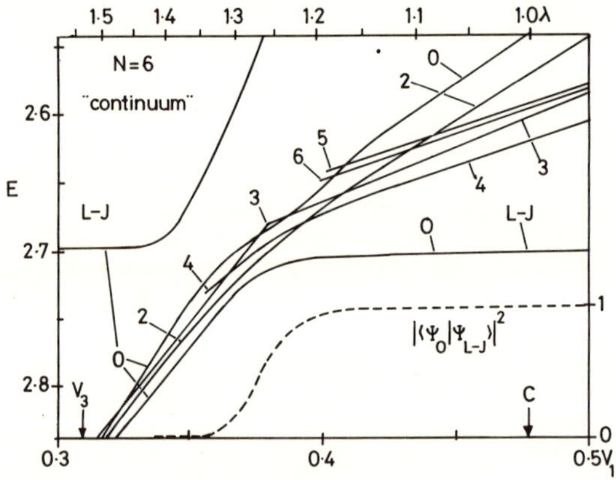

FIGURE 5.8. The energies of the low-lying states of six electrons confined to the surface of a sphere are plotted for the case $\nu = \frac{1}{3}$ labeled according to their total angular momentum. The interparticle interaction is varied from hard-core ($\lambda = 0$) through Coulomb ($\lambda = 1$) to longer-range interactions ($\lambda > 1$). The strength of the hard-core component V_1 is also shown; see equation (5.4.56). Also shown is the overlap between the ground state and the Laughlin wave function ψ_{LJ}. For the range of likely physical interactions the ground state remains qualitatively the same as the Laughlin state (after Haldane and Rezayi[23]).

TABLE 5.1. Overlap between ψ_L and Exact Ground State with a Coulomb Interaction ψ_C at $\nu = \frac{1}{3}$ and $\nu = \frac{1}{5}$ (after Fano et al.[24]).

N	3	4	5	6	7	8	9
$\nu = \frac{1}{3}$	1	0.99804	0.99906	0.99644	0.99636	0.99540	0.99406
$\nu = \frac{1}{5}$	1	0.98405	0.99743	0.94865	0.97681		

Other numerical studies of small systems at $\nu = 1/m$ have been reported by many authors for various boundary conditions. All lend support to Laughlin's interpretation at $\nu = 1/m$: all show large overlaps between the exact ground state with a Coulomb interaction and the Laughlin ground state and almost equally large overlaps between the exact excitations and those suggested by Laughlin. Table 5.1 shows as an example the results obtained by Fano et al.[24] in a spherical geometry first introduced by Haldane.

5.3.5. Other Filling Fractions—the Hierarchy

"Binding" ν^{-1} zeros directly to the particle positions, as in the Laughlin wave function, is only possible for ν^{-1} an odd integer. The theory needs generalizing if it is to explain the observation of plateaux at $\nu = \frac{2}{5}$, $\nu = \frac{2}{7}$, and so on.

A suggestion of Halperin[19,25] and Haldane[26] was to consider the properties of the excitations from ψ_L. If these were point charges, with coordinates η_a^1, they would constitute a system similar to the original one of electrons/holes, but with effective charge $q_1 = \alpha_1 e/m$ and magnetic length $l_1^2 = l_0^2 m$. The term $\alpha_1 = \pm 1$, according to whether the excitations are quasiholes or quasiparticles. One could therefore imagine a pseudo-Hamiltonian for these excitations, including some kind of repulsive interaction between them, and a pseudo-wave function $\tilde{\psi}_1$ written in the coordinates η_a^1. The η_a^1 would be the z_0 of the Laughlin excitations or the "center of mass" coordinates of the pair excitations equation (5.3.13).

Also if one imagined taking one of these particles around the perimeter of the system enclosing flux Φ, the pseudo-wave function would change to $\tilde{\psi}_1'$ [see equation (5.3.4) with $q = \alpha_1 e/m$]:

$$\tilde{\psi}_1' = \tilde{\psi}_1 e^{\alpha_1 2\pi i(\Phi/m\phi_0)} \tag{5.3.16}$$

If no other excitation is enclosed, $\Phi = mN_e$ and $\tilde{\psi}_1' = \tilde{\psi}_1$. If one other excitation is enclosed, $\tilde{\psi}_1' = \tilde{\psi}_1 e^{i\alpha_1(2\pi/m)}$. Arovas et al.[27] have argued by analogy with electrons: When one electron encircles another, the phase change is -2π corresponding to a phase change of $-\pi$ for interchange of two particles. The statistic for interchange of two fractionally charged excitations would then be

$$\tilde{\psi}_1(\eta_a^1, \eta_b^1) = \tilde{\psi}_1(\eta_b^1, \eta_a^1)e^{i\alpha_1(\pi/m)} \qquad (5.3.17)$$

This result can be incorporated into a trial ground-state pseudo–wave-function for N_1 particles:

$$\tilde{\psi}_1(\eta_1^1, \ldots, \eta_{N_1}^1) = \prod_{a<b}^{N_1} (\eta_a^1 - \eta_b^1)^{p_1+\alpha_1/m} \exp\left(-\sum_c \frac{|\eta_c|^2}{4m}\right) \qquad (5.3.18)$$

Only even p_1 are consistent with equation (5.3.17).

The excitations from this pseudo–wave function, $\psi_1^{\pm\eta_0}$, could be expected to be described by operators similar to those in equation (5.3.12) operating on $\tilde{\psi}_1$. These excitations would form the particles of the second level of a hierarchy of excitations carrying charge

$$q_2 = \pm\frac{q_1}{p_1 + \alpha_1/m} = \frac{e}{mp_1 \pm 1} \qquad (5.3.19)$$

This follows by analogy with the Laughlin fluid at $\nu = 1/m$. The ratio of the charge of the particles of "parent" fluid to that of its excitations is equal to the number of zeros in the wave function bound to the coordinates of the other particles. This is just the exponent in equation (5.3.18).

This process leads one to the hypothesis of a hierarchy of Laughlin-like fluid ground states at different filling fractions for the lowest Landau level. For a hierarchy of n levels this leads to a filling fraction, $\nu(m; p_1, p_2, \ldots, p_n)$, derived in section 5.4.6:

$$\nu(m; p_1, p_2, \ldots, p_n) = \cfrac{1}{m + \cfrac{\alpha_1}{p_1 + \cfrac{\alpha_2}{p_2 + \cfrac{\alpha_3}{p_3 + \cfrac{\cdots + \alpha_n}{p_n}}}}} \qquad (5.3.20)$$

This interpretation implies a natural order in which the ground-state fluids would become apparent as the temperature is reduced. First, one would expect to see a parent fluid at $\nu = 1/m$ or its hole counterpart at $\nu = 1 - 1/m$. These would be followed by first-level fluids at $\nu = \frac{2}{5}$ ($m = 3, \alpha_1 = -1, p_1 = 2$), $\nu = \frac{2}{7}$ ($m = 3, \alpha_1 = +1, p_1 = 2$), or $\nu = \frac{3}{5}$ and $\nu = \frac{5}{7}$. Then second-level fluids at $\nu = \frac{3}{7}$ ($m = 3, \alpha_1 = -1, p_1 = 2, \alpha_2 = -1, p_2 = 2$), and so on. This natural order seems to agree quite well with the apparent "strength" of the experimentally observed plateaux.[14]

5.3.6. Microscopic Trial Wave Functions for the Hierarchy

The hierarchical picture is only phenomenological and is clearly not on the same basis as the microscopic trial wave function (MTW) ψ_L for $\nu = 1/m$.

In an attempt to improve on the phenomenological nature of the hierarchical fluid, Morf et $al.$[28] suggested a family of MTWs, which work explicitly with the electron coordinates, but which appear to incorporate the idea of the hierarchy. Taking the "paired" excitation, equation (5.3.13), as the basis for the quasiparticle, they suggested writing, with A the antisymmetrizing operator,

$$\psi_T = A\phi_D$$

$$= A\left(\psi_L \prod_n^{N/2} (z_{2n} - z_{2n-1})^{-t} \prod_{n,n'}^{N/2} (z_{2n}z_{2n-1} + z_{2n'}z_{2n'-1} - 2Z_nZ_{n'})^s\right) \quad (5.3.21)$$

Here Z_n $[=(z_{2n} + z_{2n-1})/2]$ denotes the center-of-mass coordinate of the pair formed by particles $2n$ and $2n - 1$. To ensure that the wave function is antisymmetric with respect to interchange of particles $2n$ and $2n - 1$ in the pair n, $m - t$ must be odd (m is the exponent in ψ_L). Also $m - t \geq 1$ and $s \geq 0$. By evaluating the maximum power of z_i for any i and dividing by N, ψ_T can be seen to be a trial wave function for a filling factor $\nu = 1/(m + s/2)$.

For $t > 0$ the partially antisymmetric part ϕ_D can be thought of as a Laughlin fluid with $N/2$ quasiparticles nucleated at the coordinates Z_n. These excitations are essentially prevented from approaching each other by the term in parentheses raised to the sth power. For $t < 0$ the partially antisymmetric part ϕ_D might be thought of as ψ_L with $N/2$ quasiholes at the coordinates Z_n, although the interpretation is not so clear in this case.

Morf et $al.$[28] have made extensive studies both by Monte Carlo and CI-type calculations of these wave functions. Fano et $al.$[24] have also made similar studies. The results for the overlaps between the trial wave functions and the exact wave function with the Coulomb interaction are presented in Table 5.2 for $\nu = \frac{2}{5}$ ($m = 2$, $s = t = 1$) and $\nu = \frac{2}{7}$ ($m = 3$, $s = 1$, $t = -2$). Again the overlaps between the trial wave functions and the exact ground-state wave function with a coulomb interaction add further support to the picture of a series of Laughlin-type fluids.

5.3.7. Higher Landau Levels

It is no problem to write the equivalent wave functions to ψ_L for higher Landau levels. In the lowest Landau level the operators z_i and $\partial/\partial z_i$ have the same matrix elements as the raising and lowering operators Z_i^+ and Z_i^-

TABLE 5.2. Coulomb Energy and Overlap for the Trial Wave Function ψ_T [equation (5.3.21)] and Overlap with Exact Ground State with a Coulomb Interaction ψ_C at $\nu = \frac{2}{5}$ and $\nu = \frac{2}{7}$. (N is the particle number, energies are measured in units of $e^2/\varepsilon l_0$. After Morf *et al.*[28])

| | | $\nu = \frac{2}{5}$ | | | $\nu = \frac{2}{7}$ | |
N	$E(\psi_T)/N$	$E(\psi_C)/N$	$\langle\psi_C|\psi_T\rangle$	$E(\psi_T)/N$	$E(\psi_C)/N$	$\langle\psi_C|\psi_T\rangle$
4	−0.426104	−0.426104	1	−0.386004	−0.386012	0.999939
6	−0.427641	−0.428517	0.98840	−0.384527	−0.384626	0.997163
8	−0.428283	−0.429543	0.97712	−0.383671	−0.383811	0.996293
10	−0.428939	−0.430258	0.97154			
12	−0.429327					
∞	−0.4310	−0.4330				

defined in section 5.4.3. The obvious generalization of ψ_L to the nth Landau level is[29]

$$\psi_L^n = \prod_{i<j} (Z_i^+ - Z_j^+)^m |n, 0\rangle \tag{5.3.22}$$

where $|n, 0\rangle$ is the symmetric product of single-particle states in the nth Landau level with $L_{z_i} = 0$.

With $n = 1$ when any two particles i and j approach each other, the norm of the wave function $|\psi_L^1|^2$ rises only as[30]

$$|\psi_L^1|^2 \sim r_{ij}^{2(m-2)}(L_2^{m-2}(r_{ij}^2/2))^2 \tag{5.3.23}$$

For *both* $m = 1$ and $m = 3$, $|\psi_L^1|^2$ increases as r_{ij}^2.

For $m = 5$, $|\psi_L^1|^2$ increases as r_{ij}^6. Therefore ψ_L is the exact nondegenerate ground state for a hard-core interaction at $\nu = 1/m$ only for $m = 5, 7, \ldots$, but not for $m = 3$; see Ref. 31. One might therefore expect to find an approximate analogy between states at filling fractions in the first Landau level, $\frac{1}{5} < \nu < \frac{1}{3}$, and states in the lowest Landau level at $\frac{1}{3} < \nu < 1$. This is supported by the numerical calculations of d'Ambrumenil and Reynolds[30] on small systems. As an example Figure 5.9 shows the equivalent curve to Figure 5.7 for the first Landau level at filling fraction of the first Landau level $\nu = \frac{1}{5}$. As in the lowest Landau level, there is a well-defined excitation dispersion and gap.

Results of calculations for small systems have led Haldane[29] and d'Ambrumenil and Reynolds[30] to conclude that at a filling fraction of the $n = 1$ Landau level $\nu = \frac{1}{3}$ the system is close to the borderline between having an incompressible Laughlin-like fluid ground state and compressibility.

d'Ambrumenil and Reynolds[30] also found evidence that there should be a fractional quantized Hall state at a filling fraction of the $n = 1$ Landau level $\nu = \frac{2}{7}$. This is again compatible with the premise that there should be at least an approximate analogy between the states in the $n = 1$ Landau

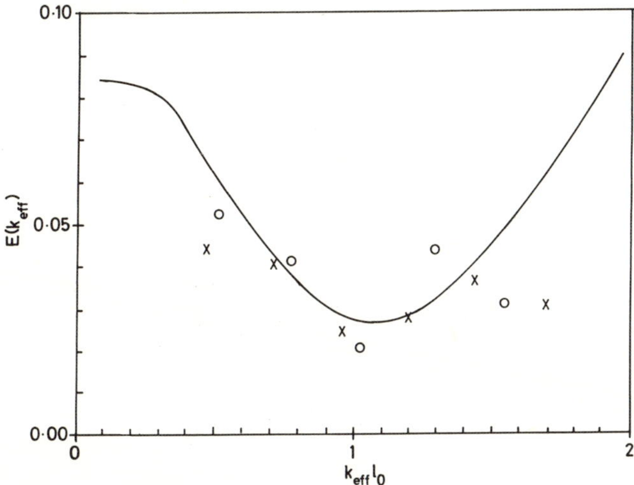

FIGURE 5.9. The excitation energies at $\nu_1^{-1} = 5$ for six and seven particles. The energies are defined as the total energy difference between the lowest energy state for the various L and the exact ground state. The solid line represents the spectrum predicted by MacDonald and Girvin[31] on the basis of the Feynman–Bijl ansatz. Energies are in units of $e^2/\varepsilon l_0$ (after d'Ambrumenil and Reynolds[30]).

level at filling fractions $\frac{1}{5} < \nu < \frac{1}{3}$ and those in the lowest ($n = 0$) Landau level at filling fractions $\frac{1}{3} < \nu < 1$.

The observations of Clark et $al.$[15] are, however, still only partially understood.

5.3.8. Ring Exchange

An alternative approach to understanding the FQHE has been suggested by Kivelson et $al.$[32] (KKAS), which does not seek to describe the systems with explicit wave functions.

KKAS found from an approximate expansion scheme for the free energy that at certain rational filling fractions there were large contributions to the free energy from coherent exchanges of particles round large rings. These ring exchanges (CRE) add coherently when the ratio of particles to flux quanta is rational with odd denominator, and preferentially stabilize the system at the corresponding filling fraction.

Initially their calculations seemed to imply that the ground state would have to have the kind of long-range order generally associated with a solid. However, Baskaran[33] later argued that CRE would stabilize any incompressible state, including ψ_L.

5.4. DICTIONARY OF STANDARD RESULTS (MORE OR LESS)

5.4.1. Hamiltonian and Energy Spectrum

The Hamiltonian for a particle of charge q, with coordinates (x, y, z) and effective mass m^* is given by

$$H_0' = \frac{(\mathbf{p} - q\mathbf{A})^2}{2m^*} = \frac{\boldsymbol{\pi}^2}{2m^*} \tag{5.4.1}$$

where $\mathbf{A}$ is the vector potential and $\mathbf{p}$ is the momentum. The Hamiltonian will also include items describing the particle–impurity interaction H_i, the particle–lattice interaction $H_{\text{p-l}}$, the lattice dynamics important in the magnetophonon effect H_L, and the particle–particle interaction $H_{\text{p-p}}$.

It is usual to separate the Hamiltonian H_0' into a part describing motion parallel to the magnetic field and a part describing motion perpendicular to the magnetic field. If the magnetic field is taken to be parallel to the z-direction, H_0' separates and the eigenfunctions can be written as

$$\Psi = \psi(x, y)\phi(z) \tag{5.4.2}$$

with

$$H_0'\Psi = (E_z + H_0)\psi \tag{5.4.3}$$

where H_0 operates on the x and y coordinates only. In a three-dimensional jellium model $\phi(z)$ would be just a plane wave with $E_z = \hbar^2 k_z^2/2m^*$. In a semiconductor heterostructure $\phi(z)$ would be the envelope wave function of the subband.

It is often helpful to write H_0 in so-called magnetic units. Introduce the cyclotron frequency $\omega_c = eB/m^*$ and the magnetic length l_0, $l_0^2 = \hbar/eB$, where e is the magnitude of the electron charge, and write

$$x' = x/l_0$$
$$y' = y/l_0$$
$$\nabla' = l_0\nabla \tag{5.4.4}$$
$$\mathbf{A}' = \mathbf{A}/l_0 B$$

Primed operators have been used to denote operators and variables in magnetic units.

Then

$$H_0 = (\hbar\omega_c/2)\boldsymbol{\pi}'^2 \tag{5.4.5}$$

with

$$\boldsymbol{\pi}' = \left(\frac{1}{i}\nabla' - \frac{q}{e}\mathbf{A}'\right) = \frac{l_0}{\hbar}\boldsymbol{\pi} \tag{5.4.6}$$

The components π_α of the conjugate momentum $\boldsymbol{\pi}$ do not commute with one another in the presence of a magnetic field:

$$[\pi_\alpha, \pi_\beta] = iq\hbar\varepsilon^{\alpha\beta\gamma}B^\gamma \tag{5.4.7}$$

With B parallel to the z-direction, this reduces to the simple relation

$$[\pi_x, \pi_y] = iq\hbar B \tag{5.4.8}$$

The eigenfunctions of H_0, which are known as the cyclotron orbits, have the well-known energy spectrum

$$E(n) = (n + \tfrac{1}{2})\hbar\omega_c \tag{5.4.9}$$

where n is known as the Landau level index. One may also show that the density of states per unit area in each Landau level ρ_0 is given by

$$\rho_0 = \frac{B}{h/e} = \frac{1}{2\pi l_0^2} \tag{5.4.10}$$

Therefore, N_A, the total number of states in each Landau level, is

$$N_A = \rho_0 A = \frac{\Phi}{h/e} \tag{5.4.11}$$

where Φ is the total flux threading the system of area A. Noting that h/e is just the magnetic flux quantum, we may say that each flux quantum threading a system gives rise to one cyclotron orbit per Landau level.

5.4.2. Gauge Choice

The form of the eigenfunctions of H_0 depend on the choice of gauge for $\mathbf{A}$. The two most popular choices are the Landau gauge and the symmetric gauge. In the Landau gauge an orthonormal basis can be easily constructed which makes translational invariance parallel to some direction, the x-direction, say, explicit. In the symmetric gauge the construction of an orthonormal basis of eigenstates of angular momentum is simple.

In the Landau gauge,

$$\mathbf{A} = (0, Bx, 0) \tag{5.4.12}$$

One may write the cyclotron orbits ψ as

$$\psi = |nk\rangle \sim e^{iky'}H_n(x' - k(q/e))e^{-(x'-k(q/e))^2} \tag{5.4.13}$$

where the H_n are the Hermite polynomials[37] and, again, primes denote the use of magnetic units.

In the symmetric gauge,

$$\mathbf{A} = \left(-B\frac{y}{2}, B\frac{x}{2}, 0\right) \tag{5.4.14}$$

$$\psi = |nm\rangle = A_{n,m}\left(x' + i\frac{q}{e}y'\right)^m e^{-r'^2/4} L_n^m\left(\frac{r'^2}{2}\right) \tag{5.4.15}$$

$$A_{n,m} = \left(2^m \frac{(m+n)!}{n!}\right)^{-1/2} \tag{5.4.16}$$

where $r'^2 = x'^2 + y'^2$ and the L_m^n are the Laguerre polynomials.[34] The $|nm\rangle$ are eigenstates of the angular momentum L_z defined in equation (5.4.31).

However, there is no unique choice for the cyclotron orbits in any particular gauge, nor is there any unique choice of gauge. All results for physical quantities must always be derivable independent of gauge. However, it is often possible to exploit the freedom in the choice of gauge and basis to derive certain results more easily, so I note here the form of a gauge transformation.

Some observable O with matrix elements $\langle \psi O \psi \rangle$ will transform to O' under the gauge transformation

$$\psi' = e^{i(q/e)\theta(x,y)}\psi \tag{5.4.17}$$

with

$$O' = e^{i(q/e)\theta(x,y)} O e^{-i(q/e)\theta(x,y)} \tag{5.4.18}$$

which has the same matrix elements in the transformed basis ψ' as does O in the original basis.

In particular, the operator π becomes

$$\pi_1 = \left(\frac{\hbar}{i}\nabla - q\mathbf{A}_1\right) \tag{5.4.19}$$

where

$$\mathbf{A}_1 = \mathbf{A} + \frac{\hbar}{e}\nabla\theta \tag{5.4.20}$$

As an example, the transformation from the symmetric gauge to the Landau gauge is as follows:

$$\frac{\hbar}{e}\nabla\theta = \mathbf{A}_{\text{Landau}} - \mathbf{A}_{\text{Symmetric}} = \left(B\frac{y}{2}, B\frac{x}{2}, 0\right) \tag{5.4.21}$$

$$\theta = \frac{B}{e/\hbar}\frac{xy}{2} = \frac{x'y'}{2} \tag{5.4.22}$$

So an eigenfunction of angular momentum, $|nm\rangle$, in the symmetric gauge becomes, after the gauge transformation,

$$|\widetilde{nm}\rangle = A_{n,m}\left(x' + i\frac{q}{e}y'\right)^m e^{-r'^2/4} L_m^n\left(\frac{r'^2}{2}\right) \exp\left(i\frac{q}{e}\frac{x'y'}{2}\right) \quad (5.4.23)$$

which is an eigenstate of H_0 in the Landau gauge, as may easily be checked by acting on the state $|\widetilde{nm}\rangle$ with H_0 explicitly.

5.4.3. Conserved Momenta, Magnetic Translations, and Rotations

In the presence of a magnetic field, the quantities $\boldsymbol{\pi}$ and $\mathbf{J} = \mathbf{r} \times \boldsymbol{\pi}$ are not conserved. However, the conserved quantities corresponding to momentum and angular momentum are easily constructed.

The time-dependence of $\boldsymbol{\pi}$ in a magnetic field is

$$\frac{d\boldsymbol{\pi}}{dt} = q\left(\frac{d\mathbf{r}}{dt} \times \mathbf{B}\right) \quad (5.4.24)$$

In a constant magnetic field we may define the conserved quantity

$$\boldsymbol{\pi}^B = \boldsymbol{\pi} - q\mathbf{r} \times \mathbf{B} \quad (5.4.25)$$

The $\boldsymbol{\pi}^B$ can then be used to define magnetic translation operators $T(\mathbf{a})$ with,[35]

$$T(\mathbf{a}) = \exp\left(-i\frac{\boldsymbol{\pi}^B \cdot \mathbf{a}}{\hbar}\right) \quad (5.4.26)$$

which commute with the Hamiltonian. In the absence of a magnetic field, $T(\mathbf{a})$ reduces to the standard translation operator.

However, for two nonparallel vectors $\mathbf{a}, \mathbf{b}$ magnetic translation operators $T(\mathbf{a})$ and $T(\mathbf{b})$ only commute up to a phase factor:

$$T(\mathbf{a})\,T(\mathbf{b}) = \exp\left(-i\frac{\boldsymbol{\pi}^B \cdot \mathbf{a}}{\hbar}\right)\exp\left(-i\frac{\boldsymbol{\pi}^B \cdot \mathbf{b}}{\hbar}\right)$$

$$= \exp\left(-i\frac{\boldsymbol{\pi}^B \cdot (\mathbf{a} + \mathbf{b})}{\hbar}\right)\exp\left(i\frac{q\mathbf{B} \cdot (\mathbf{a} \times \mathbf{b})}{2\hbar}\right)$$

$$= T(\mathbf{a} + \mathbf{b})\exp\left(i2\pi\frac{q}{e}\frac{\mathbf{B} \cdot (\mathbf{a} \times \mathbf{b})}{2\phi_0}\right)$$

$$= T(\mathbf{a} + \mathbf{b})\exp\left(i2\pi\frac{q}{e}\frac{\Phi}{\phi_0}\right) \quad (5.4.27)$$

The phase factor Φ/ϕ_0 is just the flux in units of $\phi_0 = h/e$ threading the triangle formed by the vectors $\mathbf{a}, \mathbf{b}$, and $-(\mathbf{a} + \mathbf{b})$.

The angular momentum **J**,

$$\mathbf{J} = \mathbf{r} \times \boldsymbol{\pi} = \hbar \mathbf{r}' \times \boldsymbol{\pi}' \qquad (5.4.28)$$

is also time-dependent. One finds

$$\frac{d\mathbf{J}}{dt} = \mathbf{r} \times \frac{d\boldsymbol{\pi}}{dt} = \mathbf{r} \times q\left(\frac{d\mathbf{r}}{dt} \times \mathbf{B}\right)$$

$$= q\left((\mathbf{r} \cdot \mathbf{B})\frac{d\mathbf{r}}{dt} - \mathbf{B}\left(\mathbf{r} \cdot \frac{d\mathbf{r}}{dt}\right)\right) \qquad (5.4.29)$$

For particles confined to a plane $\mathbf{r} \cdot \mathbf{z} = c$, c constant, and with **B** parallel to the z-direction, we may take $c = 0$ and define a conserved quantity L_z, given by

$$L_z = \frac{q}{e}(\mathbf{r} \times \boldsymbol{\pi})_z + eB\frac{r^2}{2} + n\hbar \qquad (5.4.30)$$

which in magnetic units is

$$L_z = \hbar\left(\frac{q}{e}(\mathbf{r}' \times \boldsymbol{\pi}')_z + \frac{r'^2}{2} + n\right) \qquad (5.4.31)$$

where n is the Landau level index. Characterization of eigenstates of H_0 by angular momentum is useful when discussing the fractional quantum Hall effect. Its inclusion in the definition of L_z ensures that L_z is positive. Operating with L_z on equation (5.4.16) shows that the $|nm\rangle$ are indeed eigenstates of L_z.

One may also define the so-called guiding center coordinate **R** with components (X, Y),

$$\mathbf{R} = \mathbf{r} - \frac{q}{e}\frac{l_0^2}{\hbar}\mathbf{z} \times \boldsymbol{\pi} \qquad (5.4.32)$$

which classically is just the center of the cyclotron motion in the plane perpendicular to the magnetic field. In the absence of any electric fields the quantity **R** is a constant of the motion. In a constant magnetic field the components of **R**, R_α, have the following commutation relations:

$$[R_\alpha, R_\beta] = -il_0^2(q/e)\varepsilon^{\alpha\beta\gamma}\Omega^\gamma \qquad (5.4.33)$$

or with the magnetic field parallel to the z-direction

$$[X, Y] = -i\frac{q}{e}l_0^2 = -i\frac{\hbar}{qB} \qquad (5.4.34)$$

As the magnetic field is made stronger, it becomes an increasingly accurate approximation to treat the quantities **R** as classical variables, for the commutator $[X, Y]$ tends to zero as B is increased. This has been the basis for an intuitively appealing description of the role of localized states in the quantum Hall effect.[36]

In magnetic units, $X' = X/l_0$, $Y' = Y/l_0$, equation (5.4.34) becomes

$$[X', Y'] = -i(q/e) \tag{5.4.35}$$

For the case of motion in the plane, $\mathbf{z} \cdot \mathbf{r} = 0$, one may also write

$$L_z = \left(\frac{eBr^2}{2} + \frac{q}{e}(\mathbf{r} \times \boldsymbol{\pi})_z + n\hbar \right) = \frac{eB}{2}\mathbf{R}^2 - \frac{\hbar}{2} \tag{5.4.36}$$

which in magnetic units can be written as

$$L_z/\hbar = (\mathbf{R}'^2 - 1)/2 \tag{5.4.37}$$

One may use the components X' and Y' to form operators $Z^{\pm}$,

$$Z^{\pm} = \frac{X' \pm i(q/e)Y'}{\sqrt{2}} \tag{5.4.38}$$

and then

$$[L_z, Z^{\pm}] = \pm \hbar L_z \tag{5.4.39}$$

The operators $Z^{\pm}$ are raising and lowering operators for the eigenstates of the angular momentum L_z defined in equations (5.4.31) and (5.4.37).

5.4.4. The Single-Particle Green's Function

In the presence of an electric field the single-particle Green's function for electrons confined to two dimensions, $G_0(\mathbf{r}_1, \mathbf{r}_2, \varepsilon)$, has simple asymptotic forms for $|\mathbf{r}_1 - \mathbf{r}_2| \gg l_0$.[8]

Assume that the electric field is applied parallel to the x-direction. Then in the Landau gauge H_0 is given for electrons ($q = -e$) by

$$H_0 = \frac{p_x^2 + (p_y + eBx)^2}{2m^*} + eEx \tag{5.4.40}$$

which in magnetic units has the simple form

$$H_0 = \frac{\hbar\omega_c}{2}\left(-\frac{\partial^2}{\partial x'} + \left(\frac{1}{i}\frac{\partial}{\partial y'} + x' \right)^2 + 2vx' \right) \tag{5.4.41}$$

with $v = eEl_0/\hbar\omega_c$.

The gauge H_0 has eigenstates ψ_{nk} with

$$\psi_{nk} = A_{nk}e^{i(k-v)y'}H_n(x' + k)e^{-(x'+k)^2/2}$$

$$A_{nk} = (2^n n! \sqrt{\pi})^{-1/2} \tag{5.4.42}$$

where H_n is the nth Hermite polynomial.[34] The ψ_{nk} are normalized for a system of length L' $(=L/l_0)$ in the y-direction. The eigenenergies ε_{nk} measured from $(\hbar\omega_c/2)(1 + v^2)$ are

$$\varepsilon_{nk} = n - kv \tag{5.4.43}$$

where k is a continuous real variable and n takes the values $0, 1, \ldots$. Imagining that the system is bounded at $x = \pm x_0$, the energies ε_{nk} form bands bounded by ε^u and ε^l with

$$\varepsilon^u = n + vx_0 \qquad \varepsilon^l = n - vx_0 \tag{5.4.44}$$

The Green's function $G_0(\mathbf{r}_2', \mathbf{r}_1', \varepsilon)$, with primes denoting magnetic units, is given by

$$G_0(\mathbf{r}_2', \mathbf{r}_1', \varepsilon) = \sum_{n=0}^{\infty} \int \frac{dk}{2\pi} \frac{\psi_{nk}(\mathbf{r}_2')\psi_{nk}^*(\mathbf{r}_1')}{\varepsilon - \varepsilon_{nk} + i\delta} \tag{5.4.45}$$

which can be written as

$$G_0(\mathbf{r}_2', \mathbf{r}_1', \varepsilon) = \frac{X}{2\pi^{3/2}} \sum_{n=0}^{\infty} 2^{-n}(n!)^{-1} \int dk$$
$$\cdot \frac{H_{nk}(x_2' + k)H_{nk}(x_1' + k)e^{-(k+\alpha/2)^2}}{\varepsilon - \varepsilon_{nk} + i\delta} \tag{5.4.46}$$

with

$$X = \exp\left(-i\left(v + \frac{x_1' + x_2'}{2}\right)(y_1' - y_2') - \frac{(r_2' - r_1')^2}{4}\right) \tag{5.4.47}$$

and

$$\alpha = x_1' + x_2' - i(y_2' - y_1') \tag{5.4.48}$$

If ε lies in one of the bands $m - kv$, this integral can be evaluated by integrating around the contour with the real axis as one edge and the line $\text{Im } k = (y_2' - y_1')/2$. The contributions from off the real axis and all contributions with $n \neq m$ contribute only exponentially small terms in the limit $|y_1' - y_2'| \to \infty$, $|x_1' - x_2'| \to \infty$. The asymptotic form for G is determined by the pole at $k = k_0$, where $\varepsilon = m - kv_0$. With the electric field parallel to the positive x-direction, as in equation (5.4.41), this only contributes for $y_2 < y_1$, so that,

$$\lim|\mathbf{r}_2' - \mathbf{r}_1'| \to \infty$$

$$G_0(\mathbf{r}_2', \mathbf{r}_1', \varepsilon) = \frac{-i}{v\sqrt{\pi}\,2^m m!}\,\theta(y_1' - y_2')H_m(x_2' + k_0)H_m(x_1' + k_0)$$
$$\cdot \exp\left(-\frac{(x_2' + k_0)^2}{2} - \frac{(x_1' + k_0)^2}{2} + i(k_0 - v)(y_2' - y_1')\right) \tag{5.4.49}$$

where θ denotes the Heaviside function. Comparing with equation (5.4.42) shows that G_0 is of the form

$$G_0(\mathbf{r}_2', \mathbf{r}_1', \varepsilon) \sim i\theta(y_1' - y_2')\psi_{nk_0}(\mathbf{r}_2')\psi_{nk_0}^*(\mathbf{r}_1') \tag{5.4.50}$$

The factoring of G_0 in equation (5.4.50) implies that only "forward scattering" of a particle in state, ψ_{nk_0}, is possible for electrons with energies ε within one of the energy continua [equation (5.4.44)]. If the system included impurities close, say, to the origin, one could insert G_0 into Dyson's equation and solve for the asymptotic form for ψ, the eigenstate of the perturbed system. Again, with an electric field parallel to the positive x-direction one would find

$$y' \gg 0 \qquad \psi(\mathbf{r}', \varepsilon = m - vk_0) \sim \psi_{mk_0}(\mathbf{r}')$$
$$y' \ll 0 \qquad \psi(\mathbf{r}', \varepsilon = m - vk_0) \sim \psi_{mk_0}(\mathbf{r}')e^{-i\delta(\varepsilon)} \tag{5.4.51}$$

The impurity potential simply gives rise to the additional phase shift $\delta(\varepsilon)$.

5.4.5. Exactness of Laughlin's Wave Function

The sense in which Laughlin's wave function ψ_L [equation (5.3.9)] is exact is as follows.[26] Any interparticle interaction of the form $v_{ij} = v(|\mathbf{r}_i - \mathbf{r}_j|)$ conserves the relative angular momentum of the particles i and j, L_z^{ij} (see Section 5.4.3). The interaction in the lowest Landau level may therefore be written as

$$v_{ij} = \sum_{m=1,3,\ldots} V_m P_m(i,j) \tag{5.4.52}$$

Here, $P_m(i,j)$ projects out from some wave function $\psi(\mathbf{r}_1,\ldots,\mathbf{r}_N)$, those components with relative angular momentum m for particles i and j.

For any pair i and j the two-particle wave function $\psi(\mathbf{r}_i, \mathbf{r}_j)$ can be written as products of eigenstates of relative and center-of-mass angular momenta, with quantum numbers M and m, as

$$\psi(\mathbf{r}_i, \mathbf{r}_j) \sim \psi_M\left(\frac{\mathbf{r}_i + \mathbf{r}_j}{\sqrt{2}}\right)\psi_m\left(\frac{\mathbf{r}_i - \mathbf{r}_j}{\sqrt{2}}\right) \tag{5.4.53}$$

In the lowest Landau level and in the symmetric gauge ψ_m takes the simple form [as shown in equation (5.4.15)]

$$\psi_m\left(\frac{\mathbf{r}_i - \mathbf{r}_j}{\sqrt{2}}\right) \sim (z_i - z_j)^m e^{-|z_i|^2 + |z_j|^2/4} \tag{5.4.54}$$

where $z_i = x_i + i(q/e)y_i$. Clearly Laughlin's wave function ψ_L can be written for any particles i and j as

$$\psi_L \sim (z_i - z_j)^m e^{-|z_i|^2 + |z_j|^2/4} P(z_i, z_j) \tag{5.4.55}$$

with $P(z_i, z_j)$ some polynomial.

At $\nu = 1/m$, ψ_L is the only wave function that contains no relative angular momentum component less than m. If one were to define a hard-core interaction $V^{h.c.}$,

$$V^{h.c.}_m = 1, \qquad m < n$$
$$V^{h.c.}_m = 0, \qquad m \geq n$$

$$(5.4.56)$$

then, at $\nu = 1/n$, ψ_L would be the exact nondegenerate ground state.

5.4.6. The Hierarchy

The following is based on the works of Haldane[26] and Halperin.[25]

Suppose a hierarchical fluid with n levels describes the ground state at some filling factor $\nu(m; p_1, p_2, \ldots, p_n)$ of the lowest Landau level. The N_i quasiparticles/holes at the ith level with charge $\alpha_i q_i$, where $\alpha = \pm 1$ according to whether the excitations are holelike or particlelike, and coordinates η^i_a would have a Laughlin-like ground state:

$$\tilde{\psi}_i \sim \prod_{a<b}^{N_i} (\eta^i_a - \eta^i_b)^{p_i + \alpha_i q_i/(q_{i-1})} \exp\left(-\sum_c \frac{|\eta^i_c|^2}{4q_i}\right)$$

$$(5.4.57)$$

The exponent $p_i + \alpha_i q_i/(q_{i-1})$, with p_i an even integer, ensures a consistent statistic for interchange of two excitations. Applying the same argument as the one leading to equation (5.3.17) shows that the change in phase when one quasiexcitation is taken round another is $\Delta\phi$, with

$$\Delta\phi = 2\pi\alpha_i q_i \frac{\Delta\Phi}{h/e} = 2\pi\alpha_i \frac{q_i}{q_{i-1}}$$

$$(5.4.58)$$

where $\Delta\Phi$ $(=h/q_{i-1})$ is the flux change to nucleate an excitation at level i. This is only consistent with a statistic for interchange of two quasiexcitations with coordinates η^i_a and η^i_b:

$$\tilde{\psi}_i(\eta^i_b, \eta^i_a) = \tilde{\psi}_i(\eta^i_a, \eta^i_b)e^{i\pi\alpha_i q_i/(q_{i-1})}$$

$$(5.4.59)$$

which is clearly satisfied by the particles described by ψ_i in equation (5.4.57).

The excitations from $\tilde{\psi}_i$ form the particles of the $(i+1)$th level. These carry charge

$$\alpha_{i+1}q_{i+1} = \alpha_{i+1}\frac{q_i}{p_i + \alpha_i q_i/(q_{i-1})}$$

$$(5.4.60)$$

and have a magnetic length l^2_{i+1} given by

$$l^2_{i+1} = \frac{l^2_0}{q_{i+1}} = l^2_i \frac{q_i}{q_{i+1}}$$

$$(5.4.61)$$

These ratios are determined by the exponent of the term $\eta_a^i - \eta_b^i$ in the Laughlin-like ground state proposed in equation (5.4.57).

Writing that

$$N_i = \nu(p_i, p_{i+1}, \ldots, p_n)N_{i-1} \tag{5.4.62}$$

we may use the condition that the fluid wave function describes particles uniformly distributed over the area of the system, A, to derive a recursion relation between the $\nu(p_i, p_{i+1}, \ldots, p_n)$. The area A_i of a system described by ψ_i is related to the number of zeros in ψ_i via

$$A_i = l_i^2 \left(p_i + \alpha_i \frac{q_i}{q_{i-1}} + \alpha_{i+1}\nu(p_{i+1}, \ldots, p_n) \right) N_i \tag{5.4.63}$$

and the area A_{i-1} is related to the number of zeros in $\tilde{\psi}_{i-1}$ via

$$A_{i-1} = l_{i-1}^2 \left(p_{i-1} + \alpha_{i-1} \frac{q_{i-1}}{q_{i-2}} + \alpha_i\nu(p_i, \ldots, p_n) \right) N_{i-1} \tag{5.4.64}$$

Requiring that $A_i = A_{i-1}$ and substituting for $N_i/N_{i-1} (=\nu(p_i, \ldots, p_n))$, $l_i^2/l_{i-1}^2 (=q_{i-1}/q_i)$, and $p_{i-1} + \alpha_{i-1}q_{i-1}/q_{i-2} (=q_{i-1}/q_i)$ yields the required relation:

$$\nu(p_i, \ldots, p_n) = \frac{1}{p_i + \alpha_{i+1}\nu(p_{i+1}, \ldots, p_n)} \tag{5.4.65}$$

It is straightforward to show that equation (5.4.65) has the solution for $\nu(m; p_1, p_2, \ldots, p_n)$ given by equation (5.3.20).

ACKNOWLEDGMENTS

It is a pleasure to acknowledge conversations with B. I. Halperin, G. Fano, F.-C. Zhang, and, in particular, R. Morf. I would also like to thank E. Tosatti for his hospitality at the ICTP in Trieste.

This work was partially funded by the SERC, grant no. GR/D89790.

REFERENCES

1. K. von KLITZING, G. DORDA, and M. PEPPER, *Phys. Rev. Lett.* **45**, 494 (1980).
2. D. C. TSUI, H. L. STÖRMER, and A. C. GOSSARD, *Phys. Rev. Lett.* **48**, 1559 (1982).
3. R. B. LAUGHLIN, *Phys. Rev. B* **23**, 5632 (1981).
4. T. ANDO, Y. MATSUMOTO, and Y. UEMURA, *J. Phys. Soc. Japan* **39**, 279 (1975).
5. K. von KLITZING, *Surface Science* **113**, 113 (1982).
6. H. L. STÖRMER and D. C. TSUI, *Science* **220**, 1241 (1983).
7. R. E. PRANGE, *Phys. Rev. B* **23**, 4802 (1981).
8. R. JOYNT and R. E. PRANGE, *Phys. Rev. B* **29**, 3303 (1984).

9. W. BRENIG, *Z. Phys. B* **50**, 305 (1983).

10. N. LEVINSON, *K. Dan Vrdensk. Selsk. Mat. Fys. Medd.* **25**, No. 9 (1949).

11. D. J. THOULESS, M. KOHMOTO, M. P. NIGHTINGALE, and M. den NIJS, *Phys. Rev. Lett.* **49**, 405 (1982).

12. B. I. HALPERIN, *Phys. Rev. B* **25**, 2185 (1982).

13. A. MACKINNON, private comm. (1986).

14. A. M. CHANG, P. BERGLUND, D. C. TSUI, H. L. STÖRMER, and J. C. M. HWANG, *Phys. Rev. Lett.* **53**, 997 (1984).

15. R. G. CLARK, in: *18th International Conference on Physics of Semiconductors* (Olof Engstrom, ed.), Vol. 1, p. 393, World Scientific (1986).

16. R. B. LAUGHLIN, *Phys. Rev. Lett.* **50**, 1395 (1983).

17. Q. NIU, D. J. THOULESS, and Y. S. WU, *Phys. Rev. B* **31**, 3372 (1985).

18. J. R. SCHRIEFFER, 1985 preprint "Strange Quantum Numbers in Condensed Matter and Field Theory," NSF-ITP-85-124.

19. B. I. HALPERIN, *Helv. Phys. Acta* **56**, 75 (1983).

20. J. M. CAILLOL, D. LEVESQUE, J. J. WEIS, and J. P. HANSEN, *J. Stat. Phys.* **28**, 325 (1982).

21. R. MORF and B. I. HALPERIN, *Phys. Rev. B* **33**, 2221 (1986).

22. S. M. GIRVIN, A. H. MACDONALD, and P. M. PLATZMAN, *Phys. Rev. B* **33**, 2481 (1986).

23. F. D. M. HALDANE and E. H. REZAYI, *Phys. Rev. Lett.* **54**, 237 (1985).

24. G. FANO, F. ORTOLANI, and E. COLOMBO, *Phys. Rev. B* **34**, 2670 (1986).

25. B. I. HALPERIN, *Phys. Rev. Lett.* **52**, 1583 (1984).

26. F. D. M. HALDANE, *Phys. Rev. Lett.* **51**, 605 (1983).

27. D. AROVAS, J. R. SCHRIEFFER, and F. WILCZEK, *Phys. Rev. Lett.* **53**, 722 (1984).

28. R. MORF, N. d'AMBRUMENIL, and B. I. HALPERIN, *Phys. Rev. B* **34**, 3037 (1986).

29. F. D. M. HALDANE, in: *The Quantum Hall Effect* (S. M. Girvin, ed.), Springer (1987).

30. N. d'AMBRUMENIL and A. M. REYNOLDS, *J. Phys. C* **21**, 119 (1988).

31. A. H. MACDONALD and S. M. GIRVIN, *Phys. Rev. B* **33**, 4009 (1986).

32. S. KIVELSON, C. CALLIN, D. P. AROVAS, and J. R. SCHRIEFFER, *Phys. Rev. Lett.* **56**, 873 (1986).

33. G. BASKARAN, *Phys. Rev. Lett.* **56**, 2716 (1986).

34. I. S. GRADSHTEYN and I. M. RYZHIK, *Table of Integrals, Series and Products*, Sections 8.95 and 8.97, Academic, New York (1980).

35. E. BROWN, *Phys. Rev.* **133** 1038 (1968).

36. B. I. HALPERIN, *Scientific American*, 40 (April 1986).

Chapter 6

Green's Function Methods: Quantum Boltzmann Equation for Linear Transport

Gerald D. Mahan

6.1. INTRODUCTION

Transport is called *linear* if the current is proportional to the driving force. The forces might be electric fields F, temperature gradients ∇T, or density gradients ∇n. Examples of linear transport relations are $j = \sigma F$ or $Q = K(\nabla T)$, where the currents j (electrical) or Q (heat) are determined in terms of macroscopic coefficients σ (electrical conductivity) or K (thermal conductivity). These coefficients can be measured experimentally and calculated theoretically, which provides the meeting ground between experiment and theory. The theoretical calculation of σ and K can be done by using equilibrium Green's functions.[1-2] A set of theoretical techniques for linear transport, based on equilibrium Green's functions, have been developed and are widely used.

A very different technique for transport theory is based on nonequilibrium Green's functions. Here one derives and solves an equation of motion for a particle distribution function f.[3-15] In the semiclassical theory of Boltzmann, $f(r, t, v)$ has the three variables of position r, time t, and velocity v. Modern Green's functions methods require a four-variable distribution function $f(k, \omega; r, t)$ based on the Wigner distribution function. The correct gauge-invariant quantum Boltzmann equation (QBE) for $f(k, \omega; r, t)$ was

---

Gerald D. Mahan • Department of Physics, University of Tennessee, Knoxville, Tennessee, 37830, USA, *and* Solid State Division, Oak Ridge National Laboratory, Oak Ridge, Tennessee, 37381-6024, USA.

Quantum Transport in Semiconductors, edited by David K. Ferry and Carlo Jacoboni. Plenum Press, New York, 1991.

derived only recently by Mahan and Hansch.[12–14] For linear transport the QBE predicts results for σ which are identical to the equilibrium methods. The emphasis in the present volume is on hot-electron transport, which is very nonlinear. Then one must use the nonequilibrium methods based on the BE or the QBE. This chapter introduces the nonequilibrium methods.

6.2. TIME-DEPENDENT GREEN'S FUNCTIONS

6.2.1. Six Green's Functions

For nonequilibrium phenomena, it is necessary to define six different Green's functions. It is possible to employ fewer than six since they are not independent, but using six simplifies the notation. They are all correlation functions which (in the Heisenberg representation) relate the field operator $\psi(x)$ of the particle at one point $x_1 = (\mathbf{r}_1 t_1)$ in space-time to the conjugate field operator $\psi^+(x_2)$ at another point $x_2 = (\mathbf{r}_2 t_2)$. The six functions are the advanced G^a, retarded G^r, time-ordered G_t, antitime ordered $G_{\bar{t}}$, and $G^<$ and $G^>$, which have no names.

$$G^>(x_1 x_2) = -i\langle \psi(x_1)\psi^+(x_2)\rangle$$

$$G^<(x_1 x_2) = i\langle \psi^+(x_2)\psi(x_1)\rangle$$

$$G_t(x_1 x_2) = \theta(t_1 - t_2)G^>(x_1 x_2) + \theta(t_2 - t_1)G^<(x_1 x_2)$$

$$G_{\bar{t}}(x_1 x_2) = \theta(t_2 - t_1)G^>(x_1 x_2) + \theta(t_1 - t_2)G^<(x_1 x_2) \tag{6.2.1}$$

$$G^r(x_1 x_2) = G_t - G^< = G^> - G_{\bar{t}}$$

$$G^a(x_1 x_2) = G_t - G^> = G^< - G_{\bar{t}}$$

where $\theta(t)$ is the step function, which is zero for $t < 0$ and one for $t > 0$. These expressions have a somewhat different meaning for equilibrium than for nonequilibrium.

For equilibrium, the brackets imply a thermodynamic average over all possible states of the system. For homogeneous systems in equilibrium, the Green's functions depend only upon the difference of their arguments $(x_1, x_2) = (x_1 - x_2)$. Then the most simple, and useful, quantities are the Fourier transform of these quantities

$$G(k, \omega) = \int d^3 r\, e^{-ik\cdot r} \int dt\, e^{i\omega t} G(r, t) \tag{6.2.2}$$

Explicit expressions for these quantities are given below.

For systems out of equilibrium, the definitions in (6.2.1) have a different meaning. The bracket no longer signifies thermodynamic averaging, since that concept is now ill-defined. Instead, the bracket signifies the need to average over the available states of the system for the nonequilibrium distributions. Furthermore, the arguments of the Green's functions (x_1, x_2) are *not* usually given as the difference $(x_1 - x_2)$. For example, nonequilibrium could be caused by transients, which make the Green's functions depend upon (t_1, t_2) rather than $(t_1 - t_2)$.

The above set of equations, with the indicated sign change between $G^>$ and $G^<$, is valid for fermions, such as electrons. For boson particles, there is no sign change.

Often the Hamiltonian H can be solved exactly in terms of single-particle eigenfunctions $\psi_\lambda(r_1)$ and eigenvalues ε_λ. Two examples are the electron in a magnetic field or a free particle. Then it is useful to have the expressions for the Green's functions in terms of these eigenfunctions. They are derived by expanding the field operators in terms of the eigenfunctions and creation C_λ^+ and destruction C_λ operators ($\hbar = 1$)

$$\psi(x_1) = \sum_\lambda C_\lambda \psi_\lambda(r_1) e^{-i\varepsilon_\lambda t_1}$$

$$\psi^+(x_2) = \sum_\lambda C_\lambda^+ \psi_\lambda^*(r_2) e^{+i\varepsilon_\lambda t_2}$$

(6.2.3)

The Green's functions in (6.2.1) may be evaluated with the occupation factor $n_\lambda = \langle C_\lambda^+ C_\lambda \rangle$ and $t = t_1 - t_2$, as follows:

$$G^>(x_1, x_2) = -i \sum_\lambda (1 - n_\lambda)\psi_\lambda(r_1)\psi_\lambda^*(r_2)e^{-i\varepsilon_\lambda t}$$

$$G^<(x_1, x_2) = i \sum_\lambda n_\lambda \psi_\lambda(r_1)\psi_\lambda^*(r_2)e^{-i\varepsilon_\lambda t}$$

$$G_t(x_1, x_2) = -i \sum_\lambda [\theta(t) - n_\lambda]\psi_\lambda(r_1)\psi_\lambda^*(r_2)e^{-i\varepsilon_\lambda t}$$

$$G_{\bar{t}}(x_1, x_2) = -i \sum_\lambda [\theta(-t) - n_\lambda]\psi_\lambda(r_1)\psi_\lambda^*(r_2)e^{-i\varepsilon_\lambda t}$$

(6.2.4)

$$G^r(x_1, x_2) = -i\theta(t) \sum_\lambda \psi_\lambda(r_1)\psi_\lambda^*(r_2)e^{-i\varepsilon_\lambda t}$$

$$G^a(x_1, x_2) = i\theta(-t) \sum_\lambda \psi_\lambda(r_1)\psi_\lambda^*(r_2)e^{-i\varepsilon_\lambda t}$$

and $t = t_1 - t_2$.

The starting point for any calculation, at least conceptually, is the behavior of the Green's functions for systems without interactions. Then the wave functions are those for plane wave or noninteracting Bloch states (if such can be defined). The quantum number λ becomes the wave vector k and a spin index σ, which is usually not written. The eigenvalue combination is $\psi_k(r_1)\psi_k^*(r_2) = \exp[ik'(r_1 - r_2)]$. The subscript 0 on the Green's functions means to use those for a noninteracting system in equilibrium. Fourier-transforming the r-variable to k as in (6.2.2) gives the free-particle

Green's functions $G_0(k, t)$. For fermions of band energy ε_k and occupation number $n_k = n_F(\varepsilon_k)$ they are

$$G_{t0}(k, t) = -i[\theta(t) - n_k]\exp(-i\varepsilon_k t)$$

$$G_{\bar{t}0}(k, t) = -i[\theta(-t) - n_k]\exp(-i\varepsilon_k t)$$

$$G_0^<(k, t) = in_k\exp(-i\varepsilon_k t)$$

$$G_0^>(k, t) = -i(1 - n_k)\exp(-i\varepsilon_k t) \tag{6.2.5}$$

$$G_0^r(k, t) = -i\theta(t)\exp(-i\varepsilon_k t)$$

$$G_0^a(k, t) = i\theta(-t)\exp(-i\varepsilon_k t)$$

The t-variable can be Fourier-transformed, which gives the noninteracting Green's function of frequency. The quantity η is infinitesimal.

$$G_0^r(k, \omega) = \frac{1}{\omega - \varepsilon_k + i\eta}$$

$$G_0^a(k, \omega) = \frac{1}{\omega - \varepsilon_k - i\eta}$$

$$G_0^<(k, \omega) = 2\pi in_k\delta(\omega - \varepsilon_k) \tag{6.2.6}$$

$$G_0^>(k, \omega) = -2\pi i(1 - n_k)\delta(\omega - \varepsilon_k)$$

$$G_{t0} = G_0^r + G_0^<$$

$$G_{\bar{t}0} = -G_0^a + G_0^<$$

Note that expressions such as $G^r = G_t - G^<$ are obeyed for interacting and noninteracting functions. They are obeyed for both cases of arguments (k, t) and (k, ω).

The QBE is an equation of motion for the nonequilibrium Green's functions $G^<$. It is necessary to first know the equation of motion for the noninteraction Green's functions, which are now derived. If the noninteracting Hamiltonian is $H_0 = \sum \varepsilon_k C_k^+ C_k$, then these noninteracting Green's functions obey the transient equation, which can be derived from equation (6.2.5),

$$\left(i\frac{\partial}{\partial t} - \varepsilon_k\right)\begin{cases} G_{t0} = \delta(t) \\[6pt] G_{\bar{t}0} = -\delta(t) \\[6pt] G_0^< = 0 \\[6pt] G_0^> = 0 \\[6pt] G_0^r = \delta(t) \\[6pt] G_0^a = +\delta(t) \end{cases} \tag{6.2.7}$$

The delta functions $\delta(t - t')$ come from the derivative of the step function $\theta(t - t')$.

The above Green's functions are suitable for particles such as electrons or holes in semiconductors. Another type of Green's function is needed for boson fields such as phonons or photons. For phonons let $Q(x)$ be the displacement operator from equilibrium of the ions in the solid at position $x = (r, t)$ in space-time. The phonon Green's functions are defined as

$$D^>(x_1 x_2) = -i\langle|Q(x_1)Q(x_2)|\rangle$$
$$D^<(x_1 x_2) = -i\langle|Q(x_2)Q(x_1)|\rangle$$
$$D_t(x_1 x_2) = \theta(t_1 - t_2)D^>(x_1 x_2) + \theta(t_2 - t_1)D^<(x_1 x_2)$$
$$D_{\bar{t}}(x_1 x_2) = \theta(t_2 - t_1)D^>(x_1 x_2) + \theta(t_1 - t_2)D^<(x_1 x_2)$$
$$D^r(x_1 - x_2) = D_t - D^< = D^> - D_{\bar{t}}\theta(t_1 - t_2)[D^> - D^<]$$
$$D^a(x_1 - x_2) = D_t - D^> = D^< - D_{\bar{t}} - \theta(t_2 - t_1)[D^> - D^<]$$

$$(6.2.8)$$

These expressions are similar to those in (6.2.1) for particles. The main difference is that $D^<$ and $D^>$ have the same sign, since no sign change is made when interchanging the positions of boson operators. Also the displacement operator is Hermitian $[Q^+ = Q]$, which introduces some redundancy, such as $D^<(x_1 x_2) = D^>(x_2 x_1)$.

The displacement operators Q are usually represented in terms of phonon raising (a^+) and lowering (a) operators. The usual representation is

$$Q(x) = a_q e^{iq \cdot x} + a_q^+ e^{-iq \cdot x}$$

where $q \cdot x = \mathbf{q} \cdot \mathbf{r} - \omega_q t$, and ω_q is the phonon frequency. In this representation, the phonon Green's functions in equilibrium are expressed in terms of the phonon occupation number $N_q = \langle a_q^+ a_q \rangle$, which equals $1/[\exp(\beta\omega_q) - 1]$ in equilibrium ($\beta = 1/kt$, $\hbar = 1$).

$$D^>(q, t) = -i\{(N_q + 1)e^{-i\omega_q t} + N_q e^{i\omega_q t}\}$$
$$D^<(q, t) = -i\{(N_q + 1)e^{i\omega_q t} + N_q e^{-i\omega_q t}\}$$
$$D^r(q, t) = -2\theta(t)\sin(\omega_q t)$$

$$(6.2.9)$$

The other three phonon Green's functions are found easily from these three.

6.2.2. Time Loops and the *S*-Matrix Expansion

The Green's functions G were defined in equation (6.2.1). Explicit expressions were given for the Green's functions of one-particle systems in equation (6.2.4), and for plane-wave systems in equations (6.2.5) and (6.2.6). The treatment of interacting and nonequilibrium systems will require that we learn how to find their Green's functions. Usually the Green's functions cannot be found exactly, and approximate expressions are obtained using versions of perturbation theory. These techniques are briefly reviewed in this section.

First define the *interaction representation*, which is a fancy name for writing the Hamiltonian H as

$$H = H_0 + V$$

where H_0 can be solved exactly while V contains the other terms which cannot be solved exactly. Usually H_0 is the Hamiltonian for free particles, where band structure is included crudely by using an effective mass m instead of the actual free-electron mass m_e. For electron systems, energy is related to the chemical potential μ so that

$$H_0 - \mu N = \sum_k \varepsilon_k C_k^+ C_k$$

$$\varepsilon_k = k^2/2m - \mu \qquad (6.2.10)$$

where N is the operator $\sum_k C_k^+ C_k$. The Green's functions for the Hamiltonian H_0 were given in the previous section. How does one include the interactions V? Define an operator $U(t)$ as

$$U(t) = e^{itH_0}e^{-itH}$$

Note that the two exponential operators cannot be combined by simple algebra, since $\exp(A)\exp(B) \neq \exp(A+B)$ if the two operators do not commute. We assume that they do not commute, since if H_0 commutes with V the Hamiltonian can usually be solved exactly, and there is no need for the present approach.

An expression for $U(t)$ is obtained by taking its time derivative. One must be careful about the order of operators, since they do not commute:

$$dU(t)/dt = \exp(itH_0)[iH_0 - iH]\exp(-itH)$$

$$= -i\exp(itH_0)V\exp(-itH)$$

$$= -i\bar{V}(t)U(t) \qquad (6.2.11)$$

where

$$\bar{V}(t) = \exp(itH_0)V\exp(-itH_0)$$

The full-time dependence of $V(t)$ is $\exp(itH)V\exp(-itH)$, which is not known since the eigenstates of H are not known. In the interaction representation, we use H_0 to govern the time dependence of operators, and put an overbar over these operators to remind us that we are in the interaction representation. Equation (6.2.11) is solved by integrating the time variable, starting from $t = 0$ where $U(0) = 1$:

$$U(t) = 1 - i\int_0^t dt_1 \, \bar{V}(t_1)U(t_1)$$

This equation is solved by iteration, which results in an infinite series of terms. The nth term has n nested time integrals

$$U(t) = \sum_n^\infty (-i)^n \int_0^{t_1} dt_1 \int_0^{t_1} dt_2 \cdots \int_0^{t_{n-1}} dt_n \; \bar{V}(t_1)\bar{V}(t_2) \cdots \bar{V}(t_n) \qquad (6.2.12)$$

The $n = 0$ term is one. This series can be represented in an economical notation by introducing the time-ordering operator T. The operator T in front of a bunch of functions of time is an instruction to order the functions according to their time ordering with the earliest times to the right. Then the above series can be rewritten as

$$U(t) = T \exp\left\{-i \int_0^t dt_1 \; \bar{V}(t_1)\right\}$$

The exponential of an operator is interpreted as a power series, and the time ordering rearranges each term to give back the expression (6.2.12).

These formal manipulations are useful when solving for the time dependence of the Green's functions. Examine any one of them, say $G^<$, which is

$$G^<(x_1, x_2) = i\langle|e^{iHt_2}\psi^+(r_2)e^{iH(t_1-t_2)}\psi(r_1)e^{-iHt_1}|\rangle$$

We go to the interaction representation by introducing the factors of $1 = \exp(-itH_0)\exp(+itH_0)$ before and after each field operator:

$$G^<(x_1, x_2) = i\langle|U^+(t_2)\bar{\psi}^+(x_2)U(t_2)U^+(t_1)\bar{\psi}(x_1)U(t_1)|\rangle \qquad (6.2.13)$$

The overbar over the field operators is a reminder that their time dependence is to be evaluated in the interaction representation.

The S-matrix is defined as

$$S(t_2, t_1) = U(t_2)U^+(t_1)$$
$$= T \exp\left[-i \int_{t_1}^{t_2} dt' \; \bar{V}(t')\right] \qquad (6.2.14)$$

Equation (6.2.13) is now easy to interpret. The symbol $|\rangle$ stands for the ground state of the system, which is taken at time $t = 0$. The operation by $U(t)$ advances this state in time to t, so $|t\rangle = U(t)|\rangle$ is the state at time t. Thus, we can express the Green's function as

$$G^<(x_1, x_2) = i\langle t_2|\bar{\psi}^+(x_2)S(t_2, t_1)\bar{\psi}(t_1)|t_1\rangle \qquad (6.2.15)$$

The field operator $\bar{\psi}(t_1)$ acts upon the eigenstate at t_1 and creates some sort of excitation in the system. The S-matrix describes how the system evolves in time from t_1 to t_2. The other field operator destroys this excitation at t_2, and the eigenstate at t_2 measures the amount of correlation left in the initial excitation.

Several factors in equation (6.2.15) are now known. The S-matrix is given by (6.2.14), and the field operators are expressions such as (6.2.3), which are also known. However, equation (6.2.15) contains the eigenstates $|t\rangle$ which are not known. They are related to the exact ground state at $t = 0$ $\{|0\rangle = |\rangle\}$, which is initially unknown. One of the goals of Green's function theory is to find the ground state of the interacting system.

Solid-state Green's functions evolved historically from field theory which came from scattering theory. In scattering theory the Hamiltonian H_0 was for free particles, while V came from the interaction between the free particles and a scattering target. One conceived of a beam of free particles impinging on a target. The potential $V(t)$ acts only while the particles are at the target, so it is "on" for only a short time. The potential vanishes in both limits of $t \to \pm\infty$. The exact ground state is obtained by starting from the noninteracting ground state $|\rangle_0$ at $t = \pm\infty$, which are eigenstates of H_0, and using the S-matrix to describe how this state evolves to the interacting ground state

$$|t\rangle = S(t, -\infty)|\rangle_0$$

$$\langle t| = {}_0\langle|S(\infty, t)$$

which gives for the Green's function

$$G^<(x_1, x_2) = i{}_0\langle|S(\infty, t_2)\bar{\psi}^+(x_2)S(t_2, t_1)\bar{\psi}(x_1)S(t_1, -\infty)|\rangle_0$$

$$= i{}_0\langle|TS(\infty, -\infty)\bar{\psi}^+(x_2)\bar{\psi}(x_1)|\rangle_0 \qquad (6.2.16)$$

The second expression is equivalent to the first, since the time-ordering operator T will break up the S-matrix into the proper three parts according to the time arguments.

The above procedure is correct in scattering theory. However, in solid-state many-body theory, it is incorrect because the interactions do not go away at long time. If the potential V represents an interaction such as electron–phonon scattering, then the state at $t = +\infty$ is very different from the one at $t = -\infty$. At $t = -\infty$ one starts with the noninteracting electron gas and uses the S-matrix for the electron–phonon interaction to describe how the system evolves in time. At large time, the interactions lead to a very different state than the initial one without interactions. In a metal, the long-time state could be a superconductor, which is often caused by the electron–phonon interaction. The interactions do not shut off if we leave them on for a long time. The state at $t \to +\infty$ is unknown and cannot be used as the basis state of perturbation theory. This means that the S-matrix cannot be carried to $t = +\infty$ and terminated.

The procedure for avoiding this dilemma was proposed by Blandin, Noutier, and Hone.[11] They suggested that the integration path in the S-matrix be a time-loop which returns the system back to the noninteracting state at $t \to -\infty$. The integration path is shown in Figure 6.2.1 to go from $-\infty$ to τ and then back to $-\infty$. The value of τ is large and approaches $+\infty$. The advantage of this method is that the state $_0\langle|$ is now the initial noninteracting state, which is known as the eigenstate of H_0. The variable s denotes the integration variable along the time loop, and the time-ordering operator now orders according to the earliest values of s; those on the return loop are all later than those on the outward path.

$$S = T_s \exp\left[-i \int_{\text{loop}} ds\, \bar{V}(s)\right] \qquad (6.2.17)$$

$$G^<(x_1, x_2) = i_0\langle| T_s S \bar{\psi}^+(x_2) \bar{\psi}(x_1) |\rangle_0$$

This expression is now the basis for the evaluation of all of the Green's functions in time. The expression for $G^<$ can be generalized to the other five functions.

6.2.3. Dyson's Equation

The Green's function in equations (6.2.15)–(6.2.17) can be written as

$$G^<(x_1, x_2) = i \sum_n \frac{(-i)^n}{n!} \int ds_1 \int ds_2 \cdots \int ds_n \,{}_0\langle| T_s \bar{V}(s_1)$$

$$\cdots \bar{V}(s_n) \bar{\psi}^+(x_2) \bar{\psi}(x_1) |\rangle_0 \qquad (6.2.18)$$

where all s-integrals are over the time loop. We must learn to evaluate expressions of this type. The general evaluation is based upon a series of theorems which will be stated but not proven. Details and proofs are found in standard textbooks.[1]

The interaction potential V often has the form

$$\bar{V}(t) = \int d^3r\, \bar{\psi}^+(x) \bar{\psi}(x)\, \bar{W}(x) \qquad (6.2.19)$$

where $W(x)$ has a different form for scattering of the particle by impurities or phonons. A different form for V is used for particle–particle interactions.

When $\bar{V}(t)$ has the form in equation (6.2.19), the nth term in the series for $G^<$ has $n+1$ factors of ψ^+ and $n+1$ factors of ψ. The expectation value $_0\langle| \cdots |\rangle_0$ of all of these factors is found by *pairing* the operators: each ψ^+ is paired with one ψ. Each pair is separately time-ordered and averaged, so that the nth term gives a product of $n+1$ averages of factors such as

$$_0\langle| T_s \bar{\psi}^+(r_i, s_i) \bar{\psi}(r_j, s_j) |\rangle_0$$

FIGURE 6.1. Time-loop path for the *S*-matrix. The time integrals start and return to minus infinity.

These expressions have the same form as the Green's functions defined in equation (6.2.1). The correspondence becomes exact when we consider how to evaluate the time ordering. If both s_i and s_j are in the top loop, the above average is just the time-ordered Green's function. If they are both in the return loop, the above expression is the antitime-ordered Green's function. If one s-variable is in the top loop and the other is the bottom loop, then the T_s operator makes this expression either $G^<$ or $G^>$. These relationships are shown in Figure 6.2. Thus the n term in the Green's function expansion is a product of $n + 1$ factors, where each factor is one of the four Green's functions defined in Figure 6.2.

Suppose that the interaction V is the electron–phonon interaction. The interaction $W(x)$ is then proportional to the operator $Q(x)$ for the ion displacements

$$W(r, t) = \int dr' \, \phi(r - r')Q(r', t) \tag{6.2.20}$$

The interaction term $\phi(r)$ is described below. Relativistic retardation is neglected, and this interaction is taken to be instantaneous. The time-ordered product in (6.2.18) has n factors of the phonon displacement Q. They are also paired into phonon Green's functions using the definitions in equation (6.2.8). These pairings produce one of the four phonon Green's functions: time-ordered, antitime-ordered, $D^<$, or $D^>$ according to whether the s-arguments are on the top or bottom time loop.

There are many terms in the series for the Green's function. For the nth term in the S-matrix expansion, one takes all possible pairings of ψ^+ and ψ operators and all possible pairings of Q operators. The nth term alone has many combinations and produces many terms. Feynman diagrams are a useful way to characterize a term in the expansion. One draws a picture which represents each term. Green's functions $G(x_i x_j)$ are represented by straight solid lines. These lines represent trajectories in time between t_i and t_j. The phonon Green's functions are represented by dashed lines, which are usually curved to prevent overlap with the particle lines.

As an example, consider the case of $n = 2$. There are three Green's function lines. They are shown in Figure 6.3 as the solid lines. The dashed line represents the phonon Green's function. There are only two possible

FIGURE 6.2. Four Green's functions are defined according to whether both time arguments (s_1, s_2) are on the same or different legs of the time loop.

FIGURE 6.3. Feynman diagrams for the $n = 2$ self-energy of the electron from the electron–phonon interaction. The dashed lines represent phonon Green's functions, and the solid line represents the electron Green's function. The second diagram equals zero since it involves phonons with zero wave vector.

different S-matrix terms for $n = 2$. The Feynman diagram is drawn so that the phonon lines do not start or end at the two end points. A single phonon (dashed) phonon line must start or end at each internal electron point; these points are where one electron Green's function ends and the next one begins.

The next example is $n = 4$. There are five Green's functions lines. They are shown in Figure 6.4 as the five segments of the straight solid line. The five segments are connected, since there is a ψ^+ and ψ at each time. One of these (ψ^+) starts one segment while ψ terminates the other segment. One phonon operator Q is associated with each interior point which starts or stops a segment. A dashed line represents Q: the phonon starts or ends there. For $n = 4$, the four interior points produce two phonon lines, and the diagram can be drawn several different ways if the solid lines are connected as shown in Figure 6.4a, Figure 6.4b shows other Feynman diagrams where the solid lines are not connected to the others. Each term in Figure 6.4a, b represents a valid term in the S-matrix expansion. The disconnected diagrams of Figure 6.4b give zero and are neglected. They vanish because contributions cancel from the two parts of the time loop. One only keeps the connected terms. In writing out the various terms in the S-matrix, each connected diagram occurs more than once. These identical drawings give identical contributions, which cancel the $n!$ factor in the nth term in the Green's function (6.2.18). Correct counting shows that one keeps each different connected term once.

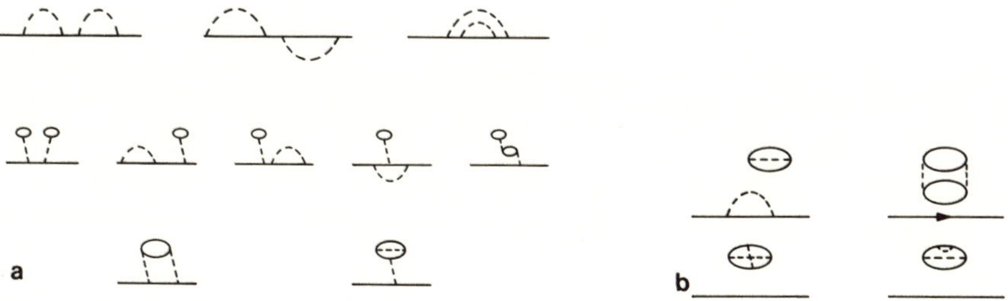

FIGURE 6.4. Feynman diagrams for the $n = 4$ self-energy of the electron from the electron–phonon interaction: (a) connected diagrams; (b) disconnected diagrams.

There are parts of the connected diagram where segments are connected only by a single Green's function, by a single solid line. Other parts are connected by several solid and dashed lines. *Self-energy functions* $\Sigma(x_1 x_2)$ are defined as the segments connected by several lines. Some of them are shown in Figure 6.5. Each Feynman diagram can be written as products of Green's functions and self-energy functions. The self-energy functions are also distinguished according to labels such as t (time-ordered), $\bar{t}$ (antitime-ordered), $\langle, \rangle$, r (retarded), and a (advanced). Note that in Figure 6.5, terms from Figure 6.4a, which represent higher-order products of those terms in Figure 6.3, do not appear as part of the self-energy.

Craig[10] introduced a useful notation which expressed four of these Green's functions as the elements of a 2×2 matrix. The self-energy terms are also a matrix.

$$\tilde{G} = \begin{bmatrix} G_t & -G^< \\ G^> & -G_{\bar{t}} \end{bmatrix}, \qquad \tilde{\Sigma} = \begin{bmatrix} \Sigma_t & -\Sigma^< \\ \Sigma^> & -\Sigma_{\bar{t}} \end{bmatrix} \tag{6.2.21}$$

The actual form for the self-energy functions is discussed in the following sections. For electrons in solids, important self-energy contributions are provided by the electron's interaction with other electrons, ions in the lattice, impurities, and phonons. For phonons, the important interactions are with other phonons, electrons, impurities, and boundaries.

For systems either in equilibrium or nonequilibrium, *Dyson's equation* is most easily expressed by using the matrix notation

$$\tilde{G}(x_1 x_2) = \tilde{G}_0(x_1 - x_2)$$
$$+ \int_{-\infty}^{\infty} dx_3 \int_{-\infty}^{\infty} dx_4 \, \tilde{G}_0(x_1 - x_3)\tilde{\Sigma}(x_3 x_4)\tilde{G}(x_4 x_2) \tag{6.2.22}$$

This expression appears rather formidable. However, some simple expressions can be obtained for the respective Green's functions. First write the above equation in a notation where the product of two functions implies an integration over the four-variable dx, which condenses the same equation to

$$\tilde{G} = \tilde{G}_0 + \tilde{G}_0 \tilde{\Sigma} \tilde{G}$$

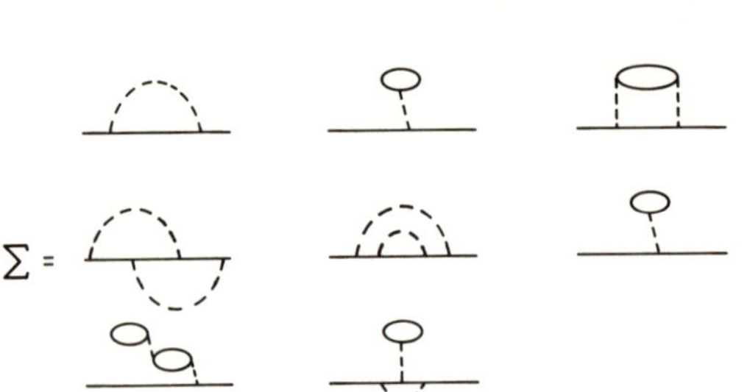

FIGURE 6.5. The Feynman diagrams for the self-energy of an electron from the electron–phonon interaction. Only connected diagrams contribute.

Then the equations are iterated. The following exact expressions are derived for the equations obeyed by the various Green's functions, using the same product notation:

$$G^r = G_0^r[1 + \Sigma^r G^r]$$

$$G^a = G_0^a[1 + \Sigma^a G^a]$$

$$G^{\lessgtr} = [1 + G^r\Sigma^r]G_0^{\lessgtr}[1 + \Sigma^a G^a] + G^r\Sigma^{\lessgtr}G^a \qquad (6.2.23)$$

$$G_t = [1 + G^r\Sigma^r]G_{0t}[1 + \Sigma^a G^r] + G^r\Sigma_{\bar{t}}G^a$$

$$G_{\bar{t}} = [1 + G^r\Sigma^r]G_{0\bar{t}}[1 + \Sigma^a G^a] + G^r\Sigma_t G^a$$

These equations represent multiple integrals in d^3r and dt.

They simplify considerably for homogeneous systems in the steady state, where the arguments of the Green's functions and self-energies depend only on $x_1 - x_2$. If the equations are Fourier-transformed, then all quantities depend only on (k, ω). The above equations are, after Fourier-transforming, just algebraic quantities which are easily solved. Using these results in (6.2.6) and (6.2.23), it is easy to show that for homogeneous, steady systems *in equilibrium* the interacting Green's functions are

$$G^r(k, \omega) = \frac{1}{\omega - \varepsilon_k - \Sigma^r}; \qquad \sigma = \omega - \varepsilon_k - \operatorname{Re}\Sigma^r$$

$$G^a(k, \omega) = \frac{1}{\omega - \varepsilon_k - \Sigma^a}; \qquad \Sigma^a = (\Sigma^r)^*$$

$$A(k, \omega) = -2\operatorname{Im} G^r = \frac{2\Gamma}{\sigma^2 + \Gamma^2}; \qquad \Gamma = -\operatorname{Im}\Sigma^r > 0$$

$$\Sigma^< = 2in_F(\omega)\Gamma(k, \omega); \qquad n_F(\omega) = \frac{1}{e^{\beta\omega} + 1} \qquad (6.2.24)$$

$$G^< = in_F(\omega)A(k, \omega)$$

$$\Sigma^> = 2i\Gamma(1 - n_F)$$

$$G^> = -iA(1 - n_F)$$

which are the only ones needed. The notation $\Gamma = -\operatorname{Im}\Sigma^r$ will often be used. These equilibrium solutions are important because they are the basis for all transport calculations. The system is always trying to relax back toward equilibrium.

6.2.4. Electron Self-Energies

For electrons in solids, the important self-energies are from interactions with (1) other electrons; (2) static lattice (band structure); (3) dynamic lattice (phonons); and (4) impurities. The band-structure effects are usually incorporated into an effective-mass approximation. The electron–electron interactions screen the other interactions. For transport, the important scattering mechanisms which limit the conductivity are provided by impurities and phonons. In this section we discuss the electron self-energies from these latter two interactions. Electron–electron interactions are assumed to screen these interactions, and the band structure is assumed to be incorporated into an effective mass. Spin effects are ignored. This model of a solid is probably adequate for simple metals and semiconductors, but is too simple for systems with d or f electrons.

All solids contain impurities. At low temperatures, the scattering of electrons by impurities is the most important mechanism for limiting the electrical conductivity. There have been numerous theoretical calculations on the electrical conductivity, or resistivity, from impurity scattering. The present example makes two assumptions which are intended to make the solution as easy as possible: (1) the impurities are dilute, so that the simultaneous scattering from several impurities can be neglected; and (2) the impurities have no internal degrees of freedom, such as spin or vibrations, which can be altered by the electron scattering. The second assumption implies the impurity is a simple potential which elastically scatters the electron.

The impurities are randomly located in the solid, and the usual methods[1] are employed for averaging over the random distributions. The factor $W(x)$ in equation (6.2.19) is written as

$$W(x) = \sum_j V(\mathbf{r} - \mathbf{R}_j)$$

where the summation is over all of the impurities in the solid at positions R_j. This expression is rewritten in terms of the Fourier transform $v(q)$ of $V(r)$:

$$W(x) = \frac{1}{\Omega} \sum_q v(q)\rho_i(q)$$

$$\rho_i(q) = \sum_j e^{iq \cdot R_j}$$

The impurity density operator $\rho_i(q)$ must be evaluated. The usual assumption is that the positions R_j of the impurities are uncorrelated. For each term in the Green's function expansion one must average each impurity over position. Then taking **averages** of products of the impurity density gives

$$\langle \rho_i(q) \rangle = N_i \delta(q)$$

$$\langle \rho_i(q)\rho_i(q') \rangle = N_i \delta(\mathbf{q} + \mathbf{q}') + N_i(N_i - 1)\delta(q)\delta(q')$$

where N_i is the number of impurities, Ω is the volume of the solid, and

$\delta(q)$ are three-dimensional delta functions setting $q = 0$. The average impurity density is $n_i = N_i/\Omega$.

Terms proportional to one power of N_i represent the scattering from a single impurity. The delta function requires the wave vector to be conserved when scattering from a single impurity. Terms proportional to higher powers of N_i represent scattering from clusters of impurities and can usually be neglected when the impurity density is small.

The retarded self-energy to order $O(n_i)$ is

$$\Sigma^r(k, \omega) = n_i T_{kk}(\omega) \tag{6.2.25}$$

The T-matrix is the one which is energy-dependent, and hence off-shell ($W \neq \varepsilon_n$). It is obtained as the solution to an integral equation involving the impurity potential $V_{kk'} = v(k - k')$ for scattering an electron from k to k'

$$T_{kk'} = V_{kk'} + \int \frac{d^3 p}{(2\pi)^3} \frac{V_{kp} T_{pk'}}{\omega - \varepsilon_p + i\delta} \tag{6.2.26}$$

In the special case that $\omega = \varepsilon_k = \varepsilon_{k'}$ the T-matrix can be expressed in terms of the phase shifts $\delta_i(k)$ for the electron scattering from the screened impurity

$$T_{kk}(\varepsilon_k) = -\frac{4\pi}{2mk} \sum_l (2l + 1) e^{i\delta_l} \sin \delta_l \tag{6.2.27}$$

For this special case, $\Gamma(k, \varepsilon_k) = -2 \operatorname{Im} \Sigma^r(k, \varepsilon_k) = n_i v_k \sigma_T$, where $\sigma_T(k)$ is the total cross section for the electron scattering from the impurity. The T-matrix can also be evaluated when $k \neq k'$, as discussed by Mahan.[1]

The Feynman diagrams for impurity scattering are shown in Figure 6.6. The x is the impurity. The dashed lines going to it are the interactions $v(q)$. The solid lines are the electron Green's functions. Repeated scattering from the same impurity yields the sequence of diagrams in Figure 6.6. The summation of this series of terms gives the retarded self-energy in equation (6.2.25).

The next step is to express the self-energies $\Sigma^<$ and $\Sigma^>$ in terms of the Green's functions $G^<$ and $G^>$. Now it is assumed that G^r and Σ^r are known. They are usually unaffected by electric fields, and one can use the

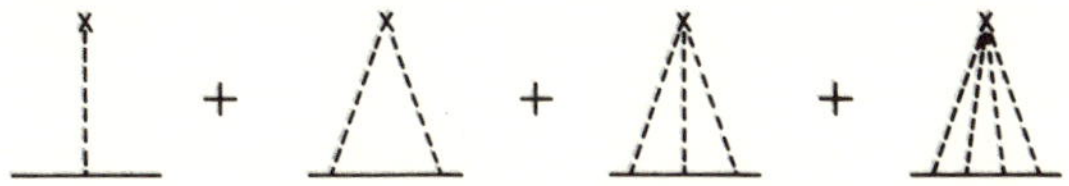

FIGURE 6.6. Feynman diagrams for the electron scattering from an impurity. The solid lines represent the electron Green's function, the dashed lines represent the potential energy between the electron and the impurity, and the x marks the impurity position. The terms represent the multiple scattering of the electron from the same impurity.

equilibrium forms for them in doing transport theory. However, $G^<$ and $G^>$ are affected by the electric field. The self-energy functions are

$$\Sigma^{\lessgtr}(k, \omega) = n_i \int \frac{d^3p}{(2\pi)^3} (T_{kp}(\omega))^2 G^{\lessgtr}(p, \omega) \qquad (6.2.28)$$

This result is derived below. The off-diagonal T-matrix is the one in equation (6.2.26) which also depends upon the energy ω.

Equation (6.2.28) is now derived. An impurity at $R = 0$ is represented by an electron potential $V(r)$, whose Fourier transform is $v(q)$. The self-energy Σ from a single scattering event is

$$\tilde{\Sigma}_0(x_1, x_2) = V(r_1)\delta^4(x_1 - x_2)\mathbf{I}$$

This self-energy is inserted into equation (6.2.22). That equation is then solved in order to find the effects of repeated scattering from the same impurity. Equation (6.2.22) is rewritten in a symbolic notation, where the product of two functions implies an integral over dx. Repeated iteration of these equations gives the series for $\tilde{\Sigma}$.

$$\tilde{\Sigma} = \tilde{\Sigma}_0 + \tilde{\Sigma}_0 \tilde{G}_0 \tilde{\Sigma}_0 + \tilde{\Sigma}_0 \tilde{G}_0 \tilde{\Sigma}_0 \tilde{G}_0 \tilde{\Sigma}_0 + \cdots \qquad (6.2.29)$$

After summing this series, one can find the formula for the individual components. The one for Σ^r gives equation (6.2.25). The other ones we want are

$$\Sigma^{\lessgtr} = [1 + G^r\Sigma^r]\Sigma_0^{\lessgtr}[1 + \Sigma^a G^a] + \Sigma^r G^{\lessgtr}\Sigma^a \qquad (6.2.30)$$

Note the analogy with the equation for $G^<$ in equation (6.2.23). This resemblance is expected, since the series (6.2.29) for $\tilde{\Sigma}$ has the same mathematical structure as the one obtained from (6.2.22) for $\tilde{G}$.

For impurity scattering, the unperturbed self-energies $\Sigma_0^{\lessgtr}$ are zero. Thus there is only the last term in (6.2.30). Since the self-energies Σ^r and Σ^a are T-matrices, we then find for $\Sigma^>$ the result in equation (6.2.28). In this notation, the damping function $\Gamma(k, \omega)$ and other self-energy $\Sigma^>(k, \omega)$ are

$$2\Gamma(k, \omega) = n_i \int \frac{d^3p}{(2\pi)^3} (T_{pk})^2 A(p, \omega) \qquad (6.2.31)$$

$$\Sigma^> = \Sigma^< - i2\Gamma \qquad (6.2.32)$$

In order to use terms which are only first order in n_i in the self-energies, it is necessary to replace the spectral function $A(k, \omega)$ by $2\pi\delta(\omega - \varepsilon_k)$ in equation (6.2.31), which then makes this expression equal to the imaginary part of the T-matrix expression in (6.2.27). These remarks complete the analysis of the self-energy expressions from impurity scattering.

The phonons are treated in the harmonic approximation, and the electron–phonon interaction is taken as the usual term which is linear in the displacements of the ions. The phonon terms in the Hamiltonian are

$$H_p = H_{0,p} + H_{e,p}$$

$$H_{0,p} = \sum_q \omega_q (a_q^+ a_q + \tfrac{1}{2})$$

$$H_{e,p} = \sum_q M_q e^{iq'r}(a_q + a_{-q}^+)$$

(6.2.33)

The phonons are represented by the operators a_q and a_{-q}^+. The electron position is r. The matrix element M_q is different for each interaction as described below. In a system of many electrons, the electron–electron interactions significantly change the phonon parts of the Hamiltonian: the phonon energies and matrix element M_q are changed by screening. It is assumed that this renormalization procedure has already occurred in the derivation. There is no need to discuss it here since it is an equilibrium process rather than a transport process. There should also be a summation λ over the different branches of the phonon spectrum (LA, TA, LO, TO). It is left out to simplify the notation, but the summation over λ is included in the appropriate final formulas.

The retarded function is examined first. For electron scattering by phonons, the retarded self-energy does depend upon the occupation of the electrons and is therefore altered in nonequilibrium. The self-energy of the electron from phonons has contributions from events involving one phonon, two phonons, three phonons, etc. Each additional phonon makes a smaller contribution, and usually the only significant contribution is from one-phonon events. For these terms, the electron self-energy has the form

$$\Sigma^\alpha(x_1, x_2) = iG^\alpha(x_1, x_2)D^\alpha(x_1, x_2)$$

where $D^\alpha(x_1, x_2)$ is the phonon Green's function which is defined in equation (6.2.8). The index α denotes either of $\rangle, \langle, t, \bar{t}$ but not r or a. In order to find the retarded or advanced Green's functions, the above expressions are used in the combinations

$$\Sigma^r = \Sigma_t - \Sigma^< = \Sigma^> - \Sigma_{\bar{t}}$$

$$\Sigma^a = \Sigma_t - \Sigma^> = \Sigma^< - \Sigma_{\bar{t}}$$

(6.2.34)

This is not the same as $\Sigma^r = G^r D^r$ or $\Sigma^a = G^a D^a$, and these latter two expressions are incorrect, and should be

$$\Sigma^r = iG^> D^> - iG_{\bar{t}}D_{\bar{t}}$$

$$\Sigma^a = iG^< D^< - iG_{\bar{t}}D_{\bar{t}}$$

The phonon system will be assumed to be in equilibrium even when the electron systems are out of equilibrium. Of course, it is well known that this approximation is often invalid in actual experiments because of "phonon drag" upon the electron.[3] The inclusion of phonon drag entails solving a set of coupled integral equations for the combined system of electrons and phonons.[2] This additional complication is avoided here by assuming the phonons remain in equilibrium. Then the phonon Green's function depends upon the difference of the arguments $x_1 - x_2$. If the electron system is in equilibrium and if one uses the unperturbed functions in equations (6.2.5) and (6.2.9), then the retarded self-energy is

$$\Sigma^r(p, t) = -i\theta(t) \sum_q M_q^2 \{(N_q + 1 - n_{p+q})e^{-i(\varepsilon_{p+q}+\omega_q)t}$$
$$+ (N_q + n_{p+q})e^{-i(\varepsilon_{p+q}-\omega_q)t}\}$$

where N_q is the phonon occupation number, and n_{p+q} is the electron occupation number. Fourier-transforming this expression yields the standard result

$$\Sigma^r(p, \omega) = \sum_q M_q^2 \left\{ \frac{N_q + 1 - n_{p+q}}{\omega - \varepsilon_{p+q} + \omega_q + i\delta} + \frac{N_q + n_{p+q}}{\omega - \varepsilon_{p+q} + \omega_q + i\delta} \right\} \quad (6.2.35)$$

which does depend upon the occupation number of the electrons. The advanced self-energy Σ^a is the complex conjugate of this expression. The quantity Γ is the negative of the imaginary part of the retarded self-energy

$$\Gamma = -\mathrm{Im}\,\Sigma^r$$
$$= \pi \sum_q M_q^2 \{(N_q + 1 - n_{p+q})\delta(\omega - \varepsilon_{p+q} - \omega_q)$$
$$+ (N_q + n_{p+q})\delta(\omega - \varepsilon_{p+q} + \omega_q)\} \quad (6.2.36)$$

In order to solve the quantum Boltzmann equation for electrons interacting with phonons, the two necessary self-energy functions are

$$\Sigma^<(p, \omega) = \sum_q M_q^2[(N_q + 1)G^<(p + q, \omega + \omega_q)$$
$$+ N_q G^<(p + q, \omega - \omega_q)]$$
$$\Sigma^>(p, \omega) = \sum_q M_q^2[N_q G^>(p + q, \omega + \omega_q)$$
$$+ (N_q + 1)G^>(p + q, \omega - \omega_q)] \quad (6.2.37)$$

Using the equilibrium forms for $G^{<,>}$ in equation (6.2.24) gives the expressions for $\Sigma^{<,>}$ in equation (6.2.24). In transport theory the Green's functions $G^<$ and $G^>$ are for the nonequilibrium case and then one uses the above relations whenever the phonons are in equilibrium and the electrons are not.

The electron–phonon interaction is governed by the matrix element M_q. Here are summarized some of the forms for this function which are commonly used in metals and semiconductors.

In metals, the interaction is described from first principles using the potential energy V_{ei} between the electron and ion.

$$M_q = \sum_G \xi_q \cdot (\mathbf{q} + \mathbf{G}) v_{ei}(\mathbf{q} + \mathbf{G}) \left| \frac{1}{2\rho\Omega\omega_q} \right|^{1/2} \qquad (6.2.38)$$

where ξ_q is the unit vector giving the direction of the ion displacements, ρ is the density of the solid, and v_{ei} is the Fourier transform of the screened electron-ion potential. The summation over G is over the reciprocal lattice vectors of the solid.

In semiconductors, the electrons or holes are usually confined to small pockets of the Brillouin zone. Since the scattering by phonons can only change the electron's wave vector by a small amount, then only long-wavelength phonons are important. Then the approach is not to calculate the electron–phonon matrix element from first principles. Instead, a number of phenomenological interactions have been described and parameterized. These parameters are found by fitting to various experiments.

The most important electron–phonon interaction in a semiconductor is the *deformation potential*. It is derived by taking the long-wavelength limit of the formula in equation (6.2.38). At small q the summation over G is neglected except for the $G = 0$ term. The screened electron–ion interaction $v_{ei}(q)$ becomes a constant which is usually called D:

$$M_q = D\xi_q \cdot \mathbf{q} \left| \frac{1}{2\rho\Omega\omega_q} \right|^{1/2} \qquad \text{Deformation Potential (acoustical)} \quad (6.2.39)$$

For long-wavelength acoustic phonons, the LA phonons have $\mathbf{q} \| \xi$ which makes M_q nonzero. TA phonons have $\mathbf{q} \perp \xi$ which makes M_q vanish. Thus, the common lore is that deformation potential mainly couples electrons to LA phonons. However, holes in a degenerate p-band have strong coupling to both TA and LA phonons for scattering between light- and heavy-hole bands.

Deformation potential scattering is also possible between optical phonons and electrons or holes. This case is very important in silicon, where hot electrons mainly lose energy by emitting optical phonons through the deformation potential coupling. For optical phonons, one does not use the above expression. Instead one thinks of an optical phonon as occurring at finite G in equation (6.2.38). For finite G, the q dependence is unimportant, so the entire matrix element becomes a constant. Thus, one has

$$M_q = M_0 \qquad \text{Deformation Potential (optical)}$$

The deformation potential is obtained by treating $v_{ei}(q)$ as a constant. Fourier-transforming back to r-space, a constant in q-space produces a delta function $\delta^\varepsilon(\mathbf{r})$ in real space. The deformation potential is viewed as a short-ranged interaction. A local dilation or compression of the lattice produces a local fluctuation in the energy of the electron or hole.

In semiconductors there are two electron–phonon interactions which are electrostatic in origin and long-ranged. The most important is the *polar* interaction which is given in terms of the dimensionless polaron constant α. The form for the interaction at long wavelength is

$$M_q = M_0/q\sqrt{\Omega} \qquad \text{Polar}$$

$$M_0 = 4\pi\alpha\omega_0[\omega_0/2m]^{1/2} \qquad\qquad (6.2.40)$$

$$\alpha = e^2(m/2\omega_{\text{LO}})^{1/2}(1/\varepsilon_\infty - 1/\varepsilon_0)$$

where ω_0 is the frequency of the LO phonon. The interaction only couples to LO phonons at long wavelength. The factor of q^{-1} makes the interaction quite strong at small q.

The polar interaction is zero in group IV elements such as silicon and germanium. It is not negligible in 3-5 semiconductors such as GaAs and dominates the mobility of these direct gap materials. It is quite strong in 2-6 semiconductors, where values of α are near unity.

The *piezoelectric* interaction is the other long-range electrostatic interaction. Only solids which lack an inversion center can be piezoelectric, and they must also be polar. Silicon is not piezoelectric, (and the interaction is weak in most 3-5 semiconductors) so the interaction is not important for present hot electron research, except at very low temperatures. The 2-6 semiconductors ZnO and CdS have the strongest piezoelectric electron–phonon interaction of common semiconductors. This interaction will be important for hot-electron measurements in these solids. The interaction causes a Cerenkov-like effect, where an electron exceeding the speed of sound emits a continuous stream of acoustical phonons in a microscopic sonic boom. Details of the interaction are given elsewhere.[16]

6.3. QUANTUM BOLTZMANN EQUATION

6.3.1. Wigner Distribution Function

The traditional Boltzmann equation is expressed in terms of the distribution function $f(r, v, t)$. The three variables are position r, velocity v, and time t. The point of view is semiclassical since it is assumed that the position and velocity (momentum) of the particle can be defined simultaneously. In order to use this distribution for quantum systems, it is necessary to perform some coarse-grain averaging in order to remove effects due to the uncertainty principle.

If quantum effects are important, it is necessary to introduce another variable into the distribution function. This variable could either be energy E or, equivalently, a frequency $\omega = E$. The technique for doing this was first suggested by Wigner, and the resulting distribution function is often called a *Wigner distribution function* $f(k, \omega; r, t)$. We have changed the velocity variable to a wave vector $k = mv$ ($\hbar = 1$).

The distribution function is derived from the Green's function $G^<(x_1, x_2)$ defined in equation (6.2.1). The Wigner distribution function (WDF) is derived using the following series of steps. First one goes to a center-of-mass coordinate system

$$(R, T) = \tfrac{1}{2}(x_1 + x_2)$$

$$(r, t) = x_1 - x_2$$

Note that T means center-of-mass time and not temperature: the symbol β is used for the inverse of Boltzmann's constant times temperature. The notation on the Green's function is altered to these center-of-mass coordinates

$$G^<(r, t; R, T) = i\langle \psi^+(R - \tfrac{1}{2}r, T - \tfrac{1}{2}t)\psi(R + \tfrac{1}{2}r, T + \tfrac{1}{2}t)\rangle$$

The next step is to Fourier-transform the relative variables (r, t) into (k, ω).

$$G^<(k, \omega; R, T) = \int d^3r\, e^{-ik\cdot r} \int dt\, e^{i\omega t} G^<(r, t; R, T) \qquad (6.3.1)$$

The relation to the WDF is quite simple:

$$f(k, \omega; R, T) = -iG^<(k, \omega; R, T)$$

Regard this assertion as the definition of $f(k, \omega; R, T)$. This choice is made reasonable by showing that various moments of f provide the macroscopic quantities of particle density $n(R, T)$, particle current $j(R, T)$, and energy density $n_E(R, T)$.

$$n(R, T) = \int \frac{d^3k}{(2\pi)^3} \int \frac{d\omega}{2\pi} f(k, \omega; R, T) = \langle \psi^+(R, T)\psi(R, T)\rangle$$

$$j(R, T) = \int \frac{d^3k}{(2\pi)^3} \frac{k}{m} \int \frac{d\omega}{2\pi} f(k, \omega; R, T)$$

$$n_E(R, T) = \int \frac{d^3k}{(2\pi)^3} \int \frac{d\omega}{2\pi} \omega f(k, \omega; R, T) \qquad (6.3.2)$$

$$= i\left[\frac{\partial}{\partial t} \langle \psi^+(R, T - \tfrac{1}{2}t)\psi(R, T + \tfrac{1}{2}t)\rangle \right]_{t=0}$$

$$= \langle \psi^+(R, T) H \psi(R, T)\rangle$$

The technique for solving nonequilibrium problems is very simple. We write an equation of motion for $G^<(k, \omega; R, T)$. This equation is just the quantum Boltzmann equation (QBE). This equation is then solved which yields directly the Wigner distribution function $f(k, \omega; R, T)$. Macroscopic observables, as measured in various experiments, are found by taking the above moments.

The semiclassical Boltzmann distribution function $f(r, v, t)$ is found by taking the frequency integral of the WDF:

$$f(r, v, t) = \int_{-\infty}^{\infty} \frac{d\omega}{2\pi} f(mv, \omega; r, t) \tag{6.3.3}$$

One approach to deriving a Boltzmann equation (BE) for $f(r, v, t)$ is to start from the rigorous QBE and to take the frequency integral of the entire equation. Each term is evaluated as best one can in order to find the BE.

6.3.2. Quantum Boltzmann Equation with Electric Field

The quantum Boltzmann equation will be derived for a particle in an electric field. The intent is to describe interacting many-particle systems which have a current flowing in response to an electric field. The derivation will be sufficiently general to include any kind of particles and nearly any kind of interactions. Here the electric field E is introduced as a scalar potential through the interaction term

$$H_E = -eE \cdot \sum_j r_j \tag{6.3.4}$$

The same QBE is derived when the electric field is introduced as a vector potential, and the QBE is gauge-invariant.

Mahan and Hansch[12-14] first derived the QBE with electric field, following the techniques introduced by Kadanoff and Baym.[4] Our goal is an equation of motion for the interacting Green's function when it is not in equilibrium. Such an equation can be found from (6.2.22) by operating by $(i\partial/\partial t - H_0)$ on both sides of the equation. The time dependence in equation (6.2.7) is expressed compactly in the matrix notation

$$\left(i\frac{\partial}{\partial t} - \varepsilon_k \right) \tilde{G}_0(k, t) = \delta(t)\tilde{I}$$

$$\left(i\frac{\partial}{\partial t} - H_0(x) \right) \tilde{G}_0(x) = \delta^4(x)\tilde{I}$$

On the right side, this operator only acts upon $\tilde{G}_0$, for which we use the above result:

$$\left[i\frac{\partial}{\partial t_1} - H_0(x_1) \right] \tilde{G}(x_1 x_2) = \delta^4(x_1 - x_2)\tilde{I} + \int dx_3 \, \tilde{\Sigma}(x_1 x_3)\tilde{G}(x_3, x_2) \tag{6.3.5}$$

This formula provides the equation of motion for the interacting Green's function. It is the basis for the derivation of the QBE.

The structure of this equation is very interesting. On the left-hand side, only the noninteracting terms are contained in the Hamiltonian H_0. The contribution from the interactions is provided by the self-energy functions on the right-hand side.

So far nothing has been said about the electric field. How is it added to this equation of motion? It can be added to either H_0 or to the self-energy terms $\tilde{\Sigma}$, and the same answer is obtained in either case. In the first method add the scalar potential H_E in (6.3.4) to H_0, and thus have the equation of motion

$$\left[i \frac{\partial}{\partial t_1} - H(x_1) \right] \tilde{G}(x_1 x_2) = \delta^4(x_1 - x_2)\tilde{I} + \int dx_3\, \tilde{\Sigma}(x_1 x_3)\tilde{G}(x_3, x_2) \tag{6.3.6}$$

$$H = H_0 + H_E$$

In the second method, the self-energy $\tilde{\Sigma}_E$ from the electric field is a *one-point function*. By this is meant that it depends only upon a single point in space-time which can be expressed as

$$\tilde{\Sigma}_E(x_1 x_2) = -eE \cdot r\delta^4(x_1 - x_2)\tilde{I} \tag{6.3.7}$$

Of course, this produces a term $H_E\tilde{G}$ on the right which is equivalent to adding H_E to H_0 on the left. The two components of the matrix $\tilde{G}$ which are needed for the QBE are G^r and $G^<$, whose separate equations are

$$\left[i \frac{\partial}{\partial t_1} - H(x_1) \right] G^<(x_1 x_2) = \int dx_3\, [\Sigma_t(x_1 x_3)G^<(x_3 x_2) - \Sigma^<(x_1 x_3)G_{\bar{t}}(x_3 x_2)]$$

$$\tag{6.3.8}$$

$$\left[i \frac{\partial}{\partial t_1} - H(x_1) \right] G^r(x_1 x_2) = \delta(x_1 - x_2) + \int dx_3 \Sigma^r(x_1 x_3)G^r(x_3 x_2)$$

So far, the equation of motion has been derived for the variable x_1 in $\tilde{G}(x_1 x_2)$. It is useful to also have an equation of motion for the other variable x_2. Since the definition (6.2.1) of the Green's functions contain the conjugate wave function $\psi^+(x_2)$, then the equation of motion on this variable is the Hermitian conjugate of Schrödinger's equation. Furthermore, Dyson's equation can be written in an alternative form, instead of (6.2.22), with G_0 on the right in the interaction terms.

$$\tilde{G}(x_1 x_2) = \tilde{G}_0(x_1 - x_2) + \int dx_3 \int dx_4\, \tilde{G}(x_1 x_3)\tilde{\Sigma}(x_3 x_4)\tilde{G}_0(x_4, x_2)$$

Then the equation of motion on the x_2 variable, when using this form for Dyson's equation, still acts only upon $\tilde{G}_0$ on the right and produces delta functions. These steps produce the alternative equation of motion for the Green's function:

$$\left[-i\frac{\partial}{\partial t_2} - H(r_2, -p_2)\right]\tilde{G}(x_1x_2) = \delta^4(x_1 - x_2)I + \int dx_3\,\tilde{G}(x_1x_3)\tilde{\Sigma}(x_3x_2)$$

This formula provides the equation of motion for all six Green's functions. For transport theory, the two most useful Green's functions are G^r and $G^<$. Their individual equations are

$$\left[-i\frac{\partial}{\partial t_2} - H(r_2, -p_2)\right]G^<(x_1, x_2) = \int dx_3[G_t(x_1x_3)\Sigma^<(x_3, x_2)$$

$$-G^<(x_1x_3)\Sigma_{\bar{t}}(x_3, x_2)]$$

$$\left[-i\frac{\partial}{\partial t_2} - H(r_2, -p_2)\right]G^r(x_1x_2) = \delta^4(x_1 - x_2)$$

$$+ \int dx_3\,G^r(x_1x_3)\Sigma^r(x_3x_2)$$

(6.3.9)

The two sets of equations (6.3.8) and (6.3.9) are the starting point for the derivation of the QBE.

The QBE is an equation of motion for the Green's function $G^<$. In order to solve it, one must invariably first solve the equation for the retarded Green's function G^r. Since G^r is the function which is causal, and somewhat more intuitive, its equation is derived first.

The two equations for G^r in (6.3.8) and (6.3.9) are first added and then subtracted.

$$\left[i\left(\frac{\partial}{\partial t_1} - \frac{\partial}{\partial t_2}\right) - H_1 - H_2\right]G^r = 2\delta(x_1 - x_2) + \int dx_3\,[\Sigma^rG^r + G^r\Sigma^r]$$

$$\left[i\left(\frac{\partial}{\partial t_1} + \frac{\partial}{\partial t_2}\right) - H_1 + H_2\right]G^r = \int dx_3\,[\Sigma^rG^r - G^r\Sigma^r]$$

The variable arguments will not be written out in the scattering terms since this takes too much space. The actual arguments are implied by the ordering of the terms. The two equations contain time derivations which relate purely to the relative or center-of-mass motion

$$\frac{\partial}{\partial t_1} + \frac{\partial}{\partial t_2} = \frac{\partial}{\partial T}; \qquad \frac{\partial}{\partial t_1} - \frac{\partial}{\partial t_2} = 2\frac{\partial}{\partial t}$$

For particles with parabolic band dispersion, the sum and difference of the two Hamiltonians produce simple expressions in relative coordinates

$$H_1 + H_2 = -\frac{1}{m}\nabla_r^2 - \frac{1}{4m}\nabla_R^2 - 2eE\cdot R$$

$$H_1 - H_2 = -\frac{1}{m}\nabla_r\cdot\nabla_R - eE\cdot r$$

Dividing the top equation by 2, we arrive at the following two equations for the retarded Green's function

$$\left[i\frac{\partial}{\partial t} + \frac{1}{2m}\nabla_r^2 + \frac{1}{8m}\nabla_R^2 + eE\cdot R\right]G^r = \delta^4(x) + \tfrac{1}{2}\int dx_3[\Sigma^r G^r + G^r\Sigma^r]$$

$$\left[i\frac{\partial}{\partial T} + \frac{1}{m}\nabla_r\cdot\nabla_R + eE\cdot r\right]G^r = \int dx_3\,[\Sigma^r G^r - G^r\Sigma^r]$$

These two equations describe the relative and center-of-mass motion of the retarded function $G^r(r, t; R, T)$. The goal is to derive the QBE for the WDF $G^<(k, \omega; R, T)$. The retarded function which is most useful in solving the QBE has the same set of variables $(k, \omega; R, T)$. Thus, Fourier-transform the variables (r, t) to the set (k, Ω) as in (6.3.1), and the above equations become, with $q = (k, \Omega)$,

$$\left[\Omega - \varepsilon_k + \frac{1}{8m}\nabla_R^2 + eE\cdot R\right]G^r(k, \Omega; R, T)$$

$$= 1 + \tfrac{1}{2}\int dx\, e^{iq\cdot x}\int dx_3\,[\Sigma^r G^r + G^r\Sigma^r] \qquad\qquad (6.3.10)$$

$$i\left[\frac{\partial}{\partial T} + v_k\cdot\nabla_R + eE\cdot\nabla_k\right]G^r = \int dx\, e^{-iq\cdot x}\int dx_3\,[\Sigma^r G^r - G^r\Sigma^r]$$

The second of these equations has, on the left-hand side, exactly the same terms as are found in the Boltzmann equation. This similarity has caused many workers to assume that the QBE was the same as the BE. However, this is not correct as is evident from the first equation. There is the term $eE\cdot R$. This term seems to combine with Ω to produce a center-of-mass energy $\omega = \Omega + eE\cdot R$. Together they suggest that the energy of a particle depends upon its location. The energy is different at one end of a sample than at the other. However, this behavior is contrary to common sense. When there is a small electric field along the sample, and a small current flowing, then we expect the system to be uniform. There is the same particle density, current density, etc., at each point in the solid. There is no dependence upon R. So this undesirable term has to be eliminated.

   Mahan and Hansch[12-14] proposed to eliminate this R-dependence through a variable transformation to the new energy variable ω. Of course, this transformation also causes some derivatives to change:

$$\Omega + eE \cdot R \Rightarrow \omega$$

$$\nabla_R \Rightarrow \nabla_R + eE \frac{\partial}{\partial \omega}$$

$$(6.3.11)$$

These definitions, and changes, cause the equations (6.3.10) for the retarded Green's function to now have the form

$$\left[\omega - \varepsilon_k + \frac{1}{8m} \left(\nabla_R + eE \frac{\partial}{\partial \omega} \right)^2 \right] G^r(k, \omega; R, T)$$

$$= 1 + \tfrac{1}{2} \int dx\, e^{-iq \cdot x} \int dx_3 \left[\Sigma^r G^r + G^r \Sigma^r \right]$$

$$i \left[\frac{\partial}{\partial T} + v_k \cdot \nabla + eE \cdot \left(\nabla_k + v_k \frac{\partial}{\partial \omega} \right) \right] G^r$$

$$= \int dx\, e^{-iq \cdot x} \int dx_3 \left[\Sigma^r G^r - G^r \Sigma^r \right]$$

$$(6.3.12)$$

The left-hand side of these two equations are now in the form which is useful. So far no approximations have been made, and the equations are exact.

   The lower equation has a linear term in the electric field E. However, the equation is exact to all powers of E, not just to the first power. However, this equation is usually too complicated to solve because the scattering terms on the right have a complicated form. Some sort of approximation has to be introduced to simplify the right side. The approximation described below is only valid to first order in the field.

   Kadanoff and Baym[4] introduced an approximation for evaluating these scattering terms, which is called the *gradient expansion*. They assume that the center-of-mass time T is very large, and take the limit that $T \to \infty$. At large values of T, they assume that the system is approaching its asymptotic limit, so that variations with respect to T are small. Obviously the gradient expansion is not suitable for studying transients, since it is poor at small values of T. Neither is it useful for steady-state ac phenomena.[15] Indeed, the T-dependence is so poorly described in the gradient expansion, that it should not be used and the T-derivative terms should be dropped from the QBE. Nevertheless, we shall use the gradient expansion since it is applicable for homogeneous (small R-dependence) steady-state (small T-dependence) systems.

Examine just one term in the right-hand side of equation (6.3.12): The various variable arguments yield a fairly complicated expression which is derived by the following steps:

1. Change the two sets of variables (x_1, x_3), (x_3, x_2) to the center-of-mass grouping $[x_1 - x_3, (x_1 + x_3)/2]$, $[x_3 - x_2, (x_3 + x_2)/2]$.
2. Change the integration variable from x_3 to $y = x_1 - x_3$, so the variable grouping becomes $(y, x_1 - y/2)$, $[x_1 - x_2 - y, (x_1 + x_2 - y)/2]$.
3. Change to center-of-mass variables $x = (x_1 - x_3)$, $X = 1/2(x_1 + x_2)$ which produces the variable grouping $[y, X + 1/2(x - y)]$, $(x - y, X - 1/2y)$.
4. Change the integration variable x to $z = x - y$, which produces the final integral of

$$I_1 = \int dz \, e^{-iq \cdot z} \int dy \, e^{-iq \cdot y} \Sigma^r(y, X + \tfrac{1}{2}z) G^r(z, X - \tfrac{1}{2}y)$$

The center-of-mass variables are all in the form $(R + \Delta R, T + \Delta T)$, which are expanded in a Taylor series about the point (R, T). The integrals can be done for each term in the series. These integrals usually cause further derivatives

$$I_1 = \Sigma^r(q, X) G^r(q, X) + \frac{i}{2}[(\nabla_q \Sigma^r)\nabla_X G^r - \nabla_X \Sigma^r \nabla_q G^r] + \cdots$$

It is customary to retain only the first derivative terms. They can be expressed using a Poisson bracket notation

$$[\Sigma^r, G^r] = \nabla_q \Sigma \nabla_X G^r - \nabla_X \Sigma^r \nabla_q G^r$$

$$= \frac{\partial \Sigma^r}{\partial \Omega} \frac{\partial G^r}{\partial T} - \frac{\partial \Sigma^r}{\partial T} \frac{\partial G^r}{\partial \Omega} - \nabla_k \Sigma^r \cdot \nabla_R G^r + \nabla_R \Sigma^r \cdot \nabla_k G^r$$

These frequency derivatives are with respect to the Ω-variables, since the variable change $\omega = \Omega + eE \cdot R$ has not yet been made. If this step is taken now, the derivatives get altered according to equation (6.3.11) which changes the Poisson brackets to

$$[C, D] \Rightarrow [C, D] + eE \cdot \left[\left(\frac{\partial C}{\partial \omega}\right)\nabla_k D - \left(\frac{\partial D}{\partial \omega}\right)\nabla_k C \right] \tag{6.3.13}$$

The first term in the gradient expansion produces a term linear in the field E. Higher-order terms in the gradient expansion, which are neglected, produce terms in higher powers of E. The neglect of these terms is the reason the resulting equation is valid only to linear terms in the electric field. Although equation (6.3.12) is valid to all powers of field, evaluating it using the first term in the gradient expansion limits the validity to quantities linear in the field.

In order to have a rigorous theory of transport in high electric fields, a method has to be found for accurately evaluating the scattering terms in (6.3.12). Several methods have been proposed and evaluated. They will be discussed in the other chapters.

This analysis finally derives from (6.3.12) the following expression for the nonequilibrium retarded Green's function in an electric field, when G^r and Σ^r depend upon $(k, \omega; R, T)$:

$$\left[\omega - \varepsilon_k + \frac{1}{8m}\left(\nabla_R + eE\frac{\partial}{\partial\omega}\right)^2 - \Sigma^r\right]G^r = 1$$

$$i\left\{\frac{\partial}{\partial T} + v_k \cdot \nabla_R + eE \cdot \left[\left(1 - \frac{\partial\Sigma^r}{\partial\omega}\right)\nabla_k + (v_k + \nabla_k\Sigma^r)\frac{\partial}{\partial\omega}\right]\right\}G^r$$

$$= i[\Sigma^r, G^r]$$

$$\tag{6.3.14}$$

Note that the additional terms from the Poisson bracket, which are linear in the field E, have been transferred to the left of the equal sign. These set of equations were first derived by Mahan and Hansch.[12-14]

These equations simplify for nonequilibrium systems which are both homogeneous ($\nabla_R = 0$) and steady state ($\partial/\partial T = 0$). The Poisson brackets vanish, as do several terms on the left. We also ignore terms which are nonlinear in the electric field (E^2), and find for the above two equations

$$[\omega - \varepsilon_k - \Sigma^r]G^r = 1$$

$$ieE \cdot \left[\left(1 - \frac{\partial\Sigma^r}{\partial\omega}\right)\nabla_k + (v_k + \nabla_k\Sigma^r)\frac{\partial}{\partial\omega}\right]G^r = 0 \tag{6.3.15}$$

The first equation is easily solved, to yield

$$G^r(k, \omega) = \frac{1}{\omega - \varepsilon_k - \Sigma^r(k, \omega)} + O(E^2) \tag{6.3.16}$$

The retarded Green's function has no first-order term in the electric field. So when current flows in response to a weak electric field, the retarded Green's function is unchanged from its value in equilibrium. This result considerably simplifies the solution to the QBE. One can show that the solution (6.3.16) also satisfies the second equation in (6.3.15). Related quantities such as the retarded function $G^a = G^{r*}$ and the spectral function $A(k, \omega) = -2\,\mathrm{Im}[G^r(k, \omega)]$ are also unchanged to first order in the electric field. This completes the discussion of the retarded Green's function in a static electric field.

The next step is to discuss the similar equations for the Green's function $G^<$. The same steps, used to derive (5.3.16) for the retarded Green's function, are now applied to $G^<$. First add and subtract the two equations in (6.3.8) and (6.3.9). The variables are changed to center of mass, and then the relative coordinates are Fourier transformed to (k, Ω). Finally, one makes the variable change $\omega = \Omega + eE \cdot R$ in (6.3.11) which brings us to the pair of equations

$$\left[\omega - \varepsilon_k + \frac{1}{8m}\left(\nabla_R + eE\frac{\partial}{\partial\omega}\right)^2\right] G^<(k, \omega; R, T)$$

$$= \tfrac{1}{2}\int dx\, e^{-iq\cdot x}\int dx_3\, [\Sigma_t G^< - \Sigma^< G_{\bar t} + G_t \Sigma^< - G^< \Sigma_{\bar t}]$$

$$\hspace{7cm}(6.3.17)$$

$$i\left[\frac{\partial}{\partial T} + v_k \cdot \nabla_R + eE \cdot \left(\nabla_k + v_k\frac{\partial}{\partial\omega}\right)\right] G^<$$

$$= \int dx\, e^{-iq\cdot x}\int dx_3\, [\Sigma_t G^< - \Sigma^< G_{\bar t} - G_t \Sigma^< + G^< \Sigma_{\bar t}]$$

In the absence of collisions, the right-hand side of these equations vanish. Particle flow is then governed by the left-hand side, which contains the new term $eE \cdot v(\partial G^</\partial\omega)$ found by Mahan and Hansch.

The right-hand side of these expressions are evaluated using the KB gradient expansion, which is described above for the derivation of the equations for G^r. Again keeping only first order derivatives, we derive the final results, which are the quantum Boltzmann equations.

$$\left[\omega - \varepsilon_k + \frac{1}{8m}\left(\nabla_k + eE\frac{\partial}{\partial\omega}\right)^2\right] G^<$$

$$= G^< \operatorname{Re}\Sigma^r + \Sigma^< \operatorname{Re} G^r$$

$$\hspace{7cm}(6.3.18)$$

$$+ \frac{i}{4}[\Sigma^>, G^<] - \frac{i}{4}[\Sigma^<, G^>]$$

$$+ \frac{i}{4}eE \cdot \left\{\frac{\partial\Sigma^>}{\partial\omega}\nabla_k G^< - \frac{\partial G^<}{\partial\omega}\nabla_k\Sigma^< - \frac{\partial\Sigma^<}{\partial\omega}\nabla_k G^< + \frac{\partial G^>}{\partial\omega}\nabla_k\Sigma^<\right\}$$

$$i\left\{\frac{\partial}{\partial T} + v_k \cdot \nabla_R + eE \cdot \left[\left(1 - \frac{\partial\operatorname{Re}\Sigma}{\partial\omega}\right)^r \nabla_k + (v_k + \nabla_k \operatorname{Re}\Sigma^r)\frac{\partial}{\partial\omega}\right]\right\} G^<$$

$$- ieE \cdot \left[\frac{\partial\Sigma^<}{\partial\omega}\nabla_k \operatorname{Re} G^r - \frac{\partial\operatorname{Re} G^r}{\partial\omega}\nabla_k\Sigma^<\right]$$

$$= \Sigma^> G^< - \Sigma^< G^> + i[\operatorname{Re}\Sigma^r, G^<] + i[\Sigma^<, \operatorname{Re} G^r]\hspace{2cm}(6.3.19)$$

In deriving this equation, we have used some relationships such as $G_t - G_{\bar t} = 2\operatorname{Re} G^r$, $G_t + G_{\bar t} G^< + G^>$, etc.

Equation (6.3.19) is the quantum Boltzmann equation. It is rather formidable. It is also difficult to solve, since it is usually an integral equation, and sometimes nonlinear in the particle density. This happens because the self-energy functions $\Sigma^<$ and $\Sigma^>$ are also functions of $G^<$ and $G^>$. It is only valid to the first power of electric field, since we have neglected higher terms in the gradient expansion which contain higher powers of the field.

For transport in high electric fields, one needs to derive a transport equation which is valid to all powers in the field. Some other means of evaluating the scattering terms is needed, since the gradient expansion is unsuitable. Alternative techniques have been proposed, which will be discussed in the other chapters.[17-21]

The QBE also contains the functions $G^>$ and $\Sigma^>$. We can also start from the general equations (6.3.5) and derive similar equations for $G^>$. This derivation shows that the equation for $G^>$ is almost identical to the one for $G^<$. In fact, one can prove that the following identities are valid:

$$G^> = G^< - iA$$

$$\Sigma^> = \Sigma^< + i\,2\,\mathrm{Im}\,\Sigma^r = \Sigma^< - i2\Gamma$$

These relations are trivial to show for equilibrium, but they are also valid for nonequilibrium situations. These identities will be used often to simplify expressions. For example, the main scattering term in the QBE can be immediately simplified to

$$\Sigma^> G^< - \Sigma^< G^> = -i[2\Gamma G^< - \Sigma^< A]$$

This result will be employed in the calculations. The quantities $G^<$ and $\Sigma^<$ are generally proportional to the density of particles. Retarded functions are only indirectly dependent upon the density of particles—only through the self-energy function Σ^r. In the QBE, each term has one factor which is either $G^<$ or $\Sigma^<$, so each term is proportional to the density of particles. Thus this equation does have the character of a transport equation.

The QBE simplifies for the treatment of systems which are homogeneous ($\nabla_R = 0$) and steady ($\partial/\partial T = 0$). These derivatives are dropped, as well as the Poisson brackets, since the latter contain similar derivatives. Then the QBE is

$$eE \cdot \left[\left(1 - \frac{\partial\,\mathrm{Re}\,\Sigma^r}{\partial\Omega}\right)\nabla_k + (v_k + \nabla_k\,\mathrm{Re}\,\Sigma^r)\frac{\partial}{\partial\Omega}\right]G^<$$

$$- eE \cdot \left[\frac{\partial\Sigma^<}{\partial\Omega}\nabla_k\,\mathrm{Re}\,G^r - \frac{\partial\,\mathrm{Re}\,G^r}{\partial\Omega}\nabla_k\Sigma^<\right] = \Sigma^< A - 2\Gamma G^< \quad (6.3.20)$$

Next we will make a remarkable simplification of this equation. The QBE is only valid to first power in the electric field, since we have systematically ignored terms of $O(E^2)$. This fact is utilized to simplify the left-hand side of the equation. Since the field E multiplies each term, then for each term on this side we can use the Green's functions which are independent of field. Of course, these are just the equilibrium quantities in (6.2.24).

Each term on the left-hand side of equation (6.3.20) contains either the factor $n(\Omega)$ or $\partial n(\Omega)/\partial\Omega$ since the equilibrium expressions of $\Sigma^<$ and $G^<$ are each proportional to $n(\Omega)$. When all of the terms are collected which contain the factor of $n(\Omega)$, they exactly cancel. There are quite a few terms, so it takes some work to show that each term is canceled by another. Thus, there remain only the terms which contain the factor $\partial n(\Omega)/\partial\Omega$:

$$iA(k,\Omega)^2 \frac{\partial n}{\partial\Omega}\, eE \cdot \{(v_k + \nabla_k \operatorname{Re}\Sigma')\Gamma + \sigma\nabla_k\Gamma\} = \Sigma^< A - 2\Gamma G^< \quad (6.3.21)$$

The factor $A(k,\Omega)^2$ appears in each term and is also taken outside. The left-hand side of this equation now contains only known quantities, which can be calculated in equilibrium. The scattering terms remain on the right-hand side. Finding them still involves work, usually in the form of an integral equation. This final form for the QBE is exact for transport which is linear in the field and for steady-state homogeneous systems. It was found by Mahan and Hansch. It is quite analogous to the similar expression for the classical BE which is

$$-eE \cdot v_k \frac{\partial f^{(0)}}{\partial\varepsilon_k} = \Sigma^> G^< - \Sigma^< G^>$$

The classical equation has $\partial f^{(0)}/\partial\varepsilon_k$, while the Mahan–Hansch expression has $\partial n(\Omega)/\partial\Omega$.

Equation (6.3.21) is the steady-state, homogeneous form of the QBE. It should be the starting point for many transport calculations. It is exact, and is an alternative formalism to using the Kubo formalism, which is also exact. The derivation of this equation has been complicated, and has entailed some work. However, once derived, it is often the easiest starting point for deriving the transport coefficients.

6.3.3. Solutions to the QBE

The starting point for solving the QBE is equation (6.3.21). On the left of the equal sign is the factor

$$(v_k + \nabla_k \operatorname{Re}\Sigma')\Gamma + \sigma\nabla_k\Gamma$$

Neglecting terms of $O(n_i^2)$ such as $\Gamma \operatorname{Re}\Sigma'$, and assuming parabolic dispersion relations, the above expression can be shortened to

$$= v_k \Phi(k,\omega)$$

where

$$\Phi(k,\omega) = \Gamma + (\omega - \varepsilon_k)\frac{\partial\Gamma}{\partial\varepsilon_k}$$

Since this expression is multiplied by A^2, which tends to make $\omega = \varepsilon_k$, the second term in Φ is small, and only the first term is important:

$$A^2(k, \omega)\frac{\partial n_F(\omega)}{\partial \omega} eE \cdot v_k\Gamma = i2\Gamma G^< - i\Sigma^< A \qquad (6.3.22)$$

This completes the treatment of the left-hand side of the QBE.

The scattering terms on the right vanish, in equilibrium, when $\Sigma^< = 2in_F\Gamma$ and $G^< = in_F A$. Assume that the electric field causes a small deviation from equilibrium, and that this deviation is proportional to the field. This choice suggests the following ansatz for the nonequilibrium Green's function:

$$G^< = iA(k, \omega)\left[n_F(\omega) - \left(\frac{\partial n_F}{\partial \omega}\right) eE \cdot v_k\Lambda(k, \omega)\right] \qquad (6.3.23)$$

The function $\Lambda(k, \omega)$ is unknown and needs to be determined by solving the QBE. This solution produces a current which is proportional to the electric field. It is a theory for linear transport. Using this ansatsz in the self-energy function (6.2.28) gives

$$\Sigma^< = in_i \int \frac{d^3p}{(2\pi)^3} (T_{pk})^2 A[n_F - \frac{\partial n_F}{\partial \omega} eE \cdot v_p\Lambda(p, \omega)]$$

$$= 2i\Gamma n_F - in_i\frac{\partial n_F}{\partial \omega} \int \frac{d^3p}{(2\pi)^3} A(p, \omega)(T_{pk})^2 eE \cdot v_p\Lambda(p, \omega)$$

Putting these two expressions into the right-hand side of the QBE, the equilibrium terms cancel, and the remaining terms each have the common factor of $(-\partial n_F/\partial \omega)A(k, \omega)$

$$A^2\frac{\partial n_F}{\partial \omega} eE \cdot v_k\Gamma$$

$$= \left(\frac{\partial n_F}{\partial \omega}\right) A(k, \omega)\left\{2\Gamma eE \cdot v_k\Lambda - n_i \int \frac{d^3p}{(2\pi)^3}|T_{kp}|^2 A(p, \omega)eE \cdot v_p\Lambda(p, \omega)\right\}$$

After cancelling all of the common factors, there is an integral equation for the unknown function $\Lambda(k, \omega)$:

$$v_k\Lambda(k, \omega) = \frac{A}{2}v_k + \frac{n_i}{2\Gamma} \int \frac{d^3p}{(2\pi)^3}|T_{pk}|^2 A(p, \omega)v_p\Lambda(p, \omega) \qquad (6.3.24)$$

After solving equation (6.3.24) for $\Lambda(k, \omega)$, this function is used in the evaluation (6.3.2) of the current. The first term in $G^<$ in (6.3.23) gives a zero current. The second one gives a current proportional to the field E, and this proportionality defines the electrical conductivity

$$\sigma_{\mu\nu} = e^2 \int \frac{d^3k}{(2\pi)^3} \int \frac{d\omega}{2\pi} v_{k\mu}v_{k\nu}\left(-\frac{\partial n_F}{\partial \omega}\right) A(k, \omega)\Lambda(k, \omega) \qquad (6.3.25)$$

Equations (6.3.24) and (6.3.25) provide the solution for the conductivity for the QBE. Exactly the same formulas are found, starting from the Kubo formula, in the limit of dilute impurities.

Note that the second term in (6.3.23) is multiplied by n_i and is divided by $\Gamma \propto n_i$. The prefactor is independent of n_i, and this term is not necessarily smaller than the first term when the impurity concentration is small.

Equation (6.3.24) is a vector equation. The first step in the solution is to choose a coordinate system for the vectors. Let the z-direction be the direction of the wave vector k. Then define θ_0 as the angle between the field E and k, and θ is the angle between p and k. These relations are summarized by

$$k \cdot E = kE \cos(\theta_0)$$

$$k \cdot p = kp \cos(\theta)$$

$$p \cdot E = pE[\cos\theta \cos\theta_0 + \sin\theta \sin\theta_0 \cos(\phi)]$$

Take the scalar product of the vector E with each term in equation (6.3.24). In the scattering term, $v_p \cdot E$ has the angular dependence shown in the third line above. This term contains the only dependence upon the azimuthal angle ϕ, since T_{kp} depends only upon θ and $\Lambda(p, \omega)$ is independent of angle. Thus the $\int d\phi \cos\phi = 0$ eliminates this second term. Then each term in equation (6.3.24) has the same factors of $E \cos\theta_0$ which are canceled. This brings us to the scalar equation

$$v_k \Lambda(k\,\omega) = \frac{A}{2} v_k + \frac{n_i}{2\Gamma} \int \frac{d^3p}{(2\pi)^3} |T_{kp}|^2 A(p, \omega) v_p \cos\theta \Lambda(p, \omega) \quad (6.3.26)$$

The generality of this expression is realized by the observation that it is valid for both metals and semiconductors. The only requirement is that the system be isotropic, since otherwise $\Lambda(p, \omega)$ will depend upon angle.

Equation (6.3.26) is solved by making the quasiparticle approximation. This assumes that $\omega \approx \varepsilon_k$. The spectral function $A(p, \omega)$ is replaced by $2\pi\delta(\varepsilon_p - \varepsilon_k)$. The p-integral will set the magnitude of p equal to k. This means that $\Lambda(p, \omega)$ becomes $\Lambda(k, \omega)$ and can be removed from the integral. Combining (6.3.26) and (6.2.31) gives the expression for the scattering function

$$\Lambda(k, \omega) = A(k, \varepsilon_k)\Gamma(k, \varepsilon_k)\tau(k)$$

$$\frac{1}{\tau(k)} = n_i \int \frac{d^3p}{(2\pi)^3} |T_{kp}|^2 \delta(\varepsilon_k - \varepsilon_p)[1 - \cos\theta] \quad (6.3.27)$$

The quantity $\tau(k)$ is the transport lifetime of the electron in the solid. It contains the factor $1 - \cos\theta$, which gives increased importance to those scattering events at large angle θ. Large-angle scattering contributes more to the electrical resistance. Expressing the T-matrices in terms of the phase shifts as in (6.2.27) the lifetime becomes

$$\frac{1}{\tau(k)} = \frac{4\pi n_i}{mk} \sum_l l \sin^2(\delta_l - \delta_{l-1}) \tag{6.3.28}$$

where $\delta_l(k)$ are the phase shifts of the electron for scattering by the impurity.

The above expression for $\Lambda(k, \omega)$ is used in the equation (6.3.25) for the electrical conductivity. In the quasiparticle approximation, the integral over the square of the spectral function is

$$\int \frac{d\omega}{2\pi} A(k, \omega)^2 f(\omega) = \frac{f(\varepsilon_k)}{\Gamma(k, \varepsilon_k)}$$

where $f(\omega)$ is any function of ω. In isotropic systems, the angular average of $v_{k\mu} v_{k\nu}$ is $\delta_{\mu\nu} v_k^{2/3}$. Thus we get for the conductivity

$$\sigma = \frac{e^2}{3} \int \frac{d^3k}{(2\pi)^3} v_k^2 \left(-\frac{\partial}{\partial \varepsilon_k} n_F(\varepsilon_k) \right) \tau(k) \tag{6.3.29}$$

The electrical conductivity is the average (6.3.29) of the transport lifetime (6.3.28) over the occupied electron states in the solid. For a semiconductor with light doping, n_F can be approximated as a Maxwell–Boltzmann distribution. This completes the discussion of the electrical conductivity from impurity scattering.

The electron scattering by phonons is important in most solids for providing a contribution to the resistivity which depends on temperature. The temperature dependence arises from the energy of the phonon. There are two contributions: the first is provided by the phonon occupation number $N_q = \exp(\beta\omega_q) - 1$, where $\beta = 1/k_B T$ (here T is temperature) and ω_q is the phonon energy. This quantity varies significantly with temperature. The second contribution of the phonon energy is to make the electron scattering inelastic. The electron can gain or lose energy by absorbing or emitting phonons. This process makes the algebra of the QBE much harder since the energy becomes an active variable in the integral equation.

The QBE will be solved for electron scattering by phonons. The solution is valid for steady-state, homogeneous systems, and is accurate to the first order in the electric field. The starting point is again the Mahan–Hansch equation (6.3.22), with the ansatz (6.3.23) for $G^<$. The scattering terms on the right are solved using the one-phonon self-energies in equation (6.2.37). The expressions for $G^<$ and $G^> = G^< - iA$ are inserted into these self-energy expressions which produces a rather lengthy expression.

$$\Sigma^> G^< - \Sigma^< G^> = A(p, \omega) \sum_q M_q^2 [a_1 + \frac{e}{m} p \cdot E a_2 + \frac{e}{m} (p + q) \cdot E a_3 + O(E^2)]$$

where

$$a_1 = A(p + q, \omega + \omega_q)\{(N_q + 1)[1 - n(\omega)]n(\omega + \omega_q) - N_q n(\omega)$$
$$\cdot [1 - n(\omega + \omega_q)]\}$$
$$+ A(p + q, \omega - \omega_q)\{N_q n(\omega - \omega_q)[1 - n(\omega)] - (N_q + 1)n(\omega)$$
$$\cdot [1 - n(\omega - \omega_q)]\}$$

$$a_2 = -\Lambda(p, \omega)\left[-\frac{\partial n(\omega)}{\partial \omega}\right]\{A(p + q, \omega + \omega_q)[N_q + n(\omega + \omega_q)]$$
$$+ A(p + q, \omega - \omega_q)[N_q + 1 - n(\omega - \omega_q)]\}$$

$$a_3 = \Lambda(p + q, \omega + \omega_q)A(p + q, \omega + \omega_q)\left(-\frac{\partial n}{\partial \omega}(\omega + \omega_q)\right)[N_q + 1 - n(\omega)]$$
$$+ \Lambda(p + q, \omega - \omega_q)A(p + q, \omega - \omega_q)\left(-\frac{\partial n(\omega - \omega_q)}{\partial \omega}\right)[N_q + n(\omega)]$$

The term a_1 equals 0 since the expressions for the thermal occupation factors exactly cancel. The scattering should vanish in equilibrium, so this is expected. If a_1 were not zero, the equilibrium solution would be incorrect.

The term in a_2 is proportional to $\Gamma = -\text{Im} \Sigma^r$. This quantity can be written for interacting systems as

$$\Gamma(p, \omega) = \frac{1}{2}\sum_q M_q^2\{[N_q + 1 - n(\omega - \omega_q)]A(p + q, \omega - \omega_q)$$
$$+ [N_q + n(\omega + \omega_q)]A(p + q, \omega + \omega_q)\} \qquad (6.3.30)$$

This way of writing $\Gamma(p, \omega)$ should be compared with the earlier expression in equation (6.2.36). They become identical in the quasiparticle approximation where the spectral function A is approximated as a delta function $A(p, \omega) = 2\pi\delta(\varepsilon_p - \omega)$. The expression in equation (6.3.30) includes the interactions and is more accurate. It is entirely determined by equilibrium quantities and is regarded as a known function when solving the QBE for the unknown function $\Lambda(p, \omega)$.

The term in a_3 is very important in the QBE. It can be simplified a bit by rearranging its thermal factors to extract a factor of $-\partial n(\omega)/\partial \omega$. One can show that

$$\frac{\partial n(\omega + \omega_q)}{\partial \omega}[N_q + 1 - n(\omega)] = \frac{\partial n(\omega)}{\partial \omega}[N_q + n(\omega + \omega_q)]$$

$$\frac{\partial n(\omega - \omega_q)}{\partial \omega}[N_q + n(\omega)] = \frac{\partial n(\omega)}{\partial \omega}[N_q + 1 - n(\omega - \omega_q)]$$

These two expressions seem a bit surprising. However, they can be proven by only algebraic manipulations by defining $N_q = 1/(\exp(\beta\omega_q) - 1)$ and $n(\omega) = 1/(\exp(\beta\omega) + 1)$ and then evaluating the expressions on both sides of the equal sign.

Since both a_2 and a_3 are proportional to $-\partial n(\omega)/\partial\omega$, each term in the QBE has this factor and also the factor $A(p, \omega)$. After factoring out these two common terms, the QBE for electron–phonon now has the form

$$\frac{e}{m}\,p \cdot EA(p, \omega)\Phi = \frac{2e}{m}\,p \cdot E\Lambda(p, \omega)\Gamma(p, \omega) - \sum_q M_q^2 \frac{e}{m}(p + q) \cdot E$$

$$\{A(p + q, \omega + \omega_q)\Lambda(p + q, \omega + \omega_q)[N_q + n(\omega + \omega_q)]$$

$$+ A(p + q, \omega - \omega_q)\Lambda(p + q, \omega - \omega_q)$$

$$\cdot [N_q + 1 - n(\omega - \omega_q)]\} \tag{6.3.31}$$

where

$$\Phi(p, \omega) = \Gamma\left(1 + \frac{\partial}{\partial\varepsilon_k}\,\mathrm{Re}\,\Sigma^r\right) + (\omega - \varepsilon_k - \mathrm{Re}\,\Sigma^r)\frac{\partial\Gamma}{\partial\varepsilon_k}$$

Each term in (6.3.31) contains the electric field E. It can be canceled which leaves the vector integral equation

$$\mathbf{p}\Lambda(p, \omega) = \mathbf{p}\,\frac{A\Phi}{2\Gamma} + \frac{1}{2\Gamma}\sum_q M_q^2(\mathbf{p} + \mathbf{q})\{\cdots\} \tag{6.3.32}$$

where the braces have the same factors as in (6.3.31). In the last term, the factor of $\sum_q M_q^2(\mathbf{p} + \mathbf{q})f(\mathbf{p} + \mathbf{q})$ must be a vector in the **p**-direction. Thus a scalar integral equation is obtained by taking the scalar product of this equation with p:

$$\Lambda(p, \omega) = \frac{A\Phi}{2\Gamma} + \frac{1}{2\Gamma}\sum_q M_q^2 \frac{\mathbf{p} \cdot (\mathbf{p} + \mathbf{q})}{p^2}\{\cdots\} \tag{6.3.33}$$

The above form for the QBE is now an integral equation which must be solved for the unknown function $\Lambda(p, \omega)$. Its solution is then used in the conductivity formula (6.3.25).

The solution to this equation for electrons in a metal is given elsewhere.[9] Here we provide the solution for a single electron scattering from an LO phonon in a semiconductor. The matrix element M_q is given in (6.3.1), and the factor of inverse volume Ω^{-1} is used to change $\Omega^{-1} \Sigma q \to \int d^3q/(2\pi)^3$. The integration variable is changed to $\mathbf{p}' = \mathbf{p} + \mathbf{q}$ and the electron occupation number $n(\omega)$ is set equal to zero. The LO phonon frequency ω_0 and occupation number N_0 are constants. These changes make the above integral equation become

$$\Lambda(p, \omega) = \frac{A}{2} + \frac{\alpha \omega_0}{4\pi^2 \Gamma p^2}\left(\frac{\omega_0}{2m}\right)^{1/2} \int d^3p' \frac{p \cdot p'}{(\mathbf{p} - \mathbf{p}')^2}$$

$$\cdot \{N_0 A(p', \omega + \omega_0)\Lambda(p', \omega + \omega_0)$$

$$+ (N_0 + 1)A(p', \omega - \omega_0)\Lambda(p', \omega - \omega_0)\} \qquad (6.3.34)$$

The angular integrals are done first. The azimuthal angle gives $\int d\phi = 2\pi$. The polar angle integral is done by using the variable $\nu = \cos \theta$

$$\int_{-1}^{1} d\nu \frac{pp'\nu}{p^2 + p'^2 - 2pp'\nu} = f(\varepsilon_p, \varepsilon_{p'})$$

$$f(\varepsilon, \omega) = \frac{\varepsilon + \omega}{2\sqrt{\omega}\sqrt{\varepsilon}} \ln \left|\frac{\sqrt{\varepsilon} + \sqrt{\omega}}{\sqrt{\varepsilon} - \sqrt{\omega}}\right| - 1$$

The momentum variable p is changed to the energy variable ε_p, so the unknown function is $\Lambda(\varepsilon_p, \omega)$ and the p subscript is dropped. The integral equation now has the form

$$2\Gamma\Lambda(\varepsilon, \omega) = \Gamma A(\varepsilon, \omega) + \frac{\alpha\omega_0^{3/2}}{2\pi\varepsilon}\int_0^\infty d\varepsilon'\sqrt{\varepsilon'}f(\varepsilon, \varepsilon')$$

$$\cdot \{N_0 A(\varepsilon', \omega + \omega_0)\Lambda(\varepsilon', \omega + \omega_0)$$

$$+ (N_0 + 1)A(\varepsilon', \omega - \omega_0)\Lambda(\varepsilon', \omega - \omega_0)\} \qquad (6.3.35)$$

$$2\Gamma(\varepsilon, \omega) = \frac{\alpha\omega_0^{3/2}}{2\pi\sqrt{\varepsilon}}\int_0^\infty d\varepsilon' \ln \left|\frac{\sqrt{\varepsilon} + \sqrt{\varepsilon'}}{\sqrt{\varepsilon} - \sqrt{\varepsilon'}}\right| \{N_0 A(\varepsilon', \omega + \omega_0)$$

$$+ (N_0 + 1)A(\varepsilon', \omega - \omega_0)\}$$

$$\sigma = \frac{e^2 4\sqrt{2}\sqrt{m}}{3(2\pi)^3}\int_{-\infty}^\infty d\omega \left(-\frac{\partial n}{\partial \omega}\right)\int_0^\infty d\varepsilon\, \varepsilon^{3/2} A(\varepsilon, \omega)\Lambda(\varepsilon, \omega)$$

We have also added the expression for σ and Γ, where Γ is the imaginary self-energy of the electron, and σ is the conductivity from (6.3.28). The spectral function is

$$A(\varepsilon, \omega) = \frac{2\Gamma}{(\omega - \varepsilon - R)^2 + \Gamma^2}$$

$$R = \text{Re}\{\Sigma(\varepsilon, \omega)\}$$

$$\Gamma = -\text{Im}\{\Sigma(\varepsilon, \omega)\}$$

For a small density of electrons, the occupation factor $n(\omega)$ is a Maxwell–Boltzmann distribution

$$n(\omega) = n_e \left(\frac{2\pi\beta}{m}\right)^{3/2} e^{-\beta\omega}$$

where n_e is the electron density.

A proper solution of these equations is hard. One must first solve self-consistently the equation for $\Gamma(\varepsilon, \omega)$ which is complicated by the fact that A depends upon Γ. One must also find the real part of the self-energy R. After solving these equations for the equilibrium quantities Γ, R, and A, one must next solve the integral equation (6.3.35) for the scattering function $\Lambda(\varepsilon, \omega)$.

Here we give an approximate solution which is valid in the limit that the polaron coupling constant α is small. Small α makes Γ small and the spectral function can be evaluated in the quasiparticle approximation where $A(\varepsilon, \omega) = 2\pi\delta(\omega - \varepsilon)$. Note that $\Lambda(\varepsilon, \omega)$ always appears multiplied by $A(\varepsilon, \omega)$, so the delta function makes it become $\Lambda(\omega, \omega)$ which we abreviate by $\Lambda(\omega)$. Since Λ is always multiplied by A, then we really want to solve equation (6.3.35) multiplied by $A/2\Gamma$. The first term on the right becomes $A^2/2$ which is ambiguous when A is replaced by a delta function. The right way to treat the A^2 term is to replace it by the limit when $\Gamma \to 0$ of the integral

$$\lim_{\Gamma \to 0} \tfrac{1}{2} \int d\varepsilon\, A(\varepsilon, \omega)^2 = 2\Gamma \int_{-\infty}^{\infty} d\varepsilon \frac{1}{[(\varepsilon - \omega)^2 + \Gamma^2]^2} = \frac{\pi}{\Gamma}$$

where we have neglected R because α is small. Then the above expressions for Λ, σ, and $\Gamma(\omega, \omega) \equiv \Gamma(\omega)$ simplify to

$$\Lambda(\omega) = \frac{1}{2\Gamma} + \frac{\alpha\omega_0^{3/2}}{2\Gamma\omega} \{ N_0(\omega + \omega_0)^{1/2} f(\omega, \omega + \omega_0)\Lambda(\omega + \omega_0)$$

$$+ (N_0 + 1)(\omega - \omega_0)^{1/2} f(\omega, \omega - \omega_0)\Lambda(\omega - \omega_0)\} \quad (6.3.36)$$

$$2\Gamma(\omega) = \frac{\alpha\omega_0^{3/2}}{\sqrt{\omega}}$$

$$\cdot \left\{ N_0 \ln\left|\frac{(\omega + \omega_0)^{1/2} + \sqrt{\omega}}{(\omega + \omega_0)^{1/2} - \sqrt{\omega}}\right| + (N_0 + 1) \ln\left|\frac{(\omega - \omega_0)^{1/2} + \sqrt{\omega}}{(\omega - \omega_0)^{1/2} - \sqrt{\omega}}\right| \right\}$$

$$\sigma = ne\tau/m$$

$$\tau = \frac{4\beta^{5/2}}{3\sqrt{\pi}} \int_0^{\infty} d\omega\, \omega^{3/2} e^{-\beta\omega} \Lambda(\omega)$$

Expressions which contain $(\omega - \omega_0)^{1/2}$ only make sense if $\omega > \omega_0$; otherwise these terms are absent. Equation (6.3.36) is solved for $\Lambda(\omega)$ using the expressions for f and Γ. This result is used in the bottom integral which defines the average relaxation time. The equation for $\Lambda(\omega)$ has the character of a contained fraction. An approximate solution is obtained for small α and low temperature (large β) by solving equation (6.3.36) by iteration. The variable ω only has important values within $k_B T$ of the bottom of the band since the conductivity integral limits $\beta\omega \approx 1$. With a little effort one can show that the terms $\Lambda(\omega \pm l\omega_0)$ are negligible for $l > 1$ and that $f(\omega - \omega_0, \omega)$ is zero for $\omega - \omega_0 < 0$. The iteration produces the equation

$$\Lambda(\omega) = \frac{1}{2\Gamma(\omega)} \left\{ 1 + \frac{\alpha\omega_0^{3/2}}{\omega} N_0(\omega + \omega_0)^{1/2} f(\omega, \omega + \omega_0)\Lambda(\omega + \omega_0) \right\}$$

$$\Lambda(\omega + \omega_0) = \frac{1}{2\Gamma(\omega + \omega_0)} \left\{ 1 + \frac{\alpha\omega_0^{3/2}}{\omega + \omega_0} (N_0 + 1)\sqrt{\omega}\, f(\omega + \omega_0, \omega)\Lambda(\omega) \right\}$$

This equation has the obvious solution

$$\Lambda(\omega) = \frac{\{1 + O(\alpha)\}}{2\Gamma(\omega)(1 - \Xi)} \tag{6.3.37}$$

$$\Xi(\omega) = \frac{\alpha^2\omega_0^3 N_0(N_0 + 1)f(\omega, \omega + \omega_0)^2}{4\Gamma(\omega)\Gamma(\omega + \omega_0)[\omega(\omega + \omega_0)]^{1/2}} \tag{6.3.38}$$

at first it appears that the correction term Ξ is proportional to the coupling constant α. However, since $\Gamma \propto \alpha$, then Ξ is independent of α. In the limit that $\omega \to 0$ one finds for the various quantities

$$\lim_{\omega \to 0} \left| \begin{array}{ll} f(\omega, \omega + \omega_0) \to 3\omega/2\omega_0 \\[1em] \Gamma(\omega) \qquad \to \alpha\omega_0 N_0 \\[1em] \Gamma(\omega + \omega_0) \to \alpha\omega_0 \left\{ N_0 \ln\left|\frac{\sqrt{2} + 1}{\sqrt{2} - 1}\right| + (N_0 + 1)2\sqrt{\omega}/\sqrt{\omega_0} \right\} \\[1.5em] \Xi(\omega) \qquad \to \dfrac{\frac{8}{9}(\omega/\omega_0)^{3/2}}{2(\omega/\omega_0)^{1/2} + e^{-\beta\omega_0}\ln\left|\frac{\sqrt{2} + 1}{\sqrt{2} - 1}\right|} \end{array} \right.$$

where all correction terms are of $O(\omega/\omega_0)$ compared to the listed terms. Expanding the denominator in (6.3.37) $[(1 - \Xi)^{-1} \approx 1 + \Xi]$ gives for the average relaxation time

$$\tau = \frac{1}{2\alpha\omega_0 N_0}[1 + 10k_B T/9\omega_0 + O(k_B T/\omega_0)^2]$$

This completes the derivation of the polaron mobility using the quantum transport equation.

The self-consistent scattering theory produced a correction term Ξ which is not of order α, so is not small at small scattering strengths. This correction term was first found by Mahan.[22] The same feature was found above for the impurity scattering. There the multiple scattering also found corrections terms whose size was nearly independent of the scattering strength, which is proportional to n_i for impurity scattering.

ACKNOWLEDGMENTS

Research support gratefully acknowledged from the National Science Foundation Grant DMR 85-010110, and from the Distinguished Scientist Program of the University of Tennessee and Oak Ridge National Laboratory. Oak Ridge National Laboratory is administered by Martin Marietta Energy Systems for the U.S. Department of Energy.

REFERENCES

1. G. D. MAHAN, *Many-Particle Physics*, Second Edition, Plenum, New York (1990).
2. T. HOLSTEIN, *Ann. Phys.* **20**, 410 (1964).
3. J. M. ZIMAN, *Principles of the Theory of Solids*, Cambridge (1960). J. M. ZIMAN, *Electrons and Phonons*, Clarendon, Oxford (1967).
4. L. P. KADANOFF and G. BAYM, *Quantum Statistical Mechanics*, Benjamin, New York (1962).
5. J. T. DEVRESSE, ed. *Linear and Nonlinear Transport in Solids*, Plenum, New York (1976).
6. J. RAMMER and H. SMITH, *Rev. Mod. Phys.* **58**, 323 (1986).
7. D. C. LANGRETH and J. W. WILKINS, *Phys. Rev. B* **6**, 3189 (1972).
8. K. D. SCHOTTE, *Phys. Rep.* **46**, 93 (1978).
9. G. D. MAHAN, *Phys. Rep.* **145**, 251 (1987).
10. R. A. CRAIG, *J. Math. Phys.* **9**, 605 (1968).
11. A. BLANDIN, A. NOUTIER, and D. W. HONE, *J. Phys.* (*Paris*) **37**, 369 (1976).
12. G. D. MAHAN and W. HANSCH, *J. Phys. F* **13**, L47 (1983).
13. W. HANSCH and G. D. MAHAN, *Phys. Rev. B* **28**, 1902 (1983).
14. W. HANSCH and G. D. MAHAN, *Phys. Rev. B* **28**, 1886 (1983).
15. J. W. WU and G. D. MAHAN, *Phys. Rev. B* **30**, 5611 (1984).
16. G. D. MAHAN, in: *Polarons in Ionic Crystals and Polar Semiconductors* (J. Devreese, ed.) North-Holland, Amsterdam, 553–657 (1972).
17. J. R. BARKER, *J. Phys. C* **6**, 2663 (1973).
18. K. K. THORNBER, *Solid State Elec.* **21**, 259 (1978).
19. A. P. JAUHO and J. W. WILKINS, *Phys. Rev. B* **29**, 1919 (1984).
20. S. K. SARKER, *Phys. Rev. B* **32**, 743 (1985).
21. F. LIPAVSKY, V. SPICKA, and B. VELICKY, *Phys. Rev. B* **34** (in press).
22. G. D. MAHAN, *Phys. Rev.* **142**, 366 (1966).

Chapter 7

Green's Function Methods: Nonequilibrium, High-Field Transport

Antti-Pekka Jauho

7.1. CONTOUR-ORDERED GREEN'S FUNCTIONS

Contour-ordered Green functions have been discussed in a number of review articles.[1-3] Here I wish to go into more detail than what is usually done. The discussion is to some extent based on unpublished notes by Rammer.[4]

The nonequilibrium problem is formulated as follows. We consider a system evolving under the Hamiltonian

$$H = h + H'(t) \tag{7.1.1}$$

(here $h = H_0 + H_i$, H_0 is "simple" and H_i is "complicated") where the nonequilibrium part is assumed to vanish for times $t < t_0$. One often lets $t_0 \to -\infty$ at a suitable point; this procedure, however, excludes the discussion of transient phenomena. We will return to this point below. Before the perturbation is turned on the system is described by the thermal equilibrium density matrix

$$\rho(h) = \frac{\exp(-\beta h)}{\text{Tr}[\exp(-\beta h)]} \tag{7.1.2}$$

The task is to calculate the expectation value of a given observable for times $t \geq t_0$:

$$\langle O(t) \rangle = \text{Tr}[\rho(h) O_H(t)] \tag{7.1.3}$$

Antti-Pekka Jauho ● Physics Laboratory, H. C. Ørsted Institute, University of Copenhagen, DK2100 Copenhagen Ø, Denmark. *Present address*: Nordita (Nordisk Institut for Teoretisk Fysik), DK2100 Copenhagen Ø, Denmark.

Quantum Transport in Semiconductors, edited by David K. Ferry and Carlo Jacoboni. Plenum Press, New York, 1991.

Definition (7.1.3) can be generalized to two-time (or n-time) quantities (Green functions, correlation functions) in an obvious fashion.

The general plan of attack is similar to the equilibrium case. We transform the complicated time-dependence of O_H to a simpler form, namely O_{H_0}. Since there are two operators to be eliminated (the time-dependent external perturbation $H'(t)$ and the "complicated" interaction term H_i), we expect to meet more complicated transformations than in the equilibrium case. However, with suitable generalizations, it can be shown that the nonequilibrium and equilibrium formalisms can be made structurally equivalent.

The first step is to change the time-dependence of O_H to that of O_h. This is achieved by the relation

$$O_H(t) = v_h^\dagger(t, t_0) O_h(t) v_h(t, t_0) \tag{7.1.4}$$

where

$$v_h(t, t_0) = T\left[\exp\left(-i \int_{t_0}^{t} dt'\, H_h'(t')\right)\right] \tag{7.1.5}$$

and

$$H_h'(t) = \exp(ih(t - t_0)) H'(t) \exp(-ih(t - t_0)) \tag{7.1.6}$$

and T is the time-ordering operator which arranges the latest times to left. We now introduce contour-ordered quantities. Observe that (7.1.4) is equivalent to

$$O_H(t) = T_{C_t}\left[\exp\left(-i \int_{c_t} d\tau\, H_h'(\tau)\right) O_h(t)\right] \tag{7.1.7}$$

where the contour C_t is depicted in Figure 7.1.

The contour runs on the real axis (or slightly above it; if $H'(t)$ can be analytically continued, no problems can arise) from t_0 to t and back again. The meaning of the contour-ordering operator T_{C_t} is the following: the time that occurs latest on the contour is moved in front of all earlier times.

In order to get some acquaintance with the functions defined on a contour, we now explicitly demonstrate the equivalence of (7.1.4) and (7.1.7). We have

$$T_{C_t}\left[\exp\left(-i \int_{C_t} d\tau\, H_h'(\tau)\right) O_h(t)\right]$$
$$= \sum_{n=0}^{\infty} \frac{(-i)^n}{n!} \int_{C_t} d\tau_1 \cdots \int_{C_t} d\tau_n\, T_{C_t}[H_h'(\tau_1) \cdots H_h'(\tau_n) O_h(t)] \tag{7.1.8}$$

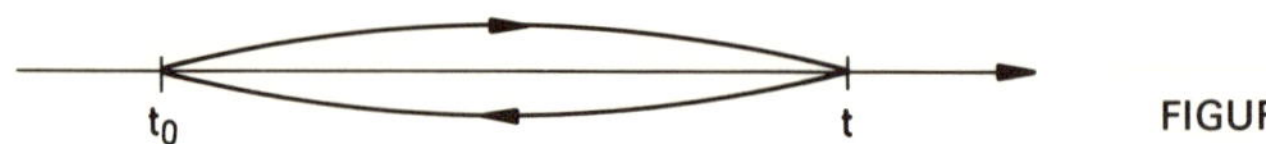

FIGURE 7.1. Contour C_t.

Now divide the contour into two branches:

$$\int_{C_t} = \int_{\to} + \int_{\leftarrow} \tag{7.1.9}$$

where $\int_{\to}$ goes from t_0 to t, and $\int_{\leftarrow}$ from t to t_0. Thus the nth-order term (7.1.8) generates 2^n terms. Let us consider one of them:

$$\int_{\to} d\tau_1 \int_{\to} d\tau_2 \int_{\leftarrow} d\tau_3 \cdots \int_{\leftarrow} d\tau_n \, T_{C_t}[H'_h(\tau_1) \cdots H'_h(\tau_n)O_h(t)]$$

$$= \int_{\leftarrow} d\tau_3 \cdots \int_{\leftarrow} d\tau_n \, T_{\leftarrow}[H'_h(\tau_3) \cdots H'_h(\tau_n)]O_h(t)$$

$$\times \int_{\to} d\tau_1 \int_{\to} d\tau_2 \, T_{\to}[H'_h(\tau_1)H'_h(\tau_2)] \tag{7.1.10}$$

We must now do some combinatorics. Out of the 2^n terms we have generated, there are $n!/[m!(n-m)!]$ terms with m $\to$'s ($m = 0, \ldots, n$ and in the above example $m = 2$). All these terms give the same contribution. Thus we can write

$$\int_{C_t} d\tau_1 \cdots \int_{C_t} d\tau_n \, T_{C_t}[H'_h(\tau_1) \cdots H'_h(\tau_n)O_h(t)]$$

$$= \sum_{m=0}^{n} \frac{n!}{m!(n-m)!} \int_{\leftarrow} d\tau_{m+1} \cdots \int_{\leftarrow} d\tau_n \, T_{\leftarrow}[H'_h(\tau_{m+1}) \cdots H'_h(\tau_n)]O_h(t)$$

$$\times \int_{\to} d\tau_1 \cdots \int_{\to} d\tau_m \, T_{\to}[H'_h(\tau_1) \cdots H'_h(\tau_m)] \tag{7.1.11}$$

Now introduce a new variable $k = n - m$; we can sum both k and m from 0 to ∞ as long as their sum equals n, and this is achieved by inserting a Kronecker delta:

$$(7.1.11) \to \sum_{m,k=0}^{\infty} \frac{n!}{m!k!} \delta_{n,k+m} \left[\int_{\leftarrow} d\tau_1 \cdots \int_{\leftarrow} d\tau_k \, T_{\leftarrow}[H'_h(\tau_1) \cdots H'_h(\tau_k)] \right] O_h(t)$$

$$\times \left[\int_{\to} d\tau_1 \cdots \int_{\to} d\tau_m \, T_{\to}[H'_h(\tau_1) \cdots H'_h(\tau_m)] \right] \tag{7.1.12}$$

We can now go back to (7.1.8). The n-sum is simple (due to the factor $\delta_{n,k+m}$), and we obtain

$$T_{C_t}\left[\exp\left[-i \int_{C_t} d\tau \, H'_h(\tau) \right] O_h(t) \right]$$

$$= \sum_{k=0}^{\infty} \frac{(-i)^k}{k!} \int_{\leftarrow} d\tau_1 \cdots \int_{\leftarrow} d\tau_k \, T_{\leftarrow}[H'_h(\tau_1) \cdots H'_h(\tau_k)]O_h(t)$$

$$\times \sum_{m=0}^{\infty} \frac{(-i)^m}{m!} \int_{\to} d\tau_1 \cdots \int_{\to} d\tau_m \, T_{\to}[H'_h(\tau_1) \cdots H'_h(\tau_m)] \tag{7.1.13}$$

But, comparing the factors multiplying $O_h(t)$ from left and right, we can identify $v^\dagger(t, t_0)$ and $v(t, t_0)$, respectively. We have thus demonstrated the equivalence of (7.1.4) and (7.1.7).

The contour ordering operator is a strong formal tool which will allow us to develop the nonequilibrium theory along lines parallel to the equilibrium theory.

We define now the contour-ordered Green function:

$$G(1, 1') \equiv -i\langle T_C[\psi_H(1)\psi_H^\dagger(1')]\rangle \tag{7.1.14}$$

where the contour C starts and ends at t_0; it runs along the real axis and passes through t_1 and t_1' once and just once (see Figure 7.2). Finally, $(1) \equiv (\mathbf{x}_1, t_1)$.

We can now repeat the analysis leading to (7.1.7). The result is

$$G(1, 1') = -i\left\langle T_C\left[\exp\left(-i\int_C d\tau\, H_h'(\tau)\right)\psi_h(1)\psi_h^\dagger(1')\right]\right\rangle \tag{7.1.15}$$

We still need one more transformation before we can apply Wick's theorem, which we need to establish the diagrammatic perturbation theory. Recall that the operator h still contains two terms, $h = H_0 + H_i$, and that Wick's theorem only works for H_0 (i.e., quadratic Hamiltonians). Thus we must replace the h-dependence by H_0-dependence. Note also that the density matrix implicit in (7.1.15) contains h, and h thus occurs in four places in (7.1.15).

We omit the details and state the final result:

$$G(1, 1') = -i\left\{ \text{Tr}\left[\rho_0 T_{C_v}\left[\exp\left(-i\int_{C_v} d\tau\, H_{H_0}^i(\tau)\right)\right.\right.\right.$$
$$\left.\left.\left. \times \exp\left(-i\int_C d\tau\, H_{H_0}'(\tau)\right)\psi_{H_0}(1)\psi_{H_0}^\dagger(1')\right]\right]\right\}$$
$$\times \left\{ \text{Tr}\left[\rho_0 T_{C_v}\left[\exp\left(-i\int_{C_v} d\tau\, H_{H_0}^i(\tau)\right)\right.\right.\right.$$
$$\left.\left.\left. \times \exp\left(-i\int_C d\tau\, H_{H_0}'(\tau)\right)\right]\right]\right\}^{-1} \tag{7.1.16}$$

where the density matrix ρ_0 is given by

$$\rho_0 = \frac{\exp(-\beta H_0)}{\text{Tr}[\exp(-\beta H_0)]} \tag{7.1.17}$$

FIGURE 7.2. Contour C.

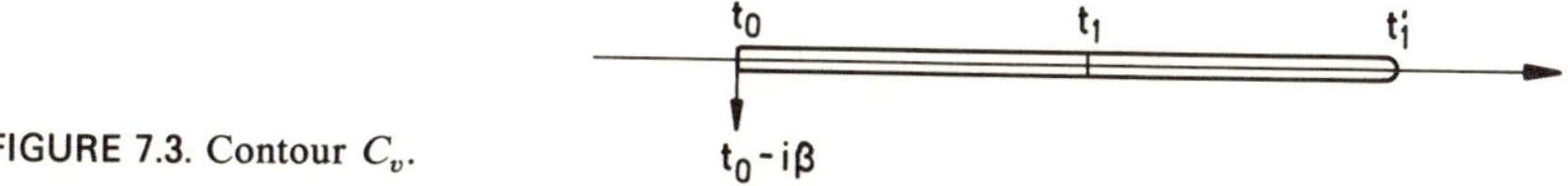

FIGURE 7.3. Contour C_v.

The contour C is defined in Figure 7.2 while the contour C_v is given in Figure 7.3.

Equation (7.1.16) is an important result. Despite its complicated appearance, it has a number of attractive features. First, it is exact. Next, all time dependence is run by the "solvable" H_0. In particular, the quadratic density matrix $\sim\exp(-\beta H_0)$ allows one to use Wick's theorem. Thus, the Feynman graphs can be constructed for the nonequilibrium problem. Once again the denominator cancels the contribution arising from the disconnected graphs.

We can summarize the main results of this section as follows. The equilibrium and nonequilibrium theories are structurally equivalent. The only difference is the replacement of real-axis integrals by contour integrals.

7.2. ANALYTIC CONTINUATION

While the result (7.1.16) is a strong formal statement, it is rather impractical in calculations unless one can replace the contour integrals by real-time integrals. This procedure is called analytic continuation, and many different formulations exist in the literature. This topic has already been discussed in Chapter 6; here we analyze in some detail the generalization of the method of Kadanoff and Baym[5] due to Langreth.[1]

As shown in Section 7.1, the contour ordered Green's function has the same Feynman diagrams as the corresponding equilibrium time-ordered Green's function. Consequently, it has the same Dyson equation:

$$G(1, 1') = G_0(1, 1') + \int d^3x_2 \int_{C_v} d\tau_2 \, G_0(1, 2)U(2)G(2, 1')$$

$$+ \int d^3x_2 \int d^3x_3 \int_{C_v} d\tau_2 \int_{C_v} d\tau_3 \, G_0(1, 2)\Sigma(2, 3)G(3, 1') \quad (7.2.1)$$

where U is the external potential and the interactions are contained in the (irreducible) self-energy $\Sigma[G]$.

The first simplification occurs if we can set $t_0 \to -\infty$. If the interactions are coupled adiabatically, the contribution from the $[t_0, t_0 - i\beta]$ piece vanishes. The information lost by this procedure is related to initial correlations. In many physical situations, for example in steady-state transport, it

appears plausible that the initial correlations have been washed out by by
the interactions when one reaches the steady state. On the contrary, if one
studies transient response, the role of initial correlations can be important.
This represents a difficult problem for which, to the best of my knowledge,
no complete treatment exists in the literature. We will briefly sketch some
of the ensuing complications in Section 7.5. Here we consider the $t_0 \rightarrow -\infty$
limit. In this limit the contours C and C_v coincide, and we can consider
only C.

Langreth theorem. In considering the Dyson equation (7.2.1), we need
terms like

$$C(t_1, t_{1'}) \equiv \int_C d\tau\, A(t_1, \tau) B(\tau, t_{1'}) \tag{7.2.2}$$

and their generalizations involving products of three (or more) terms. To
evaluate (7.2.2) we assume (for definiteness) that t_1 is one the first half of
C and that $t_{1'}$ is on the latter half (see Figure 7.2). Then, in the contour-
ordering sense, t_1 is earlier than $t_{1'}$, and we are dealing with an object which
will be dubbed a "less-than" function, i.e., $C^<$ (because the first argument
is smaller than the second, in the contour sense). Now we deform the
contour as shown in Figure 7.4. Equation (7.2.2) becomes

$$C^<(t_1, t_{1'}) = \int_{C_1} d\tau\, A(t_1, \tau) B^<(\tau, t_{1'}) + \int_{C_1'} d\tau\, A^<(t_1, \tau) B(\tau, t_{1'}) \tag{7.2.3}$$

Here, in the first term we made use of the fact that as long as the integration
variable τ is confined on the contour C_1 it is less than (in the contour sense)
$t_{1'}$. A similar argument applies to the second term. Now consider the first
term in (7.2.3) and split the integration into two parts:

$$\int_{C_1} d\tau\, A(t_1, \tau) B^<(\tau, t_{1'})$$

$$= \int_{-\infty}^{t_1} dt\, A^>(t_1, t) B^<(t, t_{1'}) + \int_{t_1}^{-\infty} dt\, A^<(t_1, t) B^<(t, t_{1'})$$

$$= \int_{-\infty}^{\infty} dt\, A_r(t_1, t) B^<(t, t_{1'}) \tag{7.2.4}$$

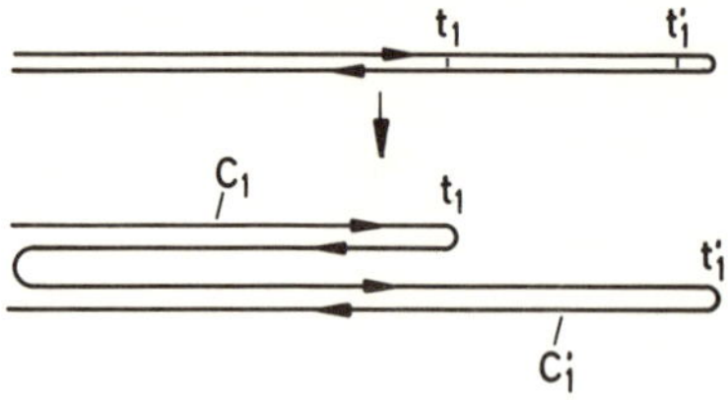

FIGURE 7.4. Deformation of contour C.

where we used the definition of the retarded function (see Chapter 6, (6.2.1)). A similar analysis can be applied to the second term; this time the advanced function is generated. The Langreth theorem can thus be summarized as follows. If on the contour $C = AB$, then on the real axis

$$C^< = A_r B^< + A^< B_a \qquad \text{and} \qquad C^> = A_r B^> + A^> B_a \qquad (7.2.5)$$

It is easy to generalize this result for three functions: If $D = ABC$ on the contour (here, as, for example, in (7.2.1) a "matrix multiplication" of the internal variables is implied), then on the real axis, one has

$$D^< = A_r B_r C^< + A_r B^< C_a + A^< B_a C_a \qquad (7.2.6)$$

and a similar equation holds for the "greater-than" functions. One often needs the retarded (or advanced) component of a series of functions defined on the contour. Using the relations

$$A_r(t, t') = \theta(t - t')(A^> - A^<)(t, t')$$
$$A_a(t, t') = \theta(t' - t)(A^< - A^>)(t, t') \qquad (7.2.7)$$

it is readily seen that

$$C_{r,a} = A_{r,a} B_{r,a} \qquad (7.2.8)$$

When considering the various terms in the diagrammatic perturbation series, one may also encounter terms where two Green's function lines run (anti)parallel [this is the case, for example, a polarization bubble (two antiparallel fermion Green's functions), or electron–phonon self-energy (parallel fermion and boson Green's functions)]; i.e., one must form the less-than or retarded components of (we consider here a structure that could originate from electron–phonon interactions; a fermion bubble can be treated in an analogous fashion)

$$C(\tau, \tau') = A(\tau, \tau') B(\tau, \tau') \qquad (7.2.9)$$

where τ is a contour variable. The required relations are

$$C^<(t, t') = A^<(t, t') B^<(t, t') \qquad (7.2.10)$$

$$C_r(t, t') = A^<(t, t') B_r(t, t') + A_r(t, t') B^<(t, t') + A_r(t, t') B_r(t, t') \qquad (7.2.11)$$

It is a useful exercise to evaluate (7.2.11) with equilibrium electron and phonon Green's functions; the result is found to coincide with the one obtained with the Matsubara technique.[6] Similar results can be derived for the greater-than and advanced functions. We will find many applications of these relations, but a word about the notation. We use the convention that suppressed time arguments imply integration over internal variables [cf. (7.2.4)–(7.2.6)], while if no integration is implied we write the arguments explicitly. Further, the reader should be cautioned that there is no unique convention as to where to put i's, therefore the present notation differs from Langreth's[1] and agrees with Refs. 2, 3, and 7.

7.3. THE KADANOFF–BAYM FORMULATION

Operate now with G_0^{-1} on the Dyson equation (7.2.1) and rearrange:

$$(G_0^{-1} - U)G = 1 + \Sigma G$$
$$G(G_0^{-1} - U) = 1 + G\Sigma \tag{7.3.1}$$

Applying the rule (7.2.5), we find (note that the delta-function term in (7.3.1) vanishes identically because the time labels are on different branches of the contour)

$$(G_0^{-1} - U)G^< = \Sigma_r G^< + \Sigma^< G_a$$
$$G^<(G_0^{-1} - U) = G_r \Sigma^< + G^< \Sigma_a \tag{7.3.2}$$

Subtract these equations from each other:

$$[G_0^{-1} - U, G^<] = \Sigma_r G^< + \Sigma^< G_a - G_r \Sigma^< - G^< \Sigma_a \tag{7.3.3}$$

We use the identity

$$A_r \equiv \tfrac{1}{2}(A_r + A_a) + \tfrac{1}{2}(A_r - A_a)$$

to rewrite the various terms on the right-hand side of (7.3.3), and we obtain

$$[G_0^{-1} - U, G^<] - [\Sigma, G^<] - [\Sigma^<, G] = -\tfrac{1}{2}\{\Sigma^<, G^>\} + \tfrac{1}{2}\{\Sigma^>, G^<\} \tag{7.3.4}$$

Here we defined $\Sigma \equiv \tfrac{1}{2}(\Sigma_r + \Sigma_a)$ and analogously for G. Equation (7.3.4) is the (generalized) Kadanoff–Baym equation (GKB).[1] In subsequent sections we will apply it to various special cases; let us make a few more general comments here.

First, to have a closed set of equations, GKB must be supplemented with the Dyson equations for $G_{r,a}$. Quite often the calculation splits in two stages; one first solves (or tries to!) for $G_{r,a}$, which then are required as inputs for the GKB equation. One can also encounter situations where the Dyson equations for $G_{r,a}$ also involve $G^{\lessgtr}$ [see (7.2.11)]. This results in extreme complications, and I am not aware of any treatment in the literature where the full consequences have been analyzed. Now let us make a few general comments about the structure of the GKB equation. The first commutator on the left-hand side corresponds to a (generalized) driving term. The next two terms are renormalization terms. Finally, the terms on the right-hand side lead to a quantum collision integral. The GKB equation is still exact and reversible. The irreversible behavior predicted by the Boltzmann equation is only introduced after certain approximations are made on the GKB.

7.4. KELDYSH FORMULATION

Almost simultaneously, and independently of Kadanoff and Baym, Keldysh[8] introduced an alternative formulation for applying diagrammatic techniques for nonequilibrium situations (of course, the Kadanoff–Baym and Keldysh techniques have their roots in Schwinger's pioneering work[9]). The Keldysh and Kadanoff–Baym methods are equivalent, and for the sake of completeness we now derive the Keldysh integral equation. Rather than following Keldysh's original treatment, we apply the Langreth theorem on the imaginary-time Dyson equation, and after some rerrangement we end up with (a variation) of the Keldysh equation. Thus, we apply (7.2.6) to (7.2.1):

$$G^< = G_0^< + G_{0,r}\Sigma_r G^< + G_{0,r}\Sigma^< G_a + G_0^< \Sigma_a G_a \qquad (7.4.1)$$

We proceed by iteration. Iterating once and regrouping the terms, we obtain

$$G^< = (1 + G_{0,r}\Sigma_r)G_0^<(1 + \Sigma_a G_a) + (G_{0,r} + G_{0,r}\Sigma_r G_{0,r})\Sigma^< G_a$$
$$+ G_{0,r}\Sigma_r G_{0,r}\Sigma_r G^< \qquad (7.4.2)$$

Inspecting (7.4.2) it is easy to convince oneself that the infinite-order iterate is

$$G^< = (1 + G_r\Sigma_r)G_0^<(1 + \Sigma_a G_a) + G_r\Sigma^< G_a \qquad (7.4.3)$$

Equation (7.4.3) is Keldysh's result (even though in the original work it was written for another function, $G_K \equiv G^< + G^>$, the difference, however, plays minor significance). The relation between the Keldysh equation and the GKB equation is analogous to the relation between an ordinary differential equation plus a boundary condition and the corresponding integral equation. It seems to be a matter of convenience as to which starting point to choose.

7.5. REMARKS ON TRANSIENT RESPONSE

A first-principles analysis of the transient response of an interacting system represents a formidable problem, and only fragments of the complete theory are beginning to emerge; in particular, we would like to mention the works of Kukharenko and Tikhodeev,[10] Hall,[11] and Danielewicz.[12] To get some idea of the complexities, we start with a simplified problem and subsequently discuss the real problem we would like to solve. The general strategy is the following: we keep the time t_0 finite and ask how far we can carry out the analysis. In the general case this means that we cannot ignore the contribution arising from the strip $[t_0, t_0 - i\beta]$.

Let us first assume that the initial-state density matrix allows a Wick decomposition (here we follow Danielewicz). In this case the transformation leading to (7.1.16) is not necessary, and we can write the following Dyson equation for $G^<$:

$$(G_0^{-1} - U)G^<(1, 1') = \int_{t_0}^{\infty} d2[\Sigma_r(1, 2)G^<(2, 1') + \Sigma^<(1, 2)G_a(2, 1')] \quad (7.5.1)$$

The general solution of (7.3.1) can be written in the form

$$G^<(t_1, t_{1'}) = G_r(t_1, t_0)G^<(t_0, t_0)G_a(t_0, t_{1'})$$
$$+ \int_{t_0}^{t_1} dt_2 \int_{t_0}^{t_{1'}} dt_3 \, G_r(t_1, t_2)\Sigma^<(t_2, t_3)G_a(t_3, t_{1'}) \quad (7.5.2)$$

Equation (7.5.2) has a very suggestive structure. Suppose one were to set the two times t_1 and $t_{1'}$ equal. Then the correlation function $G^<$ reduces to the Wigner distribution function, and the two terms in (7.5.2) can be assigned a direct interpretation. The first term describes the decay of the initial distribution, while the second one gives the feedback effect due to collisions. Note however, that the "collision" term is not diagonal in the time labels (the correlation function is contained in the self-energy), and (7.5.2) is not a closed equation for the Wigner distribution function alone. This is a major complication common to all theories, and we will discuss it in detail in Sections 7.6 and 7.9. We also note that (7.5.2) bears a striking resemblance to an equation derived long ago by Thornber,[13] using the Feynman path integral formulation. Since the two approaches are quite different, it would be very interesting to carry out a detailed comparison.

Finally, let us consider the general case where the initial-state density matrix is "complicated." There are two approaches to the problem. In the first, one numerically integrates along the imaginary part of the contour C_v, to prepare a suitable initial value for $G^<(t_0, t_0)$, and then proceeds by integrating along the real axis. This procedure requires the initial value of $G^<(t_0 - i\beta, t_0 - i\beta)$, which can be chosen to equal, say, the Fermi distribution. The second approach consists of constructing a diagrammatic technique, where one accounts for terms in which one of the time labels is located on the imaginary part of the contour. The end result of this procedure is that the Dyson equation has to be amended by two additional self-energies, Σ^c and Σ_c:

$$G = G_0 + G_0\Sigma^c G + G_0\Sigma G + G_0\Sigma_c G \quad (7.5.3)$$

The expressions for the new self-energies involve correlation matrices for which a special diagrammatic technique has to be developed. The only application of this method that I am aware of comes from nuclear matter theory,[12] where particle–particle interactions dominate. An application to semiconductors is still wanting.

7.6. RELATION TO BOLTZMANN EQUATION

The Boltzmann equation is expected to be valid for slow spatial and temporal variations. We thus need a device for separating the "fast" quantum variations from "slow" macroscopic variations. This separation is achieved in two steps: we introduce so-called Wigner (or center-of-mass) coordinates and perform a systematic gradient expansion. The lowest-order gradient approximation and the quasiparticle approximation then lead to the Boltzmann equation.

The new variables are defined by the following relations:

$$\mathbf{r} = \mathbf{x} - \mathbf{x}', \qquad \tau = t - t'$$
$$\mathbf{R} = \tfrac{1}{2}(\mathbf{x} + \mathbf{x}'), \qquad T = \tfrac{1}{2}(t + t') \tag{7.6.1}$$

The variables $\mathbf{r}$ and τ are fast and will be treated exactly (after Fourier transforms $\mathbf{r} \leftrightarrow \mathbf{p}$, $\tau \leftrightarrow \omega$), while $\mathbf{R}$ and T are macroscopic, slow variables with small gradients and, hence, are treated approximately. A word of caution is in order, however. Since one of our main objectives is to construct a high-field transport theory, we wish to treat the external field nonpeturbatively. Now a straightforward expansion in gradients of $\mathbf{R}$ may overlook some higher powers of the external field, thus leading to an inconsistent final result.[14] We will introduce in Section 7.7 a gauge-invariant formulation where such problems are avoided.

All the terms in the GKB equation (7.3.4) are (matrix) products of two functions. Expressed in terms of the new set of variables (7.6.1), these terms transform as

$$\int d^3x \, dt \, A(\mathbf{x}_1, t_1; \mathbf{x}, t) B(\mathbf{x}, t; \mathbf{x}_{1'}, t_{1'})$$

$$\rightarrow A(\mathbf{p}, \omega, \mathbf{R}, T) \exp\left[-\frac{i}{2} (\overleftarrow{\partial}_T \overrightarrow{\partial}_\omega - \overleftarrow{\partial}_\omega \overrightarrow{\partial}_T - \overleftarrow{\partial}_\mathbf{R} \cdot \overrightarrow{\partial}_\mathbf{p} + \overleftarrow{\partial}_\mathbf{p} \cdot \overrightarrow{\partial}_\mathbf{R}) \right]$$

$$\times B(\mathbf{p}, \omega; \mathbf{R}, T) \tag{7.6.2}$$

In the lowest nonvanishing order we get the following prescription for evaluating commutators and anticommutators:

$$[A, B] \rightarrow -i \left\{ \frac{\partial A}{\partial T} \frac{\partial B}{\partial \omega} - \frac{\partial A}{\partial \omega} \frac{\partial B}{\partial T} - \frac{\partial A}{\partial \mathbf{R}} \cdot \frac{\partial B}{\partial \mathbf{p}} + \frac{\partial A}{\partial \mathbf{p}} \cdot \frac{\partial B}{\partial \mathbf{R}} \right\} \tag{7.6.3}$$

$$\{A, B\} \rightarrow 2A(\mathbf{p}, \omega, \mathbf{R}, T) B(\mathbf{p}, \omega, \mathbf{R}, T)$$

We apply these rules now to the GKB equation. We discard the renormalization terms (i.e., the second and third commutators on the left-hand side) at the outset (recall we are heading toward the Boltzmann equation where such corrections are supposedly not present). One obtains

$$-i\left\{ -\frac{\partial U}{\partial T}\frac{\partial G^<}{\partial \omega} - \frac{\partial G^<}{\partial T} - \left(-\frac{\partial U}{\partial \mathbf{R}}\right)\cdot\frac{\partial G^<}{\partial \mathbf{p}} - \frac{\partial \varepsilon}{\partial \mathbf{p}}\cdot\frac{\partial G^<}{\partial \mathbf{R}} \right\}$$

$$= -\Sigma^<(\mathbf{p}, \omega, \mathbf{R}, T)G^>(\mathbf{p}, \omega, \mathbf{R}, T) + \Sigma^>(\mathbf{p}, \omega, \mathbf{R}, T)G^<(\mathbf{p}, \omega, \mathbf{R}, T)$$

$$(7.6.4)$$

Two steps are still required: we must specify the self-energy (i.e., scattering mechanisms), and we must find a connection between the correlation function $G^<$ and a distribution function. As for the self-energy we choose the dilute impurity model:

$$\Sigma^\lessgtr(\mathbf{p}, \omega, \mathbf{R}, T) = c\sum_q |V(\mathbf{p} - \mathbf{q})|^2 G^\lessgtr(\mathbf{q}, \omega, \mathbf{R}, T) \qquad (7.6.5)$$

(Here c is the concentration of impurities.) Next we eliminate $G^>$. This can be done by introducing the important concept of spectral density:

$$A(\mathbf{p}, \omega, \mathbf{R}, T) = -i[G_a(\mathbf{p}, \omega, \mathbf{R}, T) - G_r(\mathbf{p}, \omega, \mathbf{R}, T)]$$

$$= -i[G^<(\mathbf{p}, \omega, \mathbf{R}, T) - G^>(\mathbf{p}, \omega, \mathbf{R}, T)] \qquad (7.6.6)$$

Below we will examine various spectral densities in detail.

The collision integral becomes

$$\left(\frac{\partial G^<}{\partial T}\right)_{coll} = i\sum_q c|V(\mathbf{p} - \mathbf{q})|^2[G^<(\mathbf{q}, \omega, \mathbf{R}, T)A(\mathbf{p}, \omega, \mathbf{R}, T)$$

$$- A(\mathbf{q}, \omega, \mathbf{R}, T)G^<(\mathbf{p}, \omega, \mathbf{R}, T)] \quad (7.6.7)$$

The Wigner distribution function is obtained from the correlation function with the prescription

$$f^W(\mathbf{p}, \mathbf{R}, T) = -iG^<(\mathbf{p}, \tau = 0, \mathbf{R}, T) = -i\int \frac{d\omega}{2\pi} G^<(\mathbf{p}, \omega, \mathbf{R}, T) \quad (7.6.8)$$

This is an exact relation. Here our notation differs from Mahan's in Chapter 6: there the Wigner function depends on the full variable set $(\mathbf{p}, \mathbf{R}, \omega, T)$. One might call (7.6.8) the *Wigner momentum distribution*, but we call it the *Wigner function*. To avoid confusion, we always include the superscript W to distinguish our definition from Mahan's. We shall see that under certain circumstances the kinetic equation for f^W reduces to the Boltzmann equation. The goal here is to obtain a closed equation for f^W alone, and thus we must know $g^<$ in terms of f^W. The simplest way of doing this is to make the ansatz

$$G^<(\mathbf{p}, \omega, \mathbf{R}, T) = iA(\mathbf{p}, \omega, \mathbf{R}, T)f^W(\mathbf{p}, \mathbf{R}, T) \qquad (7.6.9)$$

The following remarks can be made concerning this ansatz. (i) It is consistent with (7.6.8) (because the frequency integral of A always equals unity by virtue of the equal-time (anti)commutation rules). (ii) In equilibrium the following exact relation holds:

$$G_{\text{eq}}^<(\mathbf{p}, \omega) = iA(\mathbf{p}, \omega)n_F(\omega) \qquad (7.6.10)$$

where $n_F(\omega)$ is the Fermi function. If the spectral function is sharply peaked (quasiparticle picture) around $\omega = \varepsilon_p$, then the argument in the Fermi function can be replaced by ε_p. Thus (7.6.10) has the structure (spectral density) × (equilibrium momentum distribution), and its generalization to nonequilibrium, equation (7.6.9), appears natural. (iii) It is known that in many cases, where the results can be obtained by alternative means (say by the Kubo formula), (7.6.9) leads to consistent results. (iv) However, it has been known for some time (Ref. 7, Appendix A) that for the high-field transport problem the results obtained with the GKB [with (7.6.9)] and with a density matrix formulation[15] differ slightly. Very recently Lipavsky et al.[16] have shown how (7.6.9) should be modified in order to obtain agreement with the two theories. We shall return to this point; for now it is sufficient to use (7.6.9). (v) Finally, the ansatz (7.6.9) has the advantage that the left-hand side of the GKB commutes with A, and consequently we can more A in front of all the derivatives in (7.6.4). Integrating over frequency we finally obtain

$$\left[\frac{\partial}{\partial T} + \frac{\partial \varepsilon(\mathbf{p})}{\partial \mathbf{p}} \cdot \frac{\partial}{\partial \mathbf{R}} + \left(-\frac{\partial U}{\partial \mathbf{R}}\right) \cdot \frac{\partial}{\partial \mathbf{p}}\right] f^W(\mathbf{p}, \mathbf{R}, T)$$

$$= \int \frac{d\omega}{2\pi} \sum_{\mathbf{p}} c|V(\mathbf{p} - \mathbf{p}')|^2 A(\mathbf{p}, \omega, \mathbf{R}, T)A(\mathbf{p}', \omega, \mathbf{R}, T)$$

$$\times [f^W(\mathbf{p}', \mathbf{R}, T) - f^W(\mathbf{p}, \mathbf{R}, T)] \qquad (7.6.11)$$

Thus, in the lowest-order gradient expansion the Wigner function obeys an equation which bears a close resemblance to the Boltzmann equation: the only difference is the "renormalization" of the scattering probability: $V^2 \to AV^2A$, where A is the nonequilibrium spectral density. However, as mentioned in Chapter 6, equation (7.6.11) should not be used for nonlinear fields: the gradient expansion at this stage has overlooked some terms, and in later sections we will examine in detail how these terms can be included in a consistent manner.

Finally, the Boltzmann equation is recovered if one uses the free spectral density:

$$A(\mathbf{p}, \omega, \mathbf{R}, T) \to A_0(\mathbf{p}, \omega) = 2\pi\delta(\omega - \varepsilon_\mathbf{p}) \tag{7.6.12}$$

With this choice the frequency integral becomes trivial, and one recovers the energy-conserving delta-function familiar from the impurity Boltzmann equation. A similar derivation can be given for the electron–phonon Boltzmann equation. The rest of this chapter will deal with cases where one relaxes some of the approximation made in this section: gradient approximation, free spectral density, etc.

7.7. GAUGE-INVARIANT FORMULATION

As mentioned above, care must be exercised when performing gradient approximations on GKB. The external electric field can be introduced in a variety of ways (scalar potential, vector potential, or a combination thereof), and a direct application of, say, the rules (7.6.3) leads to different transport equations for different choices of the potentials. This is clearly unacceptable, and one must find a way which avoids these problems. This is achieved by formulating the theory in a gauge-invariant manner. We adopt the following procedure. The gauge-invariant functions (denoted by a tilde) are constructed as

$$\tilde{g}(\mathbf{k}, \omega, \mathbf{R}, T) = \int d^3r \int d\tau \, \exp[iw(\omega, \tau, T, \mathbf{k}, \mathbf{r}, \mathbf{R})]g(\mathbf{r}, \tau, \mathbf{R}, T)$$

$$w(\omega, \tau, T, \mathbf{k}, \mathbf{r}, \mathbf{R}) \tag{7.7.1}$$

$$= \int_{-1/2}^{1/2} d\lambda \, \{\tau[\omega + \phi(\mathbf{R} + \lambda\mathbf{r}, T + \lambda\tau)] - \mathbf{r} \cdot [\mathbf{k} + \mathbf{A}(R + \lambda\mathbf{r}, T + \lambda\tau)]\}$$

Here we use the variables introduced in (7.6.1); $\phi(\mathbf{R}, T)$ and $\mathbf{A}(\mathbf{R}, T)$ are the scalar and vector potentials, respectively. Special cases of (7.7.1) have recently been considered by several authors.[14,17–20] While it is possible to apply (7.7.1) to the general case, we restrict ourselves here to spatially homogenous and steady fields. In this case (7.7.1) reduces to the simple rule

$$g(\mathbf{p}, t, t') = g\left(\mathbf{k} - \mathbf{E}T, T + \frac{\tau}{2}, T - \frac{\tau}{2}\right) = \tilde{g}(\mathbf{k}, \tau, T) \tag{7.7.2}$$

Application of this rule to (7.3.3) gives the following result:[21]

$$\left[i\mathbf{E} \cdot \frac{\partial}{\partial \mathbf{k}} - \left(\varepsilon\left(\mathbf{k} + \frac{\tau}{2}\right) - \varepsilon\left(\mathbf{k} - \frac{\tau}{2}\right)\right)\right] G^<(\mathbf{k}, \tau)$$

$$= \int (\Sigma_r G^< + \Sigma^< G_a - G_r \Sigma^< - G^< \Sigma_a) \tag{7.7.3}$$

where all the terms on the right-hand side have the variable structure

$$\int BC \equiv \int_{-\infty}^{\infty} du\, B\left(\mathbf{k} + \tfrac{1}{2}\mathbf{E}\left(u + \frac{\tau}{2}\right), \frac{\tau}{2} - u\right) C\left(\mathbf{k} + \tfrac{1}{2}\mathbf{E}\left(u - \frac{\tau}{2}\right), \frac{\tau}{2} + u\right)$$

$$(7.7.4)$$

This equation is exact within the model considered [steady, uniform field, and the effects of the crystal lattice are incorporated in a given dispersion relation $\varepsilon(\mathbf{k})$]. Mahan and co-workers[14,22] have developed a complete linear transport theory by expanding (7.7.3) and (7.7.4) to first order in the applied field. The nonlinear theory in its full generality has not yet been developed, and rather than pursuing this line we wish to develop an approximate theory for f^W alone. Thus, setting $\tau = 0$ in (7.7.3) leads to

$$\mathbf{E} \cdot \frac{\partial}{\partial \mathbf{k}} f^W(\mathbf{k}) = -\int (\Sigma^> G^< + G^< \Sigma^> - \Sigma^< G^> - G^> \Sigma^<)$$

$$(7.7.5)$$

$$\int BC \equiv \int_0^{\infty} du\, B(\mathbf{k} - \tfrac{1}{2}\mathbf{E}u, u)C(\mathbf{k} - \tfrac{1}{2}\mathbf{E}u, -u)$$

where we used the relations $G_r = \theta(\tau)(G^< - G^>)$, $G_a = \theta(-\tau)(G^> - G^<)$, and analogously for $\Sigma_{a,r}$. The right-hand side of (7.7.5) still involves the full correlation function, and to reduce (7.7.5) to a closed equation for f^W alone, approximations must be made. Before embarking on this, we introduce some examples of spectral densities which are needed in the subsequent developments.

7.8. SPECTRAL DENSITIES

Here we discuss a few sample spectral densities. Normally, in order to solve for the spectral density, one solves the Dyson equation for the retarded (or advanced) Green's function and then constructs the spectral density with (7.6.6). We have already encountered the free spectral density, equation (7.6.12), and we give a few other explicitly solvable cases.

7.8.1. Impurity Scattering

The retarded Green's function for electron scattering off a dilute impurity concentration (in equilibrium) is given by (see, e.g., Ref. 6)

$$G_r(\mathbf{k}, \omega) = \frac{1}{\omega - \varepsilon(\mathbf{k}) + i/2\tau}$$

$$(7.8.1)$$

where $1/\tau = c\pi\rho v^2$ (ρ is the density of states and v is the scattering potential). This results in a Lorentzian in frequency, or exponential decay in time:

$$A(\mathbf{k}, \omega) = \frac{1/\tau}{(\omega - \varepsilon(\mathbf{k}))^2 + (1/2\tau)^2}$$

$$A(\mathbf{k}, t) = \exp(-i\varepsilon(\mathbf{k})t - |t|/\tau) \tag{7.8.2}$$

7.8.2. Free Particles in a Uniform Electric Field

Consider next the Dyson equation for an electron in a uniform electric field, where the field is represented by a vector potential $[\mathbf{E}(T) = -\partial\mathbf{A}(T)/\partial T]$:

$$\left[i\frac{\partial}{\partial t} - \varepsilon(\mathbf{p} - \mathbf{A}(t))\right] G(\mathbf{p}, t, t') = \delta(t - t') \tag{7.8.3}$$

This equation is readily integrated:

$$G_r(\mathbf{p}, t, t') = -i\theta(t - t') \exp\left[-i \int_{t'}^{t} du \, \varepsilon(\mathbf{p} - \mathbf{A}(u))\right] \tag{7.8.4}$$

For simplicity, consider now steady fields. Applying the transformation (7.7.2) to (7.8.4) and forming A, we get

$$A(\mathbf{k}, \omega) = \int_{-\infty}^{\infty} d\tau \, \exp\left[i\left(\omega\tau - \int_{-\tau/2}^{\tau/2} du \, \varepsilon(\mathbf{k} - \mathbf{E}u)\right)\right] \tag{7.8.5}$$

This equation is in a manifestly gauge-invariant form, and the T-dependence has dropped out, as it should. For parabolic bands (7.8.5) can be evaluated explicitly:

$$A(\mathbf{k}, \omega) = 2 \int_{0}^{\infty} d\tau \, \cos\left(\omega\tau - \frac{k^2}{2\tau} - \frac{E^2}{24}\tau^3\right)$$

$$= \frac{2\pi}{\alpha} Ai\left(\frac{\hbar^2 k^2/2m - \hbar\omega}{\alpha}\right) \tag{7.8.6}$$

where $Ai(x)$ is the Airy function and we defined $\alpha = (\hbar^2 e^2 E^2/8m)^{1/3}$ (and reintroduced units). It is instructing to verify that (7.8.6) satisfies the frequency sum rule.

We can use (7.8.6) to evaluate the density of states. Recall first that density of states is given by

$$\rho(\omega) \equiv -\frac{1}{\pi} \text{Im} \, \text{Tr} \, G_r = \tfrac{1}{2}\sum_{\mathbf{k}} A(\mathbf{k}, \omega) \tag{7.8.7}$$

For parabolic bands we write $\rho_0(\omega) = \rho_0 \omega^{1/2}$ and convert the momentum summation to an integration over ω, to obtain

$$\frac{\rho(\omega)}{\rho_0} = \int_{-\omega/\alpha}^{\infty} dx \, (\alpha x + \omega)^{1/2} \text{Ai}(x)$$

$$= \left(\frac{1}{\pi}\right)^{1/2} \left[\text{Ai}'^2\left(-\frac{\omega}{\Theta}\right) + \frac{\omega}{\Theta} \text{Ai}^2\left(-\frac{\omega}{\Theta}\right) \right] \tag{7.8.8}$$

where we used various results for integrals involving Airy functions derived by Aspnes.[23] Here $\Theta = (e^2 E^2/2m\hbar)^{1/3}$. Equation (7.8.8) is a well-known result,[24] and it has some importance in the discussion of the Franz–Keldysh effect.

7.9. THE BARKER–FERRY EQUATION FOR HIGH-FIELD ELECTRON–PHONON TRANSPORT

We now use (7.7.5) to derive the Barker–Ferry transport equation,[15] originally obtained with a density matrix/projection operator formalism. It has been known for some time[7] that the ansatz (7.6.9) will not lead to the Barker–Ferry result: there are some minor (but annoying!) differences in the various time arguments. The reasons for this discrepancy remained obscure until the recent work of Lipavsky *et al.*,[16] who show how to modify (7.6.9) in order to obtain agreement between the different theories. We give a brief summary of Ref. 16; note, however, that our notation differs from theirs and that we have corrected a few minor misprints.

Following Lipavsky *et al.*,[16] we define the auxiliary functions $G_{r,a}^{\lessgtr}$ by

$$G_r^{\lessgtr}(t_1, t_2) = \theta(t_1 - t_2) G^{\lessgtr}(t_1, t_2)$$
$$G_a^{\lessgtr}(t_1, t_2) = \theta(t_2 - t_1) G^{\lessgtr}(t_1, t_2) \tag{7.9.1}$$

Operate now by G_r^{-1} on the first of equations (7.9.1):

$$[G_r^{-1} G_r^{<}] = \int d\bar{t} \left[\left(i\frac{\partial}{\partial t_1} - \varepsilon(\mathbf{p}) \right) \delta(t_1 - \bar{t}) - \Sigma_r(t_1, \bar{t}) \right] \theta(\bar{t} - t_2) G^{<}(\bar{t}, t_2)$$

$$= i\delta(t_1 - t_2) G^{<}(t_1, t_2)$$

$$+ \theta(t_1 - t_2) \left\{ \left[i\frac{\partial}{\partial t_1} - \varepsilon(\mathbf{p}) \right] G^{<}(t_1, t_2) \right.$$

$$\left. - \int_{t_2}^{t_1} d\bar{t} \, \Sigma_r(t_1, \bar{t}) G^{<}(\bar{t}, t_2) \right\} \tag{7.9.2}$$

The first term on the right-hand side gives directly the Wigner function [see (7.6.2)], while the term in braces still requires some work. Note first that it follows from (7.4.3) that $G_r^{-1}G^< = \Sigma^< G_a$, which reads explicity as

$$\left[i\frac{\partial}{\partial t_1} - \varepsilon(\mathbf{p})\right] G^<(t_1, t_2) - \int_{-\infty}^{t_1} d\bar{t}\, \Sigma_r(t_1, \bar{t}) G^<(\bar{t}, t_2)$$

$$= \int_{-\infty}^{t_2} d\bar{t}\, \Sigma^<(t_1, \bar{t}) G_a(\bar{t}, t_2) \tag{7.9.3}$$

We use this equation to eliminate the $G_0^{-1}G^<$ term from (7.9.2), and obtain

$$[G_r^{-1}G_r^<](t_1, t_2) = i\delta(t_1 - t_2) G^<(t_1, t_2)$$

$$+ \theta(t_1 - t_2) \int_{-\infty}^{t_2} d\bar{t}\, [\Sigma_r(t_1, \bar{t}) G^<(\tilde{\lambda}, t_2) + \Sigma^<(t_1, \bar{t}) G_a(\bar{t}, t_2)]$$

$$\tag{7.9.4}$$

Finally, operating with G_r from the left, we arrive at

$$G_r^<(t_1, t_2) = -G_r(t_1, t_2) f^W(t_2)$$

$$+ \int_{t_2}^{t_1} d\bar{t} \int_{-\infty}^{t_2} d\bar{t}'\, G_r(t_1, \bar{t})[\Sigma_r(\bar{t}, \bar{t}') G^<(\bar{t}', t_2) + \Sigma^<(\bar{t}, \bar{t}') G_a(\bar{t}', t_2)]$$

$$\tag{7.9.5}$$

The same analysis can be repeated for $G_a^<$ with the result

$$G_a^<(t_1, t_2) = f^W(t_1) G_a(t_1, t_2)$$

$$+ \int_{t_1}^{t_2} d\bar{t} \int_{-\infty}^{t_1} d\bar{t}'\, [G^<(t_1, \bar{t}') \Sigma_a(\bar{t}', \bar{t}) + G_r(t_1, \bar{t}') \Sigma^<(\bar{t}', \bar{t})] G_a(\bar{t}, t_2)$$

$$\tag{7.9.6}$$

These two equations allow, in principle, an iterative construction of $G^<(t_1, t_2)$ from its time-diagonal component, i.e., the Wigner function f^W. The first term gives the modified ansatz, while the iterative corrections represent an expansion in terms of the various (short) relaxation times in the system.

In passing we note that in the density matrix formulation one encounters a similar structure: the nondiagonal components are expressed in terms of the diagonal component. It would be interesting to make this analog more quantitative.

For further discussion of (7.9.5)–(7.9.6) we refer to Ref. 16; here we analyze the first term in some detail. The two terms originating from (7.8.13) and (7.8.14) can be combined:

$$G^<(\mathbf{p}, t_1, t_2) = -G_r(\mathbf{p}, t_1, t_2)f^W(\mathbf{p}, t_2) + f^W(\mathbf{p}, t_1)G_a(\mathbf{p}, t_1, t_2) \quad (7.9.7)$$

From now on we consider, for simplicity, only static fields (the generalization to time-varying fields is straightforward). In this case, and in terms of the gauge-invariant variables, (7.9.7) can be written as

$$G^<(\mathbf{k}, \tau) = -iA(\mathbf{k}, \tau)f^W(\mathbf{k} - \mathbf{E}|\tau|/2) \quad (7.9.8)$$

comparing (7.9.8) with (7.6.9) in the homogeneous case, we see that the only difference is the time- and field-dependent shift in the momentum label of the distribution function, and that the ansatz is made in time, and not energy, space.

We now use (7.9.8) to evaluate the kinetic equation (7.7.5). Before doing so we must specify the self-energy. Assuming that the phonon subsystem stays in thermal equilibrium and working in the lowest-order perturbation theory, we choose [see equation (6.2.37)]

$$\Sigma^{\lessgtr}(\mathbf{k}, \tau) = \sum_{\eta=\pm 1} \sum_{\mathbf{q}} |V(\mathbf{q})|^2 G^{\lessgtr}(\mathbf{k} - \mathbf{q}, \tau)(N_q + \tfrac{1}{2} + \tfrac{1}{2}\eta) \exp(\pm i\eta\omega_q\tau) \quad (7.9.9)$$

Consider the first term on the right-hand side of (7.7.5):

$$\int_0^\infty du\, \Sigma^>(\mathbf{k} - \tfrac{1}{2}\mathbf{E}u, u)G^<(\mathbf{k} - \tfrac{1}{2})\mathbf{E}u, -u)$$

$$= \sum_{\eta=\pm 1} \sum_{\mathbf{q}} \int_0^\infty du\, |V(\mathbf{q})|^2(N_q + \tfrac{1}{2} + \tfrac{1}{2}\eta)\exp(-i\eta\omega_q u)$$

$$\times A(\mathbf{k} - \tfrac{1}{2}\mathbf{E}u - \mathbf{q}, u)A(\mathbf{k} - \tfrac{1}{2}\mathbf{E}u, -u)f^W(\mathbf{k} - \mathbf{E}u)[1 - f^W(\mathbf{k} - \mathbf{q} - \mathbf{E}u)]$$

$$= \sum_{\mathbf{q}} \int_0^\infty du\, (I_+ N_q + I_-(1 + N_q))f^W(\mathbf{k} - \mathbf{E}u)(1 - f^W(\mathbf{k} - \mathbf{q} - \mathbf{E}u))$$

$$(7.9.10)$$

where

$$I_\pm = |V(\mathbf{q})|^2 \exp\left[-i\int_0^u du'\,(\varepsilon(\mathbf{k} - \mathbf{q} - \mathbf{E}u') - \varepsilon(\mathbf{k} - \mathbf{E}u')) \pm i\omega_q u\right] \quad (7.9.11)$$

In obtaining the last equation in (7.9.10) we used the spectral density (7.8.5); corrections can be calculated by using more sophisticated spectral functions.

A similar calculation can be performed with the other terms in (7.7.5), with the final result

$$
\mathbf{E} \cdot \frac{\partial}{\partial \mathbf{k}} f^{W}(\mathbf{k})
$$

$$
= -\sum_{\mathbf{q}} \int_{0}^{\infty} du \, [2 \, \mathrm{Re}\{I_{+} N_{q} + I_{-}(1 + N_{q})\}
$$

$$
\times [f^{W}(\mathbf{k} - \mathbf{E}u)(1 - f^{W}(\mathbf{k} - \mathbf{q} - \mathbf{E}u))]
$$

$$
- 2 \, \mathrm{Re}\{I_{+}(1 + N_{q}) + I_{-} N_{q}\}[f^{W}(\mathbf{k} - \mathbf{q} - \mathbf{E}u)(1 - f^{W}(\mathbf{k} - \mathbf{E}u))]]
$$

$$(7.9.12)$$

This equation coincides with the Barker–Ferry[15] in the steady-field case, apart from an exponential damping factor. It would be easy to generate this factor by adding a constant imaginary part to the single-particle energies, thus simulating the finite lifetime due to collisions. However, due to the lack of a detailed model we will refrain from doing so.

7.10. STEADY-STATE TRANSPORT EQUATION TRANSFORMED INTO A NUMERICALLY TRACTABLE FORM

The numerical evaluation of the Barker–Ferry (BF) equation represents a formidable task, and a complete analysis is still awaiting. Here we examine a recent scheme suggested by the Modena group and the present author,[21,25,32] which allows the application of standard Monte Carlo methodology.[26] When comparing the BF equation to the conventional Boltzmann equation, the main mathematical complication arises from the additional time integral on the right-hand side. As a result, the distribution function and quantum mechanical scattering rates are intertwined in a complicated way (memory effects), and the isolation of simple corrections to the Boltzmann equation becomes difficult. The idea in our work is to transform the BF equation into a form which resembles the BE as much as possible, thus allowing an unambiguous isolation of the leading quantum corrections. For simplicity, we consider nondegenerate conditions ($f^{W} \ll 1$ and $1 - f^{W} \simeq 1$).

We write the BF equation as

$$
\mathbf{E} \cdot \frac{\partial}{\partial \mathbf{k}} f^{W}(\mathbf{k}) = -\sum_{\mathbf{k}'} \int_{0}^{\infty} du [P(\mathbf{k} - \tfrac{1}{2}\mathbf{E}u, \mathbf{k}' - \tfrac{1}{2}\mathbf{E}u, u) f^{W}(\mathbf{k} - \mathbf{E}u)
$$

$$
- P(\mathbf{k}' - \tfrac{1}{2}\mathbf{E}u, \mathbf{k} - \tfrac{1}{2}\mathbf{E}u, u) f^{W}(\mathbf{k}' - \mathbf{E}u)] \quad (7.10.1)
$$

where the transition term is given by

$$P(\mathbf{k}_1, \mathbf{k}_2, u) = \sum_{\eta = \pm 1} |V(q)|^2 (N_q + \tfrac{1}{2} + \tfrac{1}{2}\eta) 2 \operatorname{Re}[A(\mathbf{k}_1, u) A(\mathbf{k}_2, -u) e^{-i\eta\omega_q u}]$$

$$(7.10.2)$$

and $q = |\mathbf{k} - \mathbf{k}'|$. Equation (7.10.1) has recently been analyzed in Ref. 19 in another context and is in fact, equivalent (if the free-field-dependent spectral densities are employed) to an equation reported by Barker, who has analyzed the field effects on P while suppressing the field-dependent shift in the momentum labels appearing in the distribution function. We follow another strategy: integration of both sides of (7.10.1) and a change of variables allows one to write it as[19]

$$f^W(\mathbf{k}) = -\sum_{\mathbf{k}'} \int_0^\infty dt\, [\, W(\mathbf{k} - \mathbf{E}t, \mathbf{k}' - \mathbf{E}t, t) f^W(\mathbf{k} - \mathbf{E}t)$$

$$- W(\mathbf{k}' - \mathbf{E}t, \mathbf{k} - \mathbf{E}t, t) f^W(\mathbf{k}' - \mathbf{E}t)] \qquad (7.10.3)$$

where

$$W(\mathbf{k}_1, \mathbf{k}_2, t) = \int_0^t dt'\, P(\mathbf{k}_1 + \tfrac{1}{2}\mathbf{E}t', \mathbf{k}_2 + \tfrac{1}{2}\mathbf{E}t', t') \qquad (7.10.4)$$

The time dependence in (7.10.3) falls into two distinct categories. First, the time-dependence of the momentum labels is identical to the one occurring in Boltzmann theory. Second, the finite upper bound of the t'-integration reflects the microscopic reversible character of the quantum mechanical scattering events. To obtain a scattering rate, we let $t \to \infty$ in the upper bound. This limit corresponds to completed collisions, and it quite analogous to the one encountered in connection with the Fermi golden rule.[33] Thus we conclude that in this limit the transport physics can be described with an equation whose mathematical structure is equivalent to that of the Boltzmann equation; the only difference is that the scattering probabilities are given by

$$W^{QM}(\mathbf{k}_1, \mathbf{k}_2) \equiv \lim_{t \to \infty} W(\mathbf{k}_1, \mathbf{k}_2, t)$$

$$= \sum_{\eta = \pm 1} |V(q)|^2 (N_q + \tfrac{1}{2} + \tfrac{1}{2}\eta)$$

$$\times \int_0^\infty dt'\, 2 \operatorname{Re}[A(\mathbf{k}_1 + \tfrac{1}{2}\mathbf{E}t', t') A(\mathbf{k}_2 + \tfrac{1}{2}\mathbf{E}t', -t') \exp(-i\eta\omega_q t')]$$

$$(7.10.5)$$

We observe that using free spectral densities in (7.10.5) reproduces the results obtained by Thornber[27] and Barker.[28] This equation shows explicitly how the straightforward application of the gradient approximation overlooks even some *linear* terms in the electric field [compare to 7.6.11)!].

It would be quite interesting to compare the linearized form of (7.10.5) to the transport equation of Mahan *et al.*:[14,22] this comparison should yield useful information on the range of validity of the new ansatz (7.9.8). The transformed equation (7.10.5) is Markovian (the complicated additional time dependence has been transformed away), and standard Monte Carlo methodology can be applied to it. Quantum mechanical effects (such as collisional broadening and intracollisional field effect) can be studied by using increasingly sophisticated spectral functions.

As a first step, one may want to study collisional broadening alone, suppressing the intracollisional field effect. In this case the field-dependent shifts in the momenta [see (7.10.4)] can be neglected, and one can write (7.10.5) in the following simple form:

$$W^{QM}(\mathbf{k}_1, \mathbf{k}_2)$$

$$= \sum_{\eta = \pm 1} |V(q)|^2 (N_q + \tfrac{1}{2} + \tfrac{1}{2}\eta) \int_{-\infty}^{\infty} \frac{d\omega}{2\pi} A(\mathbf{k}_1, \omega + \eta\omega_q) A(\mathbf{k}_2, \omega) \quad (7.10.6)$$

It is illustrative to consider a simple example. If the spectral density corresponds to free particles, $A(\mathbf{k}, \omega) = 2\pi\delta(\omega - \varepsilon(\mathbf{k}))$, the convolution integral over energy reduces to an energy-conserving δ-function, and the semiclassical Boltzmann equation is recovered. Result (7.10.6) indicates that the way collisional broadening has been treated by Chang *et al.*[29] should be modified [see their equation (7.1.8)]: instead of just adding the linewidths (which holds for Lorentzian spectral functions), one should evaluate the convolution integral in (7.10.6).

7.11. APPLICATION TO A SIMPLE MODEL SEMICONDUCTOR

We now consider a single-band model where the dispersion is spherical and parabolic.[21,25,32] Only nonpolar optical processes at zero temperature (spontaneous emission) are considered. The self-energy for the many-body system is taken in first-order perturbation theory. The electric field is assumed to be homogeneous and intracollisional field effects are neglected.

In spite of its simplicity, this model displays some interesting physical features. When collision broadening is neglected, it provides the two standard results: (i) at intermediate field strengths, it gives rise to the streaming motion regime; and (ii) at asymptotically high fields, it produces the quasielastic regime (for a review, see Ref. 30). Therefore, by comparing the present results with the standard ones, we shall be in a position to give a clear answer concerning the effect of collisional broadening.

Within the usual ensemble Monte Carlo technique,[26] we suggest the following procedure to account for broadening. According to our model the linewidth is given by

$$\Gamma(\omega) = -2\,\text{Im}[\Sigma^r(\omega)] = 2\gamma(\omega - \omega_0)^{1/2} \qquad (7.11.1)$$

where ω is the many-body energy, ω_0 is the optical phonon energy, and the coupling strength γ is given by

$$\gamma = (D_t K)^2 m^{*3/2}/2^{5/2}\pi\bar\rho\omega_0 \qquad (7.11.2)$$

with the usual meaning of symbols.[26]

Using (7.11.1) one obtains the following spectral density:

$$A(x_k, x) = \frac{2}{\gamma^2}\left[\frac{(x - x_0)^{1/2}\theta(x_k - x_0)\theta(x - x_0)}{(x - x_k)^2 + (x - x_0)}\right.$$

$$\left. + \pi\delta(x - x_k)\theta(x_0 - x_k)\theta(x_0 - x)\right] \qquad (7.11.3)$$

where we introduced dimensionless energies, $x = \omega/\gamma^2$, $x_k = \varepsilon_k/\gamma^2$, $x_0 = \omega_0/\gamma^2$, and the band energy is $\varepsilon_k = k^2/2m^*$. Note that $A(x_k, x)$ satisfies the required normalization constraint, $(\gamma^2/2\pi)\int_{-\infty}^{\infty} A\,dx = 1$. Figure 7.5 illustrates the shape of the spectral density for different energies (with respect to the optical phonon energy).

The carrier dynamics are assumed to follow the semiclassical equation of motion:

$$\dot{\mathbf{k}} = \mathbf{E} \qquad (7.11.4)$$

where the dot means time derivative. In the simulation the free-flight duration has been determined according to equation (7.11.1). The final **k**-state after each scattering event, taken to be isotropic, has been chosen

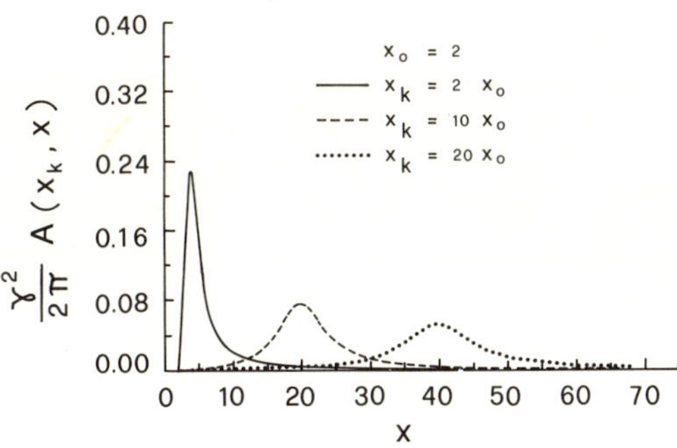

FIGURE 7.5. Spectral density as a function of the many-body energy at three different values of the kinetic energy ($x_k = 2x_0$, $x_k = 10x_0$, $x_k = 20x_0$). Energies are expressed in dimensionless units (see text).

according to the joint spectral density $K(x_i, x_f)$ given by

$$K(x_i, x_f) \equiv \frac{\gamma^2}{2\pi} \int_{-\infty}^{\infty} dx\, A(x, x_i) A(x - x_0, x_f)$$

$$= \frac{2}{\pi\gamma^2} \left\{ \int_{2x_0}^{\infty} dx\, \frac{(x - x_0)^{1/2}(x - 2x_0)^{1/2}\theta(x_i - x_0)\theta(x_f - x_0)}{[(x - x_i)^2 + (x - x_0)][(x - x_0 - x_f)^2 + (x - 2x_0)]} \right.$$

$$+ \pi \frac{x_f^{1/2}\theta(x_i - x_0)\theta(x_0 - x_f)\theta(x_f)}{(x_0 + x_f - x_i)^2 + x_f}$$

$$\left. + \pi^2 \delta(x_0 + x_f - x_i)\theta(x_0 - x_i)\theta(x_0 - x_f) \right\} \quad (7.11.5)$$

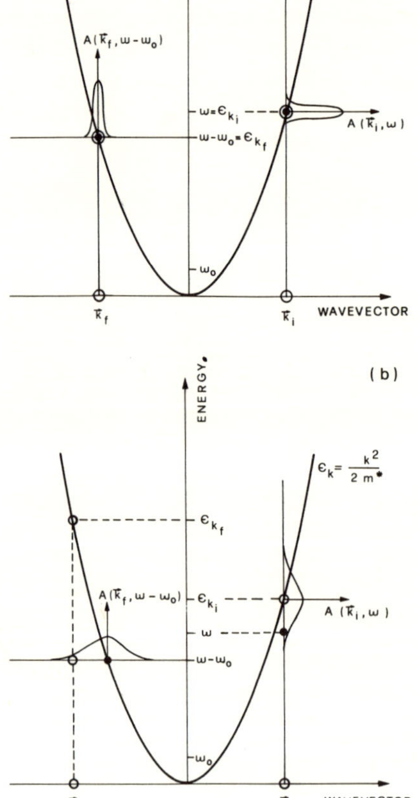

FIGURE 7.6. Schematic picture of a transition due to optical phonon emission as described in terms of spectral density. (a) Noninteracting model: the spectral densities are delta functions and $\varepsilon_{k_f} = \varepsilon_{k_i} - \omega_0$; thus kinetic energy conservation is fulfilled (ω_0 is the optical phonon energy and m^* the effective mass). (b) Interacting model: the spectral densities are broadened and $\varepsilon_{k_f} \neq \varepsilon_{k_i} - \omega_0$; thus the kinetic energy is not conserved.

Here x_i and x_f are the normalized kinetic energies of the initial and final k-states. Once again we point out that, as verified numerically, the joint spectral density satisfies the normalization requirements $(\gamma^2/2\pi)\int_{-\infty}^{\infty} K\, dx_i = (\gamma^2/2\pi)\int_{-\infty}^{\infty} K\, dx_f = 1$. In this way, energy conservation for the many-body system is strictly fulfilled. Of course, this is no longer the case for the carrier kinetic energy. Owing to the finite width, the kinetic energy is not conserved in a given scattering event, and, in particular, the final kinetic energy of a carrier may be greater than the initial one even if an emission process has occurred. Of course, for the interacting system energy conservation is strictly fulfilled. Figure 7.6 gives a schematic illustration of this effect.

We emphasize that the high-energy tail of the joint spectral density, equation (7.11.5), which is an inevitable consequence of the corresponding tail in the spectral density, equation (7.11.3), may have profound effects on the carrier dynamics as is evident in our numerical results discussed in the next section.

7.12. NUMERICAL RESULTS

The present model basically relies on three parameters, and to exploit the main features of the broadening we have assumed $m^* = 0.3$, $\omega_0 = 40$ meV, and $\gamma^2 = 1.1$ meV. These can be considered as typical values since γ^2 is usually in the range 0.1–10 meV. The results obtained with simulations using 10^4 electrons are shown in Figures 7.7 to 7.9.

In Figure 7.7 the distribution functions of the kinetic energy, with and without broadening, are compared for an electric field of 500 kV/cm. In the absence of broadening, the electron gas achieves the quasielastic regime. Accordingly, the distribution function of the kinetic energy agrees quite satisfactorily with a heated Maxwell–Boltzmann distribution. On the contrary, broadening has been found to strongly modify such a distribution. More carriers are found in low- as well as in the high-energy tails. In the presence of broadening, the kinetic energy has been found to increase significantly (about 60%); the drift velocity also increases, but to a minor extent (about 20%).

Figure 7.8 shows the distribution of **k** along the field direction as calculated with and without collision broadening. Once again it is seen that the presence of broadening is responsible for an increase of the carrier population in the low- and high-energy regions of the wave vector distribution function.

A few electrons are found to reach very high energies (above 3 keV) during the simulation. These "lucky electrons" originate from the tail of the spectral density, therefore this effect is inherent with the model of

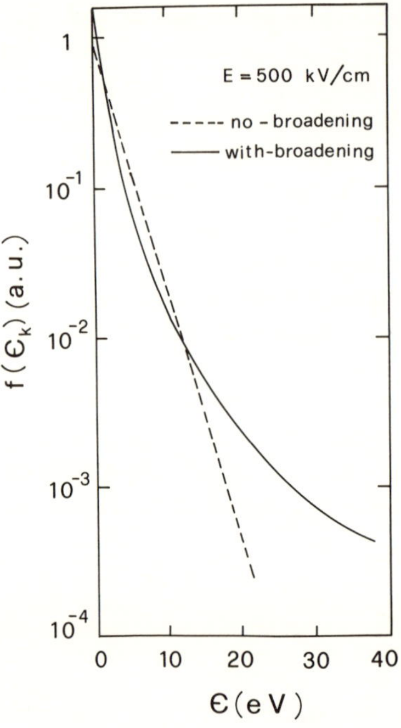

FIGURE 7.7. Energy distribution function as a function of the kinetic energy at $E = 500\,\text{kV/cm}$. Dashed (continuous) curves refer to calculations without (with) collision broadening.

broadening. Accordingly, we introduce the concept of *broadening-assisted runaway*. (The same argument can be used for the case of ionization processes which, by analogy with the previous case, are called *broadening-assisted impact ionization*.) Of course, in real cases, deviations from the simple parabolic energy spectrum and/or the presence of other scattering mechanisms (e.g., intervalley transfer, impact ionization, etc.) will modify or even forbid this kind of runaway. In any case it is worth noting that collision broadening favors not only the presence of hot electrons in the

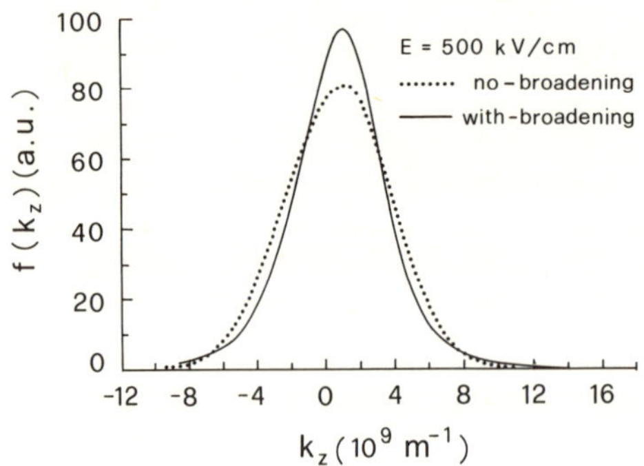

FIGURE 7.8. Wave vector distribution function for k along the field direction at $E = 500\,\text{kV/cm}$. Dotted (continuous) curves refer to calculations without (with) collision broadening.

tails of the distribution function, but it also allows for carrier runaway even if the scattering rate is a monotonically increasing function of energy.

For completeness, in Figure 7.9 we give the distribution functions of the kinetic energy at lower electric field strengths. Note that for $E = 10\,\text{kV/cm}$, when a streaming motion regime is achieved, the calculations show that the effect of collision broadening becomes negligible (see Figure 7.9b).

An important remark on the results concerns the estimators used for the calculation of average quantities. To be fully consistent, the average values should be calculated by using the field-dependent correlation function $G^<(\mathbf{k}, \omega)$. However, from the Monte Carlo simulation we know only the wave vector distribution function, which is linked to $G^<$ by equation (7.9.8). As described in section 7.9, the construction of $G^<$ starting from f is far from trivial, and therefore care should be taken in the interpretation of the kinetic mean energy and of the other average quantities obtained from the standard estimators. These quantities are, in general, different from the many-body averages whose determination will require the knowledge of $G^<(\mathbf{k}, \omega)$.

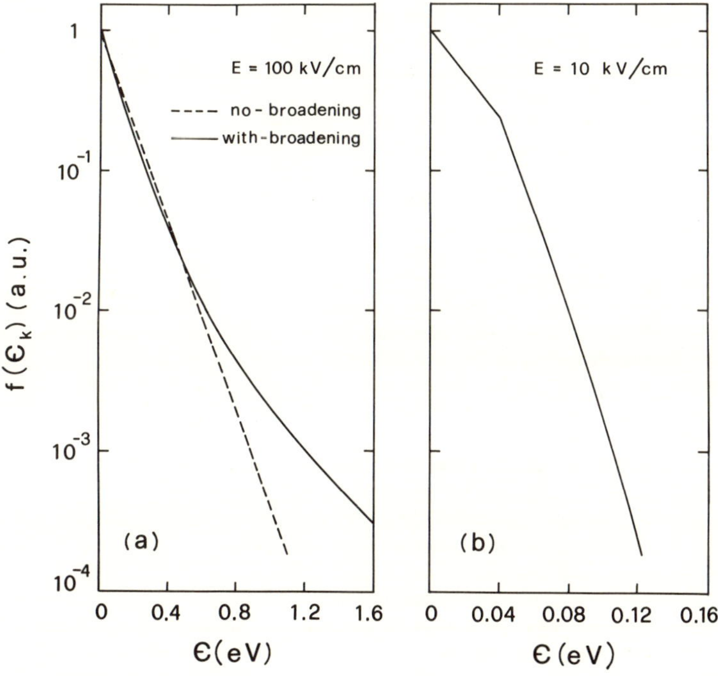

FIGURE 7.9. The same as Figure 7.7 at $E = 100\,\text{kV/cm}$ (a) and $E = 10\,\text{kV/cm}$ (b). In the latter case (b) the results without and with collisional broadening are found to be practically indistinguishable.

Our results obtained for a simple model give clear indication of an increase of the carrier population in the tails of the kinetic energy distribution function as evidenced experimentally from direct measurements of the energy distribution of hot electrons in SiO_2.[31] In particular, the spectral density relating the many-body energy and wave vector is found to be at the origin of carrier runaway even in the presence of a nonpolar scattering mechanism.

REFERENCES

1. D. C. LANGRETH, in: *Linear and Nonlinear Electron Transport in Solids* (J. T. Devreese and E. Van Doren, Plenum, New York (1976).
2. K. C. CHOU *et al.*, *Phys. Rep.* **118**, 1 (1985).
3. J. RAMMER and H. SMITH, *Rev. Mod. Phys.* **58**, 323 (1986).
4. J. RAMMER, Master's Thesis, University of Copenhagen (1980) unpublished (in Danish).
5. L. P. KADANOFF and G. BAYM, *Quantum Statistical Mechanics*, Benjamin, New York (1962).
6. G. D. MAHAN, *Many-Particle Physics*, Plenum, New York (1981).
7. A. P. JAUHO and J. W. WILKINS, *Phys. Rev. B* **29**, 1919 (1984).
8. L. V. KELDYSH, *Sov. Phys. JETP* **20**, 1018 (1965).
9. J. SCHWINGER, *J. Math. Phys.* **2**, 407 (1961).
10. Yu A. KUKHARENKO and S. G. TIKHODEEV, *Sov. Phys. JETP* **56**, 831 (1982).
11. A. G. HALL, *J. Phys. A* **8** 214 (1975).
12. P. DANIELEWICZ, *Ann. Phys. (N.Y.)* **152**, 239 (1984).
13. K. K. THORNBER, in: *Path Integrals* (G. J. Papadopoulos, ed.), Plenum, New York (1978).
14. W. HAENSCH and G. D. MAHAN, *Phys. Rev. B* **28**, 1902 (1983).
15. J. R. BARKER and D. K. FERRY, *Phys. Rev. Lett.* **42**, 1779 (1979).
16. P. LIPAVSKY *et al.*, *Phys. Rev. B* **34**, 6933 (1986).
17. P. BADZIAG, *Physica* **130**A, 565 (1985).
18. O. SERIMAA *et al.*, *Phys. Rev. A* **33**, 2913 (1986).
19. F. S. KHAN, J. H. DAVIES, and J. W. WILKINS, *Phys. Rev. B* **36**, 2578 (1987). The author is grateful to Dr. Davies for communicating these results prior to publication.
20. A. P. JAUHO, *Phys. Rev. B* **32**, 2248 (1985).
21. L. REGGIANI, P. LUGLI, and A. P. JAUHO, *Phys. Rev. B* **36**, 6602 (1987).
22. G. D. MAHAN, *Phys. Rep.* **110**, 321 (1984); **145**, 235 (1987).
23. D. E. ASPNES, *Phys. Rev.* **147**, 554 (1966).
24. See, e.g., R. KEIPER and O. ZIEP, *Phys. Stat. Sol. B* **131**, K91 (1985).
25. P. LUGLI *et al.*, *Superlattices and Microstructures* **2**, 143 (1986).
26. For a review, see L. REGGIANI and C. JACOBONI, *Rev. Mod. Phys.* **55**, 645 (1983).
27. K. K. THORNBER, *Solid State Electronics* **21**, 259 (1978).
28. J. R. BARKER, *Solid State Electronics* **21**, 267 (1978).
29. Y. C. CHANG *et al.*, *Appl. Phys. Lett.* **42**, 76 (1983).
30. L. REGGIANI, in: *Topics in Applied Physics*, vol. 58, Springer-Verlag, Heidelberg (1985).
31. W. POROD and D. K. FERRY, *Phys. Rev. Lett.* **54**, 1189 (1985); *Physica B* **134**, 137 (1985).
32. L. REGGIANI, P. LUGLI, and A. P. JAUHO, *J. Appl. Phys.* **64**, 3072 (1988).
33. Recently, P. LIPAVSKY has shown how this Markov limit can be carried out systematically (private communication).

Chapter 8

Numerical Techniques for Quantum Transport and Their Inclusion in Device Modeling

Lino Reggiani

8.1. INTRODUCTION

The aim of this chapter is to present a brief survey of the numerical techniques which, following the recent literature, have been applied to the field of quantum transport associated with nanometer electron devices. To begin, we specialize the area of interest into the three main lines which are schematically summarized in Figure 8.1. The choice of treating them separately is a matter of taste and only justified by the need of simplifying the otherwise too complicated problem we are facing. On the other hand, each line indicates a well-defined class of physical phenomena to which the reader can easily refer to the recent literature. The following sections address and introduce the physical problems related to each line and identify the numerical techniques used for their theoretical description. Finally, some concluding remarks are given.

8.2. RESONANT TUNNELING

A prototype of the resonant tunneling structure together with its current-voltage characteristic is exemplified in Figure 8.2. As a bias voltage is applied

Lino Reggiani • Dipartimento di Fisica e Centro Interuniversitario di Struttura della Materia, Università di Modena, 41100 Modena, Italy.

Quantum Transport in Semiconductors, edited by David K. Ferry and Carlo Jacoboni. Plenum Press, New York, 1991.

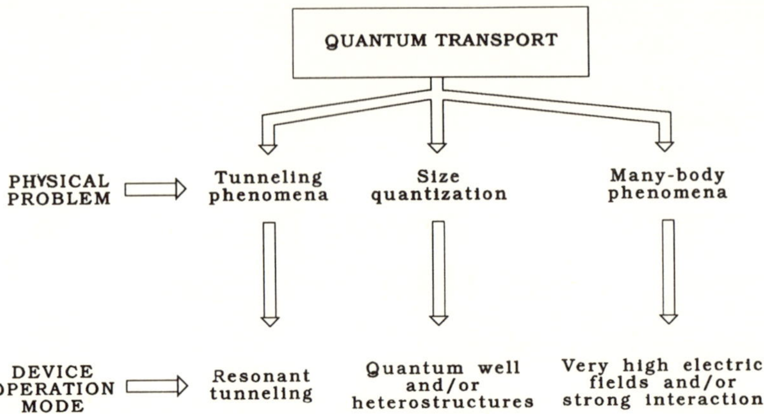

FIGURE 8.1. Schematic of the main lines of interest in the physics of quantum transport related to nanometer devices.

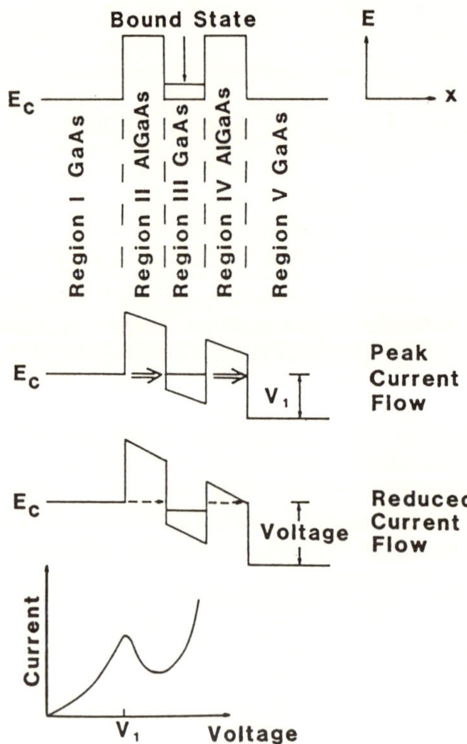

FIGURE 8.2. Schematic diagram showing resonance in a resonant tunneling device. Regions I and V are doped n-type $5 \times 10^{16}\,\text{cm}^{-3}$ and are 500 nm thick. Regions II, III, and IV are not intentionally doped and are each 5 nm thick.[1]

to the device, the resonant state in the well is pulled down in energy with respect to the more negative electrode, and the tunneling current through this state depends on the density of occupied states in the electrode. When the resonant state is pulled below the conduction band edge of the electrode, the tunneling current decreases. The device thus shows negative differential resistance which is attributed to quantum interference.

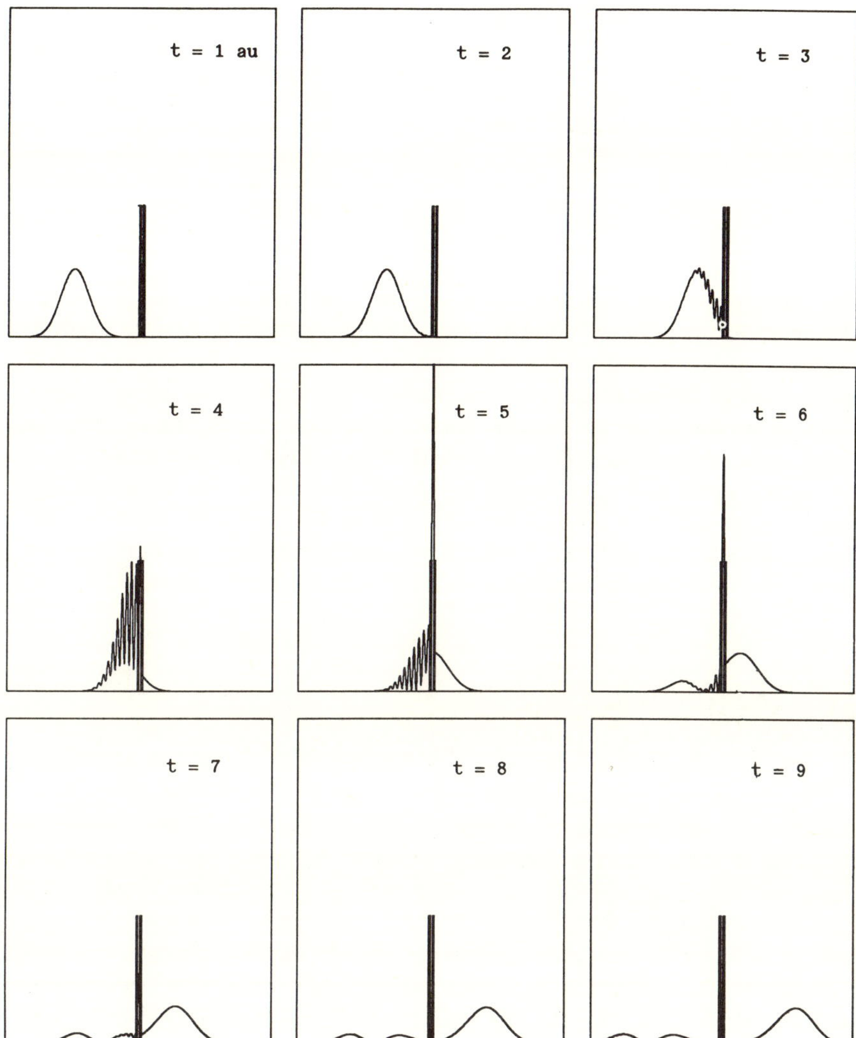

FIGURE 8.3. Temporal evolution of a Gaussian wave packet with energy equal to the resonance energy of the double barrier.

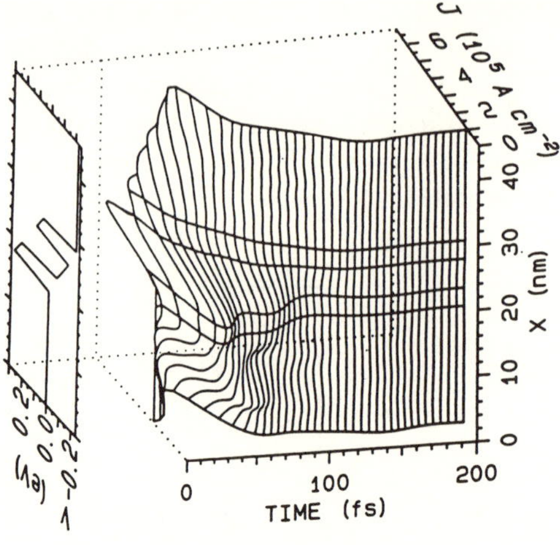

FIGURE 8.4. Transient response of the resonant diode as calculated with a Wigner distribution approach. Current density is plotted as a function of time and position within the device. The potential profile illustrates the device structure. At $t = 0$, the voltage was suddenly switched from 0.11 V (corresponding to the peak current) to 0.22 V (corresponding to the the valley current). After an initial peak, the current density approaches the lower steady-state value in 100–200 fs.[5]

The theoretical interpretation of the experimental current-voltage characteristic can be obtained using (i) the dynamical (or simulation) method;[2-4] (ii) a Wigner function approach.[5,6] In the former approach, a numerical solution of the one-dimensional Schrödinger equation is used. The time evolution of an electron wave packet (usually of initial Gaussian form) crossing the system of barriers is then visualized, as reported in Figure 8.3. In the latter approach, use is made of a solution of the Liouville–von Neumann equation in a Wigner representation. Figure 8.4 shows the results obtained with this method. Both approaches reproduce qualitatively the main features of the experimental current-voltage characteristic.

8.3. QUANTUM WELL

The prototype of the single-quantum-well structure is reported in Figure 8.5. Transport in a two-dimensional electron gas is essentially different from that in three dimensions, on account of the boundary conditions imposed by the layered geometry and its effects on scattering processes. Furthermore, attention should be paid to real space transfer due to electron transfer from GaAs to AlGaAs.

To study the transport properties in this regime, the semiclassical Boltzmann equation must be solved, taking into account the scattering in each subband of the AlGaAs–GaAs single-well potential. The Monte Carlo method, tailored with these details concerning the electron states and the scattering mechanisms, has been found to give plausible results.[7,8] Some of these are reported in Figure 8.6, where a direct comparison between the

LOW FIELD

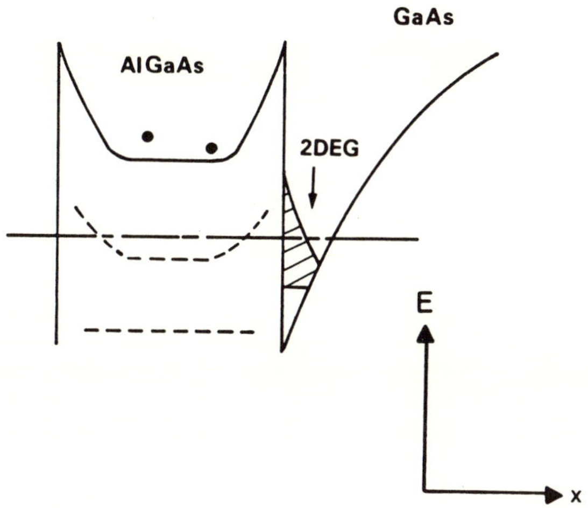

HIGH FIELD

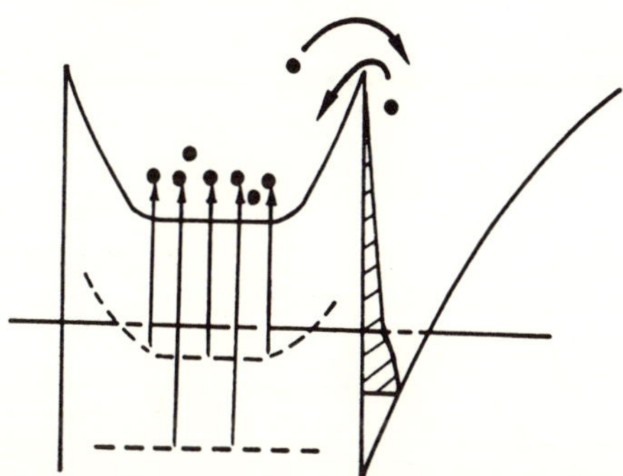

FIGURE 8.5. Schematic diagrams of hot-electron distributions and electron transfer in a heterostructure under electric field.[7] At low fields the two-dimensional electron gas at the interface remains confined in the quantum well and it represents a highly conductive channel. At high fields, hot electrons transferred from GaAs to AlGaAs can collide with electrons in deep levels, exciting them into the conduction band.

drift velocity versus electric field of the two- and three-dimensional cases can be seen. Results indicate that, for the two-dimensional case, a drift velocity response higher than in the bulk can be expected in the low-field region. In the high-field region, the velocity characteristics for the two-dimensional electron gas are very similar to those for the bulk. Experiments are found to compare favorably with theory.

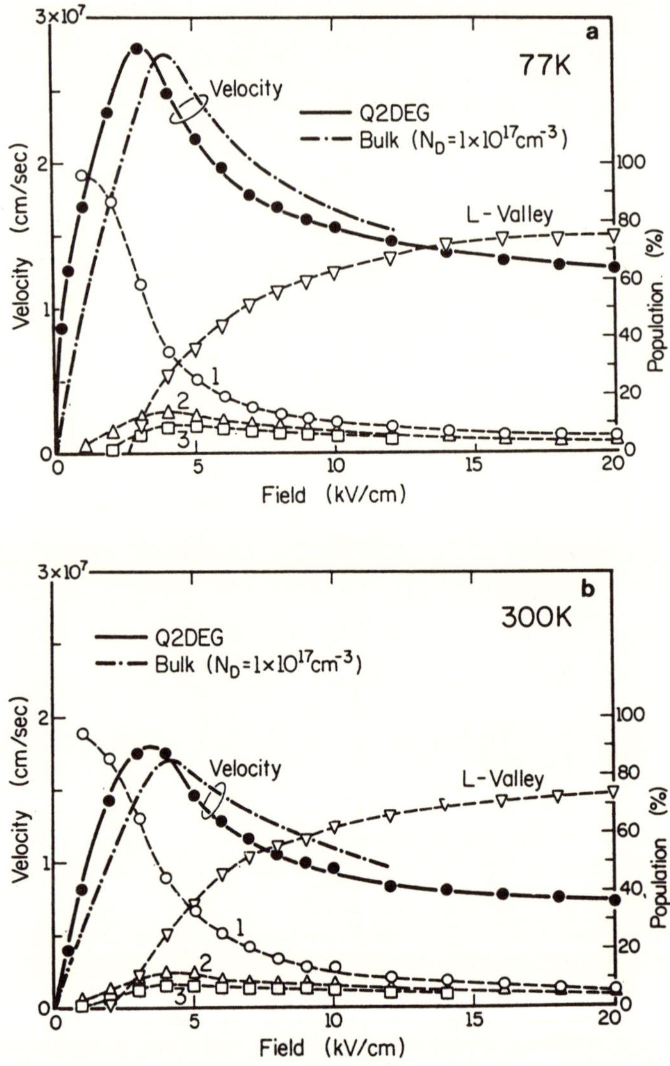

FIGURE 8.6. Steady-state drift velocity and population (1, 2, 3 refer to the subband indexes of the Γ-valley) vs. field characteristics at (a) 77 K and (b) 300 K of a two-dimensional quantum electron gas which is confined in a potential well illustrated in Figure 8.5.[8] The dashed-dotted curve represents the calculated velocity characteristics of Ref. 9 for three-dimensional transport.

8.4. MANY BODY

Values of the scattering rates above $10^{14}\,\text{s}^{-1}$ enter quite often in the analysis of transport phenomena at the high electric fields implied by submicron dimensions, especially when processes such as impact ionization and/or carrier injection into insulators are investigated. (A significant example of such a situation is the high-field properties of SiO_2.[10]) As a consequence, quantum phenomena such as collisional broadening and intracollisional field effects can no longer be neglected when a physically based microscopic interpretation is necessary. Quantum kinetic equations which include quasiparticle effects must be introduced, and their solution should be obtained numerically. Accordingly, appropriate Monte Carlo procedures have been used to this end. Three main lines of approach can be identified from the literature. Namely (i) use of more or less sophisticated spectral densities in place of an energy wave vector relationship for the charge carriers;[11] (ii) use of a Feynman path integral calculation;[12] (iii)

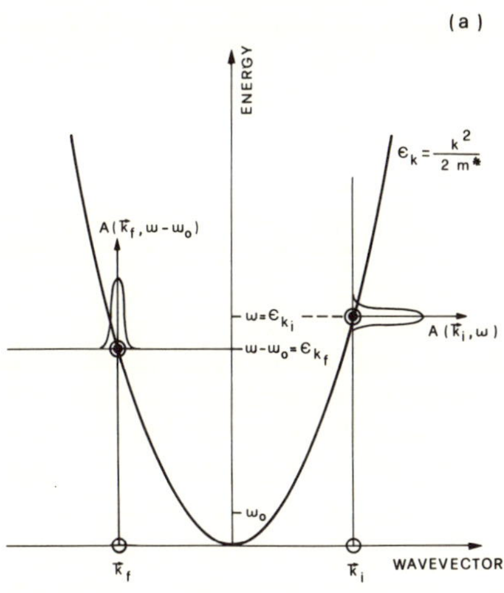

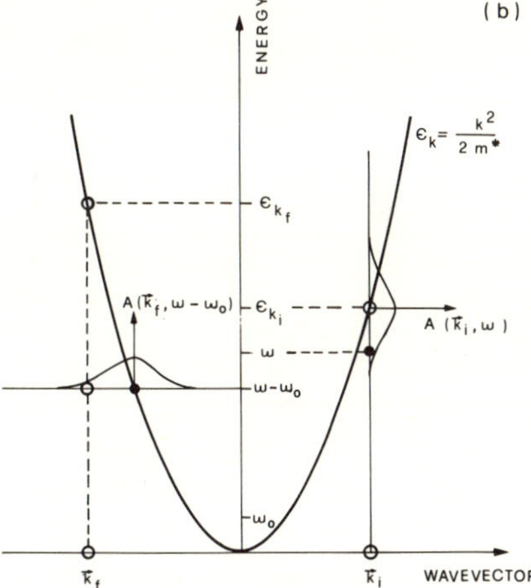

FIGURE 8.7. Schematic picture of a transition due to optical phonon emission as described in terms of spectral density. (a) Noninteracting model: the spectral densities are delta functions and $\varepsilon_{kf} = \varepsilon_{ki} - \omega_0$; thus kinetic energy conservation is fulfilled. (b) Interacting model: the spectral densities are broadened and $\varepsilon_{kf} \neq \varepsilon_{ki} - \omega_0$; thus the kinetic energy is not conserved.[14]

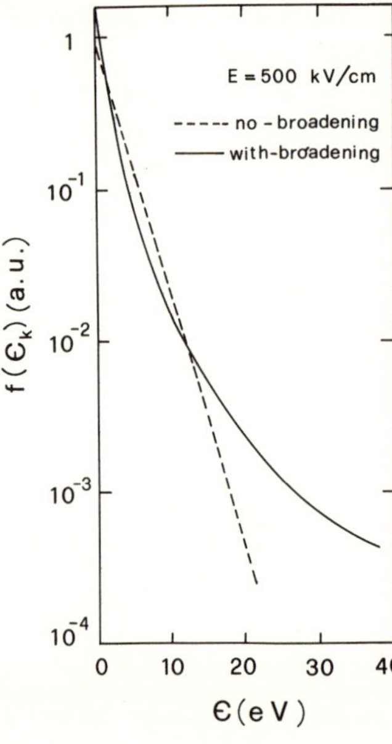

FIGURE 8.8. Energy distribution function as a function of the kinetic energy at $E = 500\,kV/cm$ for a spherical and parabolic semiconductor model which accounts for non-polar optical emission phonons only. Dashed (continuous) curves refer to calculations without (with) collision broadening.[11]

use of an iterative solution of the Liouville–von Neumann equation.[13] A schematic picture, which illustrates the main novelty associated with the inclusion of a spectral density in treating a scattering event, is reported in Figure 8.7. In all cases the comparison between the shape of the distribution functions, as obtained from the classical and the implemented approach, is physically significant. For steady-state conditions, approaches (i) and (ii) give evidence of a repopulation of both the low- and high-energy regions of the distribution function, as can be seen from Figures 8.8 and 8.9. Approach (iii) has been used only under transient conditions at very short

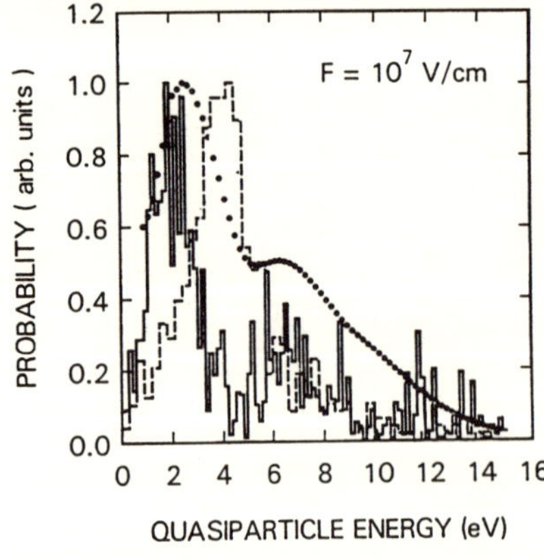

FIGURE 8.9. electron energy distributions in SiO_2 at a field of $10^4\,kV/cm$ obtained from classical Monte Carlo (dashed line) and quantum path integral (solid line) simulations and from the experimental data of Ref. 10 (dots). The minima of the distributions coincide with the minima of the density of states employed in the simulation.[12]

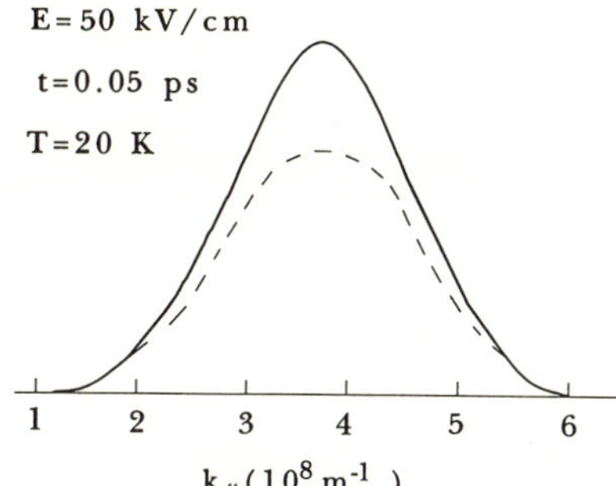

FIGURE 8.10. Electron distribution (a.u.) for **k** parallel to **E** in the quantum (dashed line) and classical (solid line) cases, respectively.[13]

times (see Figure 8.10). From this figure it is seen that, while the classical distribution still represents the ballistic shift of the initial conditions (no-scattering events), the quantum distribution appears to be more broadened. This is because of the presence of quantum interactions, which occur at short times without requiring energy conservation.

8.5. CONCLUSIONS

Attention has been drawn to three simple semiconductor structures which give evidence of quantum transport phenomena and are at the basis of advanced electronic devices. These are the resonant tunneling diode, the quantum well, and the high-field (about 10^6 V/cm) properties of SiO_2. The theoretical interpretation from first principles of the transport properties associated with these systems should start from the solution of the Liouville-von Neumann equation. This is indeed the general trend exhibited by the different efforts found in the literature. However, the inherent difficulties of such an approach oblige one to introduce simplifications and assumptions whose physical plausibility is often hard to verify. It is always doubtful whether the result comes from the approximations or from the physics. Among the numerical techniques used to provide a theoretical insight, the Monte Carlo method is found to occupy a privileged position. Because of the quantum features of statistics, particle, and fields, it is nowadays of use to call these simulation procedures *quantum Monte Carlo*. Quantum Monte Carlo is a field experiencing considerable growth in many other branches of physics, as testified by two recent conferences on the subject.[15,16]

ACKNOWLEDGMENTS

The financial support of the European Research Office (ERO) and the Computer Center of the Modena University (CICAIA) is gratefully acknowledged.

REFERENCES

1. T. J. SHEWCHUCK, P. C. CHAPIN, P. D. COLEMAN, W. KOPP, R. FISCHER, and H. MORKOC, *Appl. Phys. Lett.* **46**, 508 (1985).
2. J. R. BARKER, S. COLLINS, D. LOWE, and S. MURRAY, in *Proc. 17th I.C.P.S.* (J. D. Chadi and W. A. Harrison, eds.), p. 449, Springer-Verlag, New York (1985); J. R. BARKER, *Physica* **134B**, 22 (1985).
3. T. NAKAGAWA, N. J. KAWAI, and K. OTHA, *Superlattices and Microstructures* **1**, 217 (1985).
4. A. P. JAUHO and M. M. NIETO, *Superlattices and Microstructures* **2**, 407 (1986).
5. W. R. FRENSLEY, *J. Vac. Sci. Techn.* **B3**, 1261 (1985); W. R. FRENSLEY, *Phys. Rev, Lett.* **57**, 2853 (1986).
6. U. RAVAIOLI, M. A. OSMAN, W. POETZ, N. KLUSDAHL, and D. K. FERRY, *Physica* **134B**, 36 (1985).
7. M. INOUE, *Superlattices and Microstructures* **1**, 433 (1986).
8. K. YOKOHAMA and K. HESS, *J. Appl. Phys.* **59**, 3798 (1986); K. YOKOHAMA and K. HESS, *Phys. Rev. B* **33**, 5595 (1986).
9. J. R. RUCH AND W. FAWCETT, *J. Appl. Phys.* **41**, 3843 (1970).
10. M. V. FISCHETTI, D. J. DI MARIA, S. D. BRORSON, T. N. THEIS, and J. R. KIRTLEY, *Phys. Rev. B* **31**, 8124 (1985).
11. P. LUGLI, L. REGGIANI, and C. JACOBONI, *Superlattices and Microstructures* **2**, 143 (1986), and references therein.
12. M. V. FISCHETTI and D. J. DI MARIA, *Phys. Rev. Lett.* **55**, 2475 (1985).
13. R. BRUNETTI, C. JACOBONI, P. LUGLI and L. REGGIANI, *Proc. 18th I.C.P.S.* (O. Engstrom, ed.), p. 1527, World Scientific (1987).
14. L. REGGIANI, P. LUGLI, and A. P. JAUHO, *Phys. Rev. B* **36**, 6602 (1987).
15. M. H. KALOS (ed.), *Monte Carlo Methods in Quantum Problems*, NATO ASI Series C, **Vol. 125**, Reidel, Dordrecht (1984).
16. J. E. GUBERNATIS (ed.), *J. Stat. Phys.* **43** (1986).

Wave Packet Studies of Tunneling through Time-Modulated Semiconductor Heterostructures

Antti-Pekka Jauho

9.1. INTRODUCTION

Time series pictures of wave packets interacting with various potential structures (usually in one dimension) are a standard textbook tool in illustrating quantum mechanical concepts such as interference and tunneling.[1-4] Only a few simple cases are analytically tractable, and numerical methods are called for. In addition to their value as a pedagogical tool, the wave packets have been used during the years to address many "real" physical phenomena. Rather than giving a historical overview of the vast number of various applications,[5] we concentrate on some recent works discussing transport phenomena in submicron semiconductor structures with the help of wave packets. In a number of early papers[6-9] the wave packet transmission coefficient for a multiple-barrier structure was evaluated, and a number of physical consequences were extracted. These papers have been reviewed elsewhere,[10] and here we would like to mention a couple of very recent developments. The first is generalization to two spatial dimensions.[11,12] This is an important new development in light of many fascinating recent experiments on resistance quantization in constricted two-dimensional electron systems. In addition, a magnetic field can be included, which makes many interesting magnetotransport experiments accessible to wave packet techniques. These simulations would deserve a

Antti-Pekka Jauho ● Physics Laboratory, H. C. Ørsted Institute, University of Copenhagen, DK2100 Copenhagen Ø, Denmark. *Present address*: Nordita (Nordisk Institut for Teoretisk Fysik), DK2100 Copenhagen Ø, Denmark.

Quantum Transport in Semiconductors, edited by David K. Ferry and Carlo Jacoboni. Plenum Press, New York, 1991.

review of their own, but presently we concentrate on the second new topic: time-modulated barriers in one spatial dimension. A time modulation of the potential (the modulation can be viewed as a Gedankenexperiment or due to some real physical mechanism) introduces a "built-in" clock in the system, thus providing a reference frequency against which the tunneling process can be contrasted and from which information on the dynamics of various aspects of tunneling can be extracted.[13-17] Further, the energy exchange processes related to absorption or emission of modulation quanta bring inelastic processes within the realm of the time-dependent Schrödinger equation[18] and thus open a new area of investigation.

Before proceeding to the discussion of the various topics listed above we would like to issue a warning. By their very nature wave packet studies are model simulations, and one should be careful when attempting to make quantitative comparisons with experiments. In other words, there is always the danger of getting results that reflect the properties, say, of the chosen initial state rather than the physical system under study. For example, the transmission coefficient obtained with a wave packet simulation should not be used uncritically to calculate the current-voltage characteristics of a given barrier structure: the width, and hence momentum sharpness, of the incoming pulse should be determined by physical constraints such as contact lengths and/or scattering lengths.[8,10,19] Further, the commonly employed Gaussian minimum uncertainty packets do not form a complete orthonormal set, and therefore are only an approximate representation of the true incoming states. Below we describe how one in practice can attempt to avoid these pitfalls. Summarizing our concern: before embarking on a time-consuming (and potentially expensive) simulation, one should be prepared to face the question, what physical object is the wave packet supposed to model?

9.2. WAVE PACKETS AND TUNNELING TIMES

In recent years experimental and/or theoretical studies of tunneling in semiconductor heterostructures have attracted widespread interest. Most of the theoretical models have been time-independent, even though many experiments would, in principle, require an explicitly time-dependent treatment. As an example we mention theoretical estimation of the ultimate speed limit of a double-barrier resonant tunneling device. Phenomenological discussions often require as input an estimate of the tunneling time: the time a charge carrier spends in the classically forbidden barrier region. Several different tunneling times have been suggested;[9,13,14,20,21,22] however, due to the lack of conclusive experiments no single theoretical tunneling time has been universally accepted. (The situation might be changing, however. See Chapter 10.) Indeed, doubts have been expressed whether a unique tunneling time exists.[17,23]

Wave packet simulations, which are based on an explicit solution of the time-dependent Schrödinger equation, should be ideal for studying explicitly time-dependent situations. As a first application, we compare our numerical results to a case where analytical results are known:[13,14] a square barrier with an imposed weak harmonic perturbation. One should note that the numerical method employed can be applied with equal ease to arbitrary tunneling structures (see below) and/or time modulations (which do not have to be small). The numerical integration procedure is standard and has been discussed elsewhere.[3,8,10]

The time-modulated barrier has been used by Büttiker and Landauer[13,14] to construct a "clock" to measure tunneling times. One might wonder why an indirect method is needed, and the reason is the following. A straightforward extraction of a tunneling time by comparing the motion of the wave packet with and without the tunneling structure is complicated by the self-interference which occurs when the packet collides with the barrier system. By comparing *asymptotic* packets, for which the self-interference has died away, a time difference can be extracted, but it is not clear if one can disentangle the "waiting time" and the actual time of tunneling.[17,22]

The following points are of special importance for the numerical simulations:

1. Choice of the Initial State. Predictions for tunneling times for 25- to 50-Å-thick $Al_xGa_{1-x}As$ barriers range typically from 1 to 10 fs. These times correspond to energies of the order of 0.6 to 0.06 eV, and in order to yield quantitative results the energy resolution of the simulation must be better than the above values. This implies that the spatial width of the initial wave packet, which is inversely proportional to the momentum width and hence also determines the energy width, must be chosen accordingly. In practice we find that wave packets with half-width $\sigma \geq 1000$ Å $[\psi(t = 0) \sim \exp(-x^2/2\sigma^2)]$ yield sufficient accuracy. This, in turn, required long "contract" regions (in order to prevent the initial state overlapping with the tunnel barrier); typically we take 1-μm-long flat regions outside the barriers.

2. Discretization of the Potential. The thinnest barriers in our simulations contain only ~10 mesh points which requires some care when choosing the discretization procedure. We found out that the procedure suggested by Collins *et al.*[9] was sufficient to lead to stable results.

3. Momentum Representation. We found it extremely useful to work simultaneously in momentum and real spaces. The point is that even the slightest numerical instability immediately reflects itself in the momentum representation of the wave function: thus the "cleanness" of the momentum spectrum was an indispensable tool in judging the convergence and quality of the numerics.

4. Consistency with Static Transmission Coefficient. Another measure of the sharpness of the energy distribution of the initial state can be obtained by comparing the simulated transmission coefficient with the precisely known static value. For the parameters used in our calculations the agreement was always within a couple of percent; the error was larger for the thinner barrier presumably because the finite mesh size plays a more important role there. In addition, to make connection to the weak time-dependent perturbation considered by Büttiker and Landauer,[13,14] the amplitude of the time-dependent modulations was chosen so small ($V_1 = V_0/20$ in our simulations) that the total transmission coefficient is independent of the modulation frequency, and higher-order processes (emission/absorption of several modulation quanta) are negligible.

The physics of tunneling through a time-modulated barrier has been elucidated by Büttiker and Landauer.[13,14] The tunneling particles may absorb, or emit, modulation quanta, and thus in momentum (or energy) space the reflected and transmitted parts of the wave function consist of a main feature with the initial energy E and sidebands at $E \pm n\hbar\omega$. Below we discuss the relation of these sidebands to tunneling times.

We now turn to the numerical results. Figure 9.1 shows a temporal evolution of a typical simulation, and the sidebands at energies $E \pm \hbar\omega$ are clearly resolved. Thus a quantitative evaluation of the sideband intensities is possible (this was not the case for our preliminary data reported earlier,[24] for which we gave an erroneous interpretation).

In Figure 9.2 we show the sideband intensities obtained from our simulations and from the analytical results of Büttiker and Landauer.[13,14] As seen in the figure, the two totally independent approaches are in quantitative agreement. This serves as a stringent test for the accuracy of the numerical method and suggests its applicability to a wide range of other time-dependent phenomena.

Büttiker and Landauer (BL) have suggested using the sideband intensities for defining a tunneling time. For simplicity, let us first consider opaque barriers ($k_0 d \gg 1$, $k_0 = (2mV_0/\hbar^2)^{1/2}$, V_0 is the height of the barrier, and d is its thickness). In this case BL find that the intensities of the sidebands $T_\pm$ [$\equiv T(E \pm \hbar\omega)$, E is the energy of the incoming particle, and ω is the modulation frequency] are given by

$$T_\pm = (V_1/2\hbar\omega)^2(e^{\pm\omega md/\hbar\kappa} - 1)^2 T(E) \qquad (9.2.1)$$

where $\kappa = \sqrt{2m(V_0 - E)/\hbar^2}$, V_1 is the modulation amplitude (it is assumed that $V_1 \ll \hbar\omega$), and $T(E) = |D(E)|^2$ are the transmission coefficient and the transmission amplitude of the static barrier, respectively.

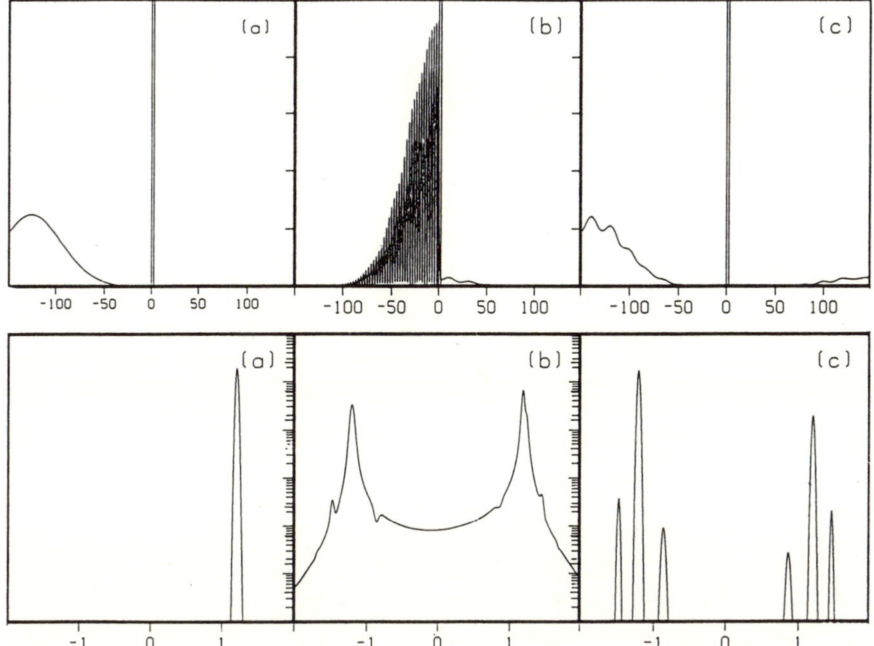

FIGURE 9.1. A Gaussian wave packet of mean energy E is shown colliding with a sinusoidally modulated square barrier, $V(x, t) = (V_0 + V_1 \sin(\omega t))\theta(x)\theta(d - x)$, where $V_0 = 0.23$ eV, $V_1 = 0.05 \times V_0$, $\hbar\omega = 0.35 \times V_0$, and $d = 50$ Å. The square modulus of the wave packet for $E = 0.72 \times V_0$ is plotted both in real space (top: linear y-axis, length unit = 22.22 Å) and in momentum space (bottom: logarithmic y-axis, momentum $\hbar k = \sqrt{2}$ corresponds to energy $E = V_0$), for three characteristic time instants during the simulation: before collision (a), "mid-collision" (b), and after collision (c). After a completed collision [frame (c)] the momentum representation of the transmitted pulse (positive momentum), and the reflected pulse (negative momentum) contains well-resolved sidebands corresponding to emission or absorption of one modulation quantum.

Next, BL define an asymmetry function

$$f(\omega) = \frac{T_+ - T_-}{T_+ + T_-} \tag{9.2.2}$$

which for the opaque barrier acquires the simple form

$$f_{\text{opaque}}(\omega) = \tanh(\omega m d / \hbar\kappa) \tag{9.2.3}$$

Thus, in this case the asymmetry function is characterized by a *single* quantity with the dimension of time, $\tau_{\text{BL}} = md/\hbar\kappa$, that separates characteristic low- and high-frequency behaviors. Therefore BL identify τ_{BL} as the traversal time for tunneling. It is interesting that for an *opaque* barrier the same traversal time appears from an analysis of field emission.[25] There

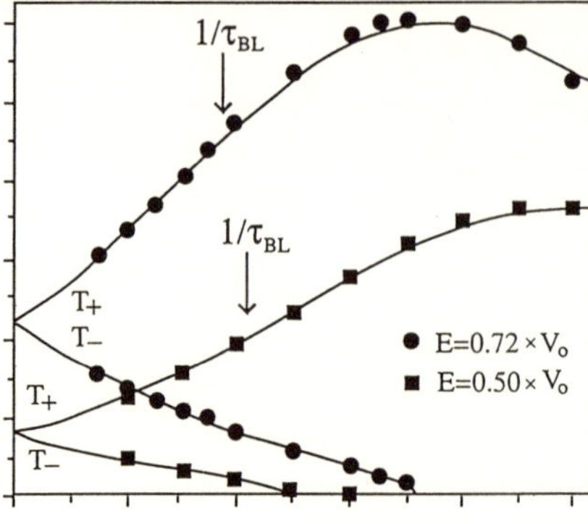

FIGURE 9.2. Sideband intensities $T_\pm$ as a function of modulation frequency obtained from a series of simulations of the kind shown in Figure 9.1. Results for two incident energies are given as squares and dots. It can be shown for Gaussian wave packets narrow in momentum space that $T_\pm = (k_\pm/k)|D_\pm|^2$ and can thus be compared directly with the analytical results of Ref. 14, which are shown as continuous lines. A quantitative agreement is found with the analytical and simulated results. Note that T_- vanishes for $\hbar\omega > E$, where $E = \hbar^2 k^2/2m$ and $\hbar k$ is the mean momentum of the incoming pulse. The barrier parameters are as in Figure 9.1.

a time-dependent field due to charge oscillations (surface plasmons) plays a similar role as the time-dependent barrier height (phonons) do in the problem discussed here.

Let us now consider general barriers. BL discuss the crossover behavior of the asymmetry function *explicitly* only in the opaque limit, but their arguments suggest the approach might have a more general validity. However, if one constructs the asymmetry function according to equation (9.2.2), and uses the results of BL for square barriers, it is seen that the resulting (complicated) expression cannot be characterized by a single quantity whose dimension is time. Even worse, for general barriers the asymmetry function can be determined only numerically, and some kind of operational procedure is called for. One may again examine the crossover from low to high frequencies as a suitable criterion. It appears obvious that (i) for low frequencies $f(\omega)$ is linear in ω, and that (ii) for high frequencies $f(\omega)$ saturates to unity. The characteristic (or crossover, in the terminology of BL) frequency could therefore be identified as the frequency where the linear low-frequency behavior meets the asymptotic limit:

$$\omega_c f'(0) = \lim_{\omega \to \infty} f(\omega) \tag{9.2.4}$$

We observe that in the opaque limit the above prescription gives the BL traversal time as the inverse of the characteristic frequency. However, this method cannot always be applied. For example, for thin barriers ($d \simeq 25$ Å) the asymmetry function has a *negative* slope at the origin, and equation (9.2.4) has no solutions (see also Ref. 17, where similar conclusions have been obtained). It is possible, of course, to construct other ad hoc procedures which do yield a critical frequency; the two points we want to make here are that (i) a simple generalization of the crossover analysis in the opaque limit is not viable, and (ii) that the connection of the tunneling time and the inverse of a characteristic frequency does not appear immediate for a general barrier.

Let us now consider the adiabatic limit. Physically, it does not appear surprising that a finite-frequency object [such as $T(E \pm \hbar\omega)$] does not directly yield intrinsic information of a static quantity (tunneling time through a static barrier), and thus this limit may provide a more direct connection. Actually the generalized traversal time introduced by Büttiker and Landauer,[14] $\tau_{\mathrm{BL}} = \hbar |d \ln D/dV|^{1/2}$, emerges from an analysis of an adiabatic limit: After first identifying a traversal time τ_{BL} for an opaque barrier from a crossover behavior at *finite* frequency of the asymmetry function of equation (9.2.2), Büttiker and Landauer notice that $T_\pm \propto \tau_{\mathrm{BL}}^2$ in the limit $\omega \to 0$. Their generalized traversal time appears when the limiting value of $T_\pm$ for an arbitrary barrier is forced to have the same quadratic form.

The adiabatic limit of the modulated barrier bears a close analogue to the Larmor clock,[26] where one extracts a tunneling time in the limit of a vanishing magnetic field. In this context it is interesting to observe that by adding the static part of the transmitted wave function, $D \exp(ikx - iEt/\hbar)$, and the one-phonon sideband terms, $D_\pm \exp(ik_\pm x - iE_\pm t/\hbar)$, the total transmitted wave function, in the limit of low frequency and amplitude of the modulation, can be written as a single term,[16]

$$\psi(x, t) = D(E, \bar{V}) \exp\{i[k(\tau_D^{\bar{V}})x - E(\tau_D^{\bar{V}})t/\hbar] + \eta\} \qquad (9.2.5)$$

This result, which is valid for a general form of the potential barrier if $t \ll 1/\omega$ and $x \ll v(k)/\omega$, is obtained by relating the sideband amplitudes $D_\pm$ to the static transmission (D) and reflection (A) amplitudes. Here

$$E(\tau_D^{\bar{V}}) = E + V_1 \omega \tau_D^{\bar{V}} \qquad (9.2.6)$$

and

$$k(\tau_D^{\bar{V}}) = k + \frac{V_1}{2E}(\omega\tau_D^{\bar{V}})k \qquad (9.2.7)$$

The complex quantity

$$\tau_D^{\bar{V}} = i\hbar \frac{d \ln D(E, \bar{V})}{d\bar{V}} \qquad (9.2.8)$$

has the dimension of time and is closely related to the complex times introduced by Sokolovski and Baskin[20] and Leavens and Aers.[27] Note that for the presently considered symmetric barrier the real and imaginary parts of equation (9.2.8) are the well-known dwell time τ_{dwell} and the so-called Büttiker time[26] τ_z, as here $\tau_d^{\bar{V}} = \tau_{\mathrm{dwell}} - i\tau_z$. The generalization of τ_{BL} to arbitrary potential barriers[14,22] is related to equation (9.2.8) as

$\tau_{\rm BL} = |\tau_D^{\bar{V}}|$. The factor η in equation (9.2.5) is quite complicated and will not be given explicitly here.

In deriving equations (9.2.5)–(9.2.8), we took the modulating potential to be $V_1 \sin(\omega t)$; i.e., $V_1 \omega t$ for $t \ll 1/\omega$. From equation (9.2.6) it is therefore tempting to interpret the modified energy and momentum of the transmitted wavefunction to be the result of an adiabatic interaction between the tunneling electron and the rising barrier during a traversal time $\tau_D^{\bar{V}}$. Such an interpretation is not meaningful, however, as $\tau_D^{\bar{V}}$ is complex and any measurable traversal time must certainly be real. Some insight into the role of $\tau_D^{\bar{V}}$ can be gained by calculating the transmitted electron density and current density. If x_0 and x are points beyond the barrier and $t_0, t \ll 1/\omega$, one finds to lowest order in ω

$$\rho(x, t) = |\psi(x, t)|^2 = \rho(x_0, t_0) \exp\left\{ \frac{-2V_1(\omega\tau_z)[(t - t_0) - (x - x_0)/v(k)]}{\hbar} \right\}$$

$$(9.2.9)$$

and

$$J(x, t) = v(k)\rho(x, t) + \frac{V_1}{2E}(\omega\tau_{\rm dwell})v(k)\rho(x_0, t_0) \qquad (9.2.10)$$

One notes that the decrease in the transmitted density with time is related to the Büttiker time τ_z, i.e., to the imginary part of $\tau_D^{\bar{V}}$. The result for the transmitted current density is a sum of two terms. The first simply reflects the change of the transmitted density with time while the second comes about because the change in density inside the barrier is associated with a change in current (to conserve charge). The latter term can also, from equation (9.2.7), be interpreted as arising from an adiabatic change of the momentum of the tunneling electron while it interacts with the barrier during the dwell time $\tau_{\rm dwell}$ given by the real part of $\tau_D^{\bar{V}}$.

The result for the transmitted current is in a sense reminiscent of the results of Büttiker's analysis[26] of the Larmor clock. There a beam of electrons, spin-polarized in the x-direction, is traveling in the y-direction impinging on a barrier. A magnetic field in the z-direction inside the barrier gives the spin of the transmitted electrons a y-component proportional to $\omega_L\tau_{\rm dwell}$, where ω_L is the Larmor frequency. In addition, Büttiker showed that the transmitted electrons also acquire a spin component in the z-direction. This is because the incoming electrons, which have no spin component in the z-direction, can be thought of as a superposition of a spin-up and spin-down state. Because of the Zeeman interaction these states have different energies, and therefore the energy dependence of the barrier transmission probability results in a nonzero z-component of the spin proportional to $\omega_L\tau_z$.

Hence, both for the Larmor clock and the time-modulated barrier systems, part of the effect on the transmitted current of electrons is due to a dependence of transmission probability on the barrier height and is characterized by the Büttiker time τ_z. Another part is due to an interaction process within the barrier during a time τ_{dwell}.

For the Larmor clock, Büttiker[26] has argued that the magnitude of the spin in the $y - z$ plane can be used to define a traversal time and thus identifies $\tau = \sqrt{\tau_z^2 + \tau_{\text{dwell}}^2}$. Though plausible, this reasoning does not appear forced, and we speculate, in light of equation (9.2.9) and equation (9.2.10), that experiments probing different aspects of the transmitted wave function equation (9.2.5) may yield different, but complementary, information about the tunneling process. A related discussion of the nonuniqueness of the tunneling time appears in Ref. 23.

9.3. RESONANT TUNNELING IN THE PRESENCE OF INELASTIC PROCESSES

Resonant tunneling in semiconductor heterostructures[28,29] is presently a very active area of research featuring both device applications and questions of basic physics. Traditionally most current-versus-voltage curves are analyzed with the Tsu-Esaki[29] tunneling formula (or its modifications[30])

$$\mathbf{J} = \frac{2e}{(2\pi)^3} \int d\mathbf{k}\, \mathbf{v}(\mathbf{k})[f_{\text{FD}}(E) - f_{\text{FD}}(E + eV)]\|T|^2 \qquad (9.3.1)$$

where f_{FD} is the Fermi–Dirac distribution, $T(E, V)$ is the transmission coefficient obtained from the solution of the static Schrödinger equation, E is the total energy of the tunneling particle, and V is the applied voltage. The use of equation (9.3.1) implies several assumptions which are of relevance to the present work: (i) use of equilibrium distribution functions (even though a biased resonant tunneling diode is manifestly in an nonequilibrium state), and (ii) scattering is not accounted for. A complete theoretical description would require a numerically tractable quantum kinetic theory for nonstationary and spatially inhomogeneous systems. No such theory exists today; for a status report of some candidate theories see Ref. 31. As a natural first step, several research groups have recently addressed partial aspects of the problem. Phenomenological theories where the main effect of inelastic collisions is to broaden the resonant transmission coefficient have been reported.[32] Wigner function simulations, with simplified collision operators, have been performed to obtain the nonequilibrium distribution function, from which the current can be extracted.[33] Several analytic calculations of the transmission coefficient in the presence of optical phonon scattering have recently been reported.[34] These calculations have been motivated by the observation of optical phonon related features in the *I-V* curves.[35]

In this section we demonstrate how wave packet simulations can be used to study energy exchange, i.e., inelastic scattering, in resonant tunneling physics. The proposed method does not overrule any of the aforementioned theoretical approaches but complements them by providing a means of obtaining additional details of various tunneling processes. The phenomenology is the same as in the single-barrier case: the incoming particles of energy E may either emit or absorb modulation quanta of energy $\hbar\omega$, and consequently the reflected and transmitted beams contain sidebands at $E \pm n\hbar\omega$. Thus there is a close analogy to interaction with a dispersionless boson field. The analogy is not complete: effects related to statistics and/or temperature are not included. In a sense the present approach is an infinite temperature calculation: absorption and emission of modulation quanta have the same probability. We do not exclude the possibility of extending the present approach to include temperature effects.

Despite this restriction, the present approach has several advantages. No heavy numerical work is required as for Wigner function simulations. The coupling to the external time modulation can be of arbitrary strength, and no truncation to low-order processes is necessary as is customarily done in the analytic calculations. Further, the shape of the heterostructure potential can be arbitrary, and the bias and energy dependence of the transmission coefficient is accounted for exactly. The part of the structure which is affected by the time modulation can be chosen at will. This allows one to analyze the relative importance of inelastic processes occurring in the barriers or in the quantum well (see below).

As an illustration of the method, we have calculated the transmission coefficient for a $50 \,\text{Å} \times 50 \,\text{Å} \times 50 \,\text{Å}$ double-barrier structure. The other parameters in the simulation were as follows: V_0 (barrier height) $= 0.23$ eV, V_1 (amplitude of harmonic modulation) $= 0.05 \times V_0$, and $\hbar\omega = 0.3 \times V_0$. The initial Gaussian wave packet had a half-width of $1000 \,\text{Å}$, which was required for sufficient energy resolution. The simulated transmission coefficient is depicted in Figure 9.3, where we show results for the static double barrier (squares), modulated barriers (circles), and modulated well (triangles). The satellite at $E_r + \hbar\omega$ (E_r is the resonant energy) is clearly visible for the case of the modulated quantum well while it is barely discernible for the modulated-barrier case. This effect can be understood as follows. The tunneling particles pass through the classically forbidden region so fat (typical estimate for a tunneling time for a 0.23-eV high and 50-Å-broad barrier at energy 0.1 eV would be a few femtoseconds) that they do not have time to absorb or emit a modulation quantum. However, once in the quantum well the particles stay there so long (semiclassically, they reflect back and forth several times before tunneling out) that a sideband

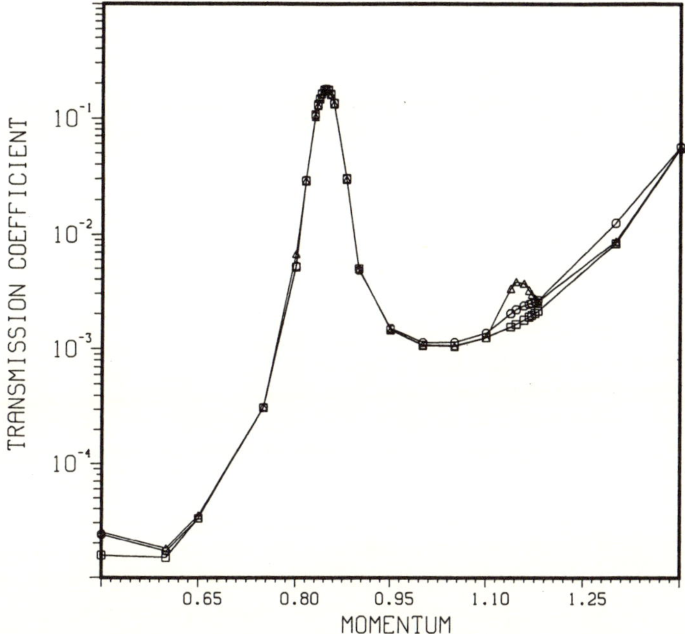

FIGURE 9.3. Transmission coefficient as a function of momentum for a 50 Å × 50 Å × 50 Å double-barrier structure. Units are chosen so that $p = \sqrt{2}$ corresponds to an energy equaling the static barrier height. Squares: static structure; circles: modulated barriers; triangles: modulated quantum well. In addition to the main resonance at $p_r = 0.864$ the satellite at $p = \sqrt{p_r^2 + 2\hbar\omega}$ is seen. The feature corresponding $p = \sqrt{p_r^2 - 2\hbar\omega}$ is weak (not shown in the figure) and best analyzed in momentum space snapshots (see Figure 9.4).

has time to form, and therefore the modulated well shows a stronger effect than modulated barriers. This seems to suggest that in experimental situations barrier phonons are of minor importance. Interfaces, however, which are probed many times (within the semiclassical picture) may have an important effect.

Further insight on the dynamics of the tunneling processes can be extracted by analyzing the wave packet in momentum space.[16] In Figure 9.4 we show the momentum space-wave function for four different energies at representative time instants during the simulation. Let us examine the transmitted part of the wave function, i.e., positive momenta (similar considerations can be made for the reflected pulse). Starting from low energies [panel (a)], $E = E_r - \hbar\omega$, we observe that the transmitted pulse does not have *any* amplitude at the incoming energy E: all its weight is located at channels $E + \hbar\omega$ and $E + 2\hbar\omega$. Thus, we have a very clear case of velocity filtering: the transmitted pulse has a velocity, or energy, different from the incoming one. The mechanism leading to this behavior is easily understood

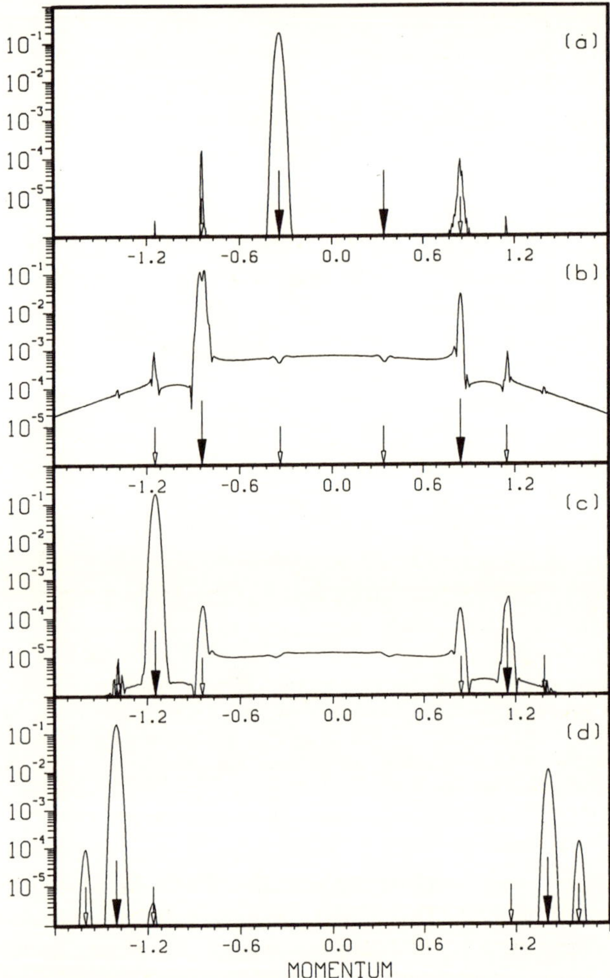

FIGURE 9.4. Square modulus of the momentum space-wave function during the simulation for four different incident energies: (a) $E = E_r - \hbar\omega$, (b) $E = E_r$, (c) $E = E_r + \hbar\omega$, (d) $E = 0.9 \times V_0$. The large solid arrows show the incident momentum, while the small open arrows indicate the locations of the $n = \pm 1$ side bands (the $n = -1$ sideband does not exist for the lowest energy).

by recalling that the transmission coefficient is very small for small energies (see Figure 9.3), and that by absorbing a modulation quantum the tunneling particles are excited to the resonant channel with an enhanced probability for transmission. For $E = E_r$ [panel (b)] the momentum space-wave function is characterized by a very broad uniform background, with a sharp peak at the incoming energy and small features corresponding to emission and absorption of modulation quanta. The background corresponds to the trapped part of the wave function which slowly leaks out from the quantum well. For $E = E_r + \hbar\omega$ [panel (c)] we also observe a large background; the interesting feature is the large feature at $E - \hbar\omega$, which corresponds to particles that have emitted a modulation quantum, and therefore are at resonance, and thus contribute strongly to the transmitted pulse. Particles that have absorbed modulation quanta, however, do not contribute significantly to the tunneling current. At even higher energies [panel (d)] the situation differs in two aspects: first, there is hardly any uniform background, which shows that tunneling is fast and dominated by nonresonant effects, and, second, the dominant sideband now occurs at $E + \hbar\omega$, which corresponds to particles that are excited above the barriers by absorbing a quantum, and therefore can contribute to the transmitted pulse.

In summary, we have described a numerical procedure which is simple and straightforward to apply, allows the simulation of inelastic processes, and gives insight to the dynamics of tunneling in the presence of inelastic effects. The method can be applied to many other tunneling problems where scattering or relaxation effects are important. Examples of potential applications include studies of stochastic time modulation, and sequential tunneling and relaxation in biased multiple quantum well systems.[36]

ACKNOWLEDGMENTS

The author is grateful to Mats Jonson, with whom the work described in the first part of this chapter was carried out, and for numerous discussions. In addition, frank and informative (but not necessarily agreeing!) correspondence with M. Büttiker and R. Landauer is gratefully acknowledged. Thanks are due to E. Hauge and J. A. Støvneng for discussions, and for sending their papers prior to publication.

REFERENCES

1. L. SCHIFF, *Quantum Mechanics*, 3rd ed., McGraw-Hill, New York (1968).
2. A. GOLDBERG *et al.*, *Am. J. Phys.* **35**, 177 (1967).
3. V. P. GUTSCHIK and M. N. NIETO, *Phys. Rev. D* **22**, 403 (1980).
4. P. SCHNUPP, *Thin Solid Films* **2**, 177 (1967).
5. The applications range from chemical physics [R. B. GERBER, R. KOSLOFF, and M. BERMAN, *Comput. Phys. Rep.* **5**, 59 (1986); R. KOSLOFF, *J. Phys. Chem.* **92**, 2087 (1988)] to foundations of quantum mechanics. [See, e.g., Y. MURAYAMA, *Phys. Lett. A* **140**, 469 (1989).]
6. J. R. BARKER, *Physica* **134B**, 22 (1985).
7. T. NAKAGAWA *et al.*, *Superlatt. Microstruct.* **1**, 217 (1985).
8. A. P. JAUHO and M. N. NIETO, *Superlatt. Microstruct.* **2**, 407 (1986).

9. S. COLLINS, D. LOWE, and J. R. BARKER, *J. Phys. C* **20**, 6233 (1987).

10. A. P. JAUHO, *Acta Polytechnica Scandinavica El-***58**, 192 (1987).

11. F. ANCILOTTO *et al.*, *Phys. Rev. B* **39**, 8322 (1989); F. ANCILOTTO *et al.*, *Phys. Rev. B* **40**, 3729 (1989).

12. N. GARCIA *et al.*, *J. Phys. Cond. Matter* **1**, 9931 (1989).

13. M. BÜTTIKER and R. LANDAUER, *Phys. Rev. Lett.* **49**, 1739 (1982).

14. M. BÜTTIKER and R. LANDAUER, *Physica Scripta* **32**, 429 (1985).

15. An early reference on numerical work is D. L. HAAVIG and R. REIFENBERGER, *Phys. Rev. B* **26**, 6408 (1982), where the motion of a wave packet whose spatial width is comparable to a time-modulated barrier, or well, was studied.

16. A. P. JAUHO and M. JONSON, *J. Phys. Cond. Matter* **1**, 9027 (1989).

17. J. A. STØVNENG and E. H. HAUGE, to appear in *Rev. Mod. Phys.*

18. A. P. JAUHO, *Phys. Rev. B* **41**, 12327 (1990).

19. D. LOWE, private communication.

20. D. SOKOLOWSKI and L. M. BASKIN, *Phys. Rev. A* **36**, 4607 (1987).

21. E. H. HAUGE *et al.*, *Phys. Rev. B* **36**, 4203 (1987); J. P. FALCK and E. H. HAUGE, *Phys. rev. B* **38**, 3287 (1988).

22. M. BÜTTIKER and R. LANDAUER, *J. Phys. C* **21**, 6207 (1988).

23. D. SOKOLOVSKI and P. HÄNGGI, *Europhys. Lett.* **7**, 7 (1988).

24. A. P. JAUHO and M. JONSON, *Superlatt. Microstruct.* **6**, 303 (1989).

25. M. JONSON, *Solid State Commun.* **33**, 743 (1980).

26. M. BÜTTIKER, *Phys. Rev. B* **27**, 6178 (1983).

27. C. R. LEAVENS and G. C. AERS, *Solid State Commun.* **63**, 1101 (1987).

28. L. ESAKI and R. TSU, *IBM J. Res. Dev.* **14**, 61 (1970).

29. R. TSU and L. ESAKI, *Appl. Phys. Lett.* **22**, 562 (1973).

30. D. D. COON and H. C. LIU, *Appl. Phys. Lett.* **47**, 172 (1985).

31. For a recent review, see A. P. JAUHO, *Solid State Electronics* **32**, 1265 (1989).

32. T. WEIL and B. VINTER, *Appl. Phys. Lett.* **50**, 410 (1987); M. JONSON and A. GRINCWAJG, *Appl. Phys. Lett.* **51**, 1729 (1987).

33. W. FRENSLEY, *Phys. Rev. B* **36**, 1570 (1987); N. C. Kluksdahl *et al.*, *Phys. Rev. B* **39**, 7720 (1989).

34. N. S. WINGREEN, K. W. JACOBSEN, and J. W. WILKINS, *Phys. Rev. Lett.* **61**, 1396 (1988); B. Y. GELFAND, S. SCHMITT-RINK, and A. F. J. LEVI, *Phys. Rev. Lett.* **62**, 1683 (1989); L. I. GLAZMAN and R. I. SHEKTER, *Zh. Eksp. Teor. Fiz.* **94**, 292 (1988) [*Sov. Phys. JETP* **67**, 163 (1988)]; M. JONSON, *Phys. Rev. B* **39**, 5924 (1989); W. CAI *et al.*, *Phys. Rev. Lett.* **63**, 418 (1989).

35. V. J. GOLDMAN, D. C. TSUI, and J. E. CUNNINGHAM, *Phys. Rev. B* **36**, 7635 (1987).

36. F. CAPASSO, K. MOHAMMED, and A. Y. CHO, *Appl. Phys. Lett.* **48**, 478 (1986).

Chapter 10

Tunneling Times in Quantum Mechanical Tunneling

M. Jonson

10.1. INTRODUCTION

In this chapter we shall be concerned with the time scale associated with quantum mechanical tunneling. More precisely, we shall discuss how long it takes for an electron to tunnel through a classically forbidden region. This *traversal time* is distinct from the lifetime of a metastable or resonant state that decays by a tunneling process. In a classical analogy[1] we may consider a gas escaping from a large vessel through a thin tube. The time scale corresponding to the lifetime of the metastable state is the average length of time a gas particle spends in the vessel. When a gas particle finally is able to escape, the much shorter time it has spent in the thin tube is the analogue of the traversal time.

The importance of time scales in tunneling was recognized early. They are not only of intrinsic interest in quantum mechanics but are clearly important in a wide range of situations where a tunneling entity interacts with additional degrees of freedom, which can adjust to the time development of the tunneling process. We shall return to discuss two examples later: tunneling in the presence of friction, which is of interest in Josephson junction circuits,[2] and electron field emission where an electron tunnels out of a metal through a potential barrier while interacting with the "image charge" at the metal surface.[3]

We consider the deceptively simple question, how long does a tunneling particle, which eventually will be transmitted, spend in the classically

---

M. Jonson • Solid State Division, Oak Ridge National Laboratory, Oak Ridge, Tennessee 37381-6024, USA. *Present address*: Institute of Theoretical Physics, Chalmers University of Technology, S-41296 Göteborg, Sweden.

Quantum Transport in Semiconductors, edited by David K. Ferry and Carlo Jacoboni. Plenum Press, New York, 1991.

inaccessible tunneling region? The first answers were given roughly 40 years ago by Bohm,[4] Wigner,[5] and Eisenbud.[6] Their answer, the *phase time*, is related to the sensitivity of the phase of a scattered wave function to a change in the particle energy. Alternative answers, now known as the *dwell time* or sojourn time and the *Larmor clock time*, were given in the 1960s by, respectively, Smith[7] and Baz',[8] without putting the question to rest. The last decade brought a revival of the interest in the time scale of tunneling, where the influential work of Büttiker and Landauer[9,10] was an important early contribution. The new wave of interest was partly due to the need to understand the dynamics of tunneling in the context of high-speed devices based on semiconductor tunneling structures.[11] Another motivation was the invention of the scanning tunneling microscope (STM) and the ensuing renewed interest in inelastic electron tunneling microscopy.[12]

In spite of the large published literature on tunneling times, no consensus has yet emerged. Nevertheless, very recent experimental and theoretical work has led to a better understanding. It is widely accepted, for instance, that the dwell time and phase times do not measure the traversal time. The reason is that the phase times, which refer to completed scattering of a wave packet with a narrow energy distribution, contain contributions from different regions of space. These include interference contributions from well outside the range of the potential, which cannot be disentangled from the contribution emanating inside the barrier. The dwell time, on the other hand, provides a well-defined measure of how long an electron spends in any given region, inside or outside the barrier. However, it does not distinguish between contributions from particles which will eventually be scattered into different channels. It does not therefore give the traversal time for electrons that eventually are transmitted. The motivation for discussing phase times and dwell time in Section 10.2 is therefore mainly pedagogical. It also serves to make contact with the discussion of tunneling times in typical quantum mechanical textbooks. These usually have something to say about the lifetime of a long-lived metastable state, but nothing about traversal times. In the context of a metastable state, it turns out that the phase time and the dwell time are useful, whereas they are not for discussing the traversal time.

In Section 10.3 we proceed to discuss tunneling times using a different philosophy than that of scattering theory. In the previously mentioned Larmor clock proposed by Baz',[8] a magnetic field was used to provide a "clock," that is, a constant frequency which could measure an interaction time. The constant frequency of the Larmor clock is due to the Larmor precession of a particle's spin in a constant magnetic field. There are several other clocks, and we shall discuss some of them: the time-modulated barrier of Büttiker and Landauer,[9,10] Büttiker's modified analysis of the Larmor clock,[13] and a clock provided by the interaction between a tunneling electron and surface polarization modes, first discussed by us.[3] In our analysis of the time-modulated barrier a complex time, equivalent to the time found by Sokolowski and Baskin[14] using a path integral technique, emerges in a natural way. This time is quite useful, although a measurable traversal time has to be a real quantity. Finally, we discuss the recent experiment by Esteve *et al.*[15] Here it is a macroscopic state of a Josephson junction rather than a particle that tunnels. It interacts with a transmission line (delay line) whose characteristic frequency can be varied *in situ* by changing its length.

Numerous papers on the subject appear in the literature. Although most technical controversies in this field have been resolved, different interpretations remain. In particular, the more philosophical questions as to whether an intrinsic tunneling time can be defined without reference to a measurement or clock, and if different clocks (interactions) could in principle give different traversal times, remain largely open. Although we will touch on several aspects of the problem of tunneling times, this is not a comprehensive review of the subject. Two recent reviews written with different viewpoints are those of Hauge and Støvneng[16] and Büttiker.[17] Together with Refs. 1, 18, and 19, they give a taste of an ongoing controversy.

10.2. LIFETIME OF A METASTABLE STATE, PHASE TIMES, AND DWELL TIME

Most textbooks on quantum mechanics have something to say about the lifetime of a metastable state. As already emphasized, this is not the main issue here, however. Instead we are concerned with the *traversal time*, that is, the time a tunneling particle spends in a classically forbidden region of space. Nevertheless, it is instructive to first review some concepts in the less controversial context of a particle decaying out of a metastable state. In particular, we shall introduce the concepts of *phase times* and *dwell time*. In doing so we shall make contact with scattering theory, as discussed in most textbooks on quantum mechanics.

10.2.1. The Decay of a Prepared State

For a start we follow Landau and Lifshitz,[20] who note that the energy eigenvalue of the Schrödinger equation, with the boundary condition that there is *only* an outgoing spherical wave in the asymptotic region, is complex:

$$\varepsilon = E_0 - \tfrac{1}{2}i\Gamma \tag{10.2.1}$$

The eigenvalue is complex because the boundary condition is complex. The corresponding eigenstate describes a prepared, unstable state. Such a state can arise, for instance, when an atom is excited by a collision. It may be able only to return to its ground state by a slow radiative process. A different example is when electrons (and holes) are photoexcited in an AlGaAs heterostructure. In such systems some electrons first relax quickly to a quasibound state in a quantum well, from which they slowly decay by tunneling through a potential barrier or by recombining with holes while sending out photons. If the prepared state is long-lived or metastable, it does not really matter for its eventual decay how it was formed; the probability of finding the electron in the metastable state is simply related to the square modulus of the wave function corresponding to the eigenvalue of (10.2.1)

$$|\Psi(t)|^2 \sim e^{-\Gamma t/\hbar} \tag{10.2.2}$$

From the time dependence of (10.2.2) it follows that Γ, the imaginary part of the complex energy eigenvalue, is related to the lifetime τ of the metastable state:

$$\tau = \hbar/\Gamma \tag{10.2.3}$$

The lifetime of a state, as indeed all times discussed here, is to be understood in the usual sense of a quantum mechanical average. For future reference, we note also that Fourier-transforming $\exp(-i\varepsilon t/\hbar)$ with respect to time, with ε given by (10.2.1), gives the energy spectrum of a metastable or quasistationary state to be quasidiscrete with a peak of width Γ around E_0.

10.2.2. The Phase Times

If we do need to describe how the metastable state is formed, the appropriate approach is scattering theory. Scattering that involves a metastable state is known as resonant scattering. Before we discuss resonant scattering, however, we shall review some general results of scattering theory. Consider the usual textbook example[20] of three-dimensional scattering in a central field $U(r)$. In this case angular momentum is not affected by the scattering, and it is useful to expand the angular part of the wave functions in spherical harmonics. It remains then to solve for the radial part of the wave function R_{kl}, where l labels angular momentum and $E_k = \hbar^2 k^2/2m$. In the asymptotic region, far outside the range of the scattering potential, the radial wave function can be written as

$$R_{kl}(r) \sim \frac{1}{r}[(-1)^{l+1}e^{-ikr} + S_l e^{ikr}] \tag{10.2.4}$$

where

$$S_l = e^{2i\delta_l} \tag{10.2.5}$$

We recognize S_l as a diagonal element of the scattering matrix $\mathbf{S}$ in the momentum representation and δ_l as the phase shift.[20] If there were no scattering potential, this would be zero and hence $S_l = 1$.

Imagine now the scattering of a wave packet Ψ_l made up of waves with given angular momentum, but with energies in a narrow interval of width σ about a mean energy E_k,

$$\Psi_l(r, t) \sim \int dk' \, A(k' - k) R_{k'l}(r) e^{-iE_{k'}t/\hbar} \qquad (10.2.6)$$

By appealing to the stationary-phase approximation, we can follow the motion of this wave packet and identify two contributions corresponding to the two terms of (10.2.4). These represent an incoming and an outgoing wave, respectively. Using the stationary-phase method to determine the position of the wave packet, we find that the time from $t_1 < 0$, when the incoming wave packet passes the position r_1 in the asymptotic region, and $t_2 > 0$, when the outgoing wave packet passes the position r_2, also in the asymptotic region, is the classic *phase time* τ^ϕ discussed by Bohm[4] and Wigner.[5] With $r_1 = r_2 = R$ as a trivial simplification, one has

$$\tau_l^\phi(R, k) = t_2 - t_1 = \frac{1}{v_k}\left(2R + \frac{2d\delta_l}{dk}\right) \qquad (10.2.7)$$

Here $v_k = \hbar^{-1}(dE/dk)$ is the group velocity of the wave packet, and we have used the fact that the normalized envelope function $A(k' - k)$ is assumed to be sharply peaked with a small width σ around $k' = k$. We emphasize again that, for (10.2.7) to be meaningful, positions r_1 and r_2 have to be asymptotically far from the scattering region so that (10.2.4) applies. In particular, r_1 and r_2 have to be much larger than σ^{-1}, the width of the wave packet. If there were no scattering, obviously the phase shift and its derivative would vanish. One can therefore define a *phase time delay* as

$$\Delta\tau_l^\phi(k) = \tau_l^\phi(R, k) - 2R/v_k \qquad (10.2.8)$$

which is independent of R as long as R is in the asymptotic region. With the scattering matrix element, the phase delay time can be written more elegantly as

$$\Delta\tau_l^\phi(k) = -i\hbar S_l^{-1} \frac{d}{dE} S_l \qquad (10.2.9)$$

For the simple "single-channel" scattering discussed so far, the scattering matrix element S_l merely gives the relative phase of the scattered and incoming waves while its modulus is unity. In a more general case there may be many scattering channels. We shall illustrate this by reviewing the scattering of a plane wave by a central-field potential.[20] Consider a wave packet that in the absence of scattering would be

$$\Psi^0(x, t) = \int dk' \, A(k' - k)\chi_{k'}^0(x) e^{-iE_{k'}t/\hbar}; \qquad \chi_k^0(x) = \frac{1}{\sqrt{v_k}} e^{ikx} \qquad (10.2.10)$$

Here we have used the conventional[20] normalization of χ_k^0 that gives a probability current density of unity, that is, one particle passing (along the x-axis) through one unit area per unit time. As in the previous discussion, $A(k' - k)$ is an envelope function sharply peaked around $k' = k$. In the presence of the scattering potential $U(r)$, the wave packet would be

$$\Psi(\mathbf{r}, t) = \int dk' \, A(k' - k)\chi_{k'}(\mathbf{r})e^{-iE_{k'}t/\hbar} \tag{10.2.11}$$

with[20]

$$\chi_k(\mathbf{r}) = \sum_{l=0}^{\infty} \frac{\sqrt{\pi(2l+1)}}{ik} Y_{l0}(\theta, \phi) R_{kl}(r) \tag{10.2.12}$$

We use the normalized spherical harmonic functions $Y_{l,m}(\theta, \phi)$ (with $m = 0$) for the angular part, although it is obvious from symmetry that there is no dependence on azimuthal angle ϕ in (10.2.12). In the *asymptotic* region R_{kl} is again given by (10.2.4) multiplied by $1/\sqrt{v_k}$ to be properly normalized to correspond to emission of one particle per unit time per unit area of incoming current.[20] With this normalization we can find the "probability" σ_l that a particle in the current has angular momentum l with respect to the origin simply by taking the modulus squared of the coefficients in the expansion of the wave function. One finds that[20,21]

$$\sigma_l = (\pi/k^2)(2l+1) \tag{10.2.13}$$

which has dimension length squared due to the choice of normalization in (10.2.10). To get a dimensionless probability p_l, we simply divide by the area A of the incoming current.

Now that we have the probabilities p_l, it is interesting to take a weighted average of the single-channel phase times over *all* available scattering channels. One defines

$$\langle \tau^\phi(R, k) \rangle_{\text{av}} \equiv \sum_{l=0}^{\infty} p_l \tau_l^\phi(R, k), \qquad p_l = \frac{\sigma_l}{A} \tag{10.2.14}$$

with $\tau_l^\phi(R, k)$ given by (10.2.7) and σ_l by (10.2.13). In the next section we shall show that this averaged phase time is equivalent to the so-called dwell time. This quantity rigorously measures the average time spent by the particle in a particular region (here $r < R$) without distinguishing between the different scattering channels in which it may eventually be found.

In contrast to the dwell time, the phase delay time of (10.2.9) is defined separately for each scattering channel (for transmission and reflection in the one-dimensional case) and is therefore a candidate for measuring the traversal time. Indeed, it has been popular to assign the total delay to the region of the potential barrier ($r < r_0$ say) and add the free-particle time $2r_0/v$ to get the traversal time.[22-26] It is now clear that this "extrapolated" phase time (in the terminology of Ref. 16) does not give the traversal time. The decisive argument is that the phase delay time, as stressed above, only applies to completed scattering events in the asymptotic region and cannot be extrapolated back to the region of the scattering potential.[16,19] Simple and striking examples, showing the interference effects that make such extrapolated phase times meaningless, have been constructed.[16,27] In addition, a number of other objections have been raised. Büttiker and Landauer[9,10,28] have criticised the phase times because they rely on relating a peak in a scattered and distorted wave packet with a peak in an incoming wave packet; "physics has no law about a peak turning into a peak."[28] In more recent work[25,26,29] the centers of gravity of wave packets, a more robust measure than peak positions, have been used without silencing all critics.[17]

Even though it has been shown[16,25,30] that the leading correction to the phase time (defined by the center-of-mass method) is of order σ, that is, linear in the width of the envelope function $A(k - k')$, it is a different question as to whether the phase time can be measured to this accuracy. Localized ideal counters,[30] measuring arrival times, will produce a distribution of phase delay times whose width is $\sim \sigma^{-1}$, reflecting the *spatial* extent of the wave packets. Because the phase delay times turn out to be equivalent to the Larmor times of Baz' in the asymptotic region,[29] it is possible to circumvent this problem; an idealized Stern–Gerlach apparatus[14,16] would give a distribution proportional to σ rather than to σ^{-1}.

10.2.3. The Dwell Time

In 1960, F. T. Smith[7] introduced a measure for the duration of a scattering event defined in terms of a wave packet $\Psi(\mathbf{r}, t)$ like (10.2.11) (but now normalized to contain one particle in a volume $\Omega = LA$) as

$$\tau_{\text{dwell}}(R, k) \equiv \int_{-\infty}^{\infty} dt \int_{r<R} d\mathbf{r} |\Psi(\mathbf{r}, t)|^2 = \frac{L}{v_k} \int_{r<R} d\mathbf{r} |\chi_k(\mathbf{r})|^2 \quad (10.2.15)$$

where the second equality is derived in Appendix 1.1. This *dwell time* can be defined for arbitrary R. It can, for instance, be defined for a small region inside or outside the barrier and does not have to refer to an asymptotically large volume. It has a clear interpretation as the average time spent by the electron in the region $r < R$. (A recent challenge[25] to this statement has been refuted.[19]) For the special case of the plane wave scattered by a central field discussed above, we show in Appendix 1.1 that

$$\tau_{\text{dwell}}(R, k) = \langle \tau^\phi(R, k) \rangle_{\text{av}} \quad (10.2.16)$$

where the weighted average of the phase times are given in (10.2.14). This relation is a special example of a more general result: the dwell time measures the time spent by a particle in a given region, *averaged over all scattering channels.* This is a quite useful measure; reflection and transmission of a particle are mutually exclusive events in the sense of Feynman and Hibbs;[31] that is, a measurement can determine, without interfering with the scattering event, whether a particle has been transmitted or reflected. Hauge and Støvneng[16] have therefore argued that the traversal and reflection times τ_T and τ_R (e.g., consider the one-dimensional case) have to obey the sum rule

$$\tau_{\text{dwell}} = T(k)\tau_T(k) + R(k)\tau_R(k) \tag{10.2.17}$$

Here $T(k)$ and $R(k)$ are the probabilities for transmission and reflection, respectively. From (10.2.14) and (10.2.16) we get an indication that the phase times obey this sum rule (they also do in a more general case). We recall, of course, that the phase times were dismissed for other reasons as measures of the traversal and reflection times in the previous section. None of the alternative *real* (as opposed to complex) quantities that have been proposed to measure the traversal and reflections time obey (10.2.17). The implication of this fact is not quite clear yet, although Hauge and Støvneng[16] conclude that none of the "times" discussed in this chapter, however useful they may be as characteristic interaction times, measure the "intrinsic" traversal time.

In a more general case than discussed above, we may have an incoming flow of electrons in channel i being scattered, possibly inelastically, into channels j. In this case the matrix element of the scattering matrix S_{ij} gives not only the relative phase but also the relative amplitude for scattering a wave from channel i into channel j. Eisenbud[6,7] found the corresponding phase delay time to be

$$\Delta\tau_{ij}^{\phi} = \text{Re}\left[-i\hbar(S_{ij})^{-1}\frac{d}{dE}S_{ij}\right] \tag{10.2.18}$$

which is a generalization of (10.2.9). Taking the real part of (10.2.18) ensures, of course, that the energy derivative is taken only with respect to the phase, not the amplitude, of S_{ij}. Smith[7] and others,[32] in their early work, showed that the dwell time defined as in (10.2.15) also is related in this more general case to a weighted average of the phase times. The weight is the probability $|S_{ij}|^2 = S_{ij}^* S_{ij}$ of being scattered from channel i to channel j. By subtracting the dwell time in the absence of scattering, we can define an *excess* dwell time $\Delta\tau_{\text{dwell},i}$ for a particle initially in channel i. One finds[7]

$$\Delta\tau_{\text{dwell},i}(R,k) = \langle\Delta\tau_{ij}^{\phi}\rangle_{\text{av over }j} = \left[-i\hbar\sum_j S_{ij}^*\frac{d}{dE}S_{ij}\right] \tag{10.2.19}$$

The right-hand side of (10.2.19) is guaranteed to be real because the scattering matrix $\mathbf{S}$ is unitary ($\mathbf{SS}^\dagger = 1$). The relation between phase and dwell times under even more general assumptions continues to be of interest to mathematical physicists.[33] We consider next the relevance of the dwell time and the phase times for a metastable state.

10.2.4. Lifetime of a Metastable State

We now return to resonant scattering and the question of the lifetime of a metastable state. For resonant scattering the phase shift can be written as[20]

$$\delta_l = \delta_l^{(0)} - \tan^{-1}\frac{\Gamma}{2(E - E_0)} \tag{10.2.20}$$

where the explicit energy dependence in the second term dominates close to resonance. One finds that the phase delay times of (10.2.9) are approximately independent of the scattering channel:

$$\Delta\tau^\phi \approx \hbar\frac{\Gamma}{(E - E_0)^2 + (\Gamma/2)^2} \tag{10.2.21}$$

Obviously this result is unchanged by averaging over scattering channels. Hence, the excess dwell time defined in the previous section is given by the same expression. When evaluated at the resonance energy $E = E_0$, (10.2.21) gives $\Delta\tau^\phi = 4\hbar/\Gamma$. However, the energy of the resonant state is not sharply defined. Fourier-transforming $\exp(i\varepsilon t)$ with ε given by (10.2.1), we find a Lorenzian line shape of the resonance. Averaging the phase delay time as a function of E with respect to this line shape, one finds

$$\Delta\tau^\phi = \int_0^\infty \frac{dE}{\pi}\,\frac{\frac{1}{2}\Gamma}{(E - E_0)^2 + (\frac{1}{2}\Gamma)^2}\,\frac{\hbar\Gamma}{(E - E_0)^2 + (\Gamma/2)^2} = \frac{2\hbar}{\Gamma} \tag{10.2.22}$$

This gives the excess time the wave packet (the electron) takes in going from and to the asymptotic regions. It is twice the time required for a prepared metastable state to decay, as given by (10.2.3).

To summarize, we can think of the lifetime of a resonant state as given either by the width of the resonant energy level, the phase delay time, or an excess dwell time. Furthermore a lifetime also follows[4] from the Heisenberg uncertainty principle; if the electron spends a limited time τ in the scattering region, its energy is only determined to a finite precision, that is, to within Γ. All answers seem to converge to the same result.

The lifetime (or decay rate) of a resonant state can be measured. Recently decay rates of electrons trapped in metastable quantum well states in $GaAs/Al_xGa_{1-x}As$ double-barrier heterostructures were measured using a time-resolved photoluminescence technique.[34] Electrons trapped in the conduction band of the GaAs quantum well can decay either by recombination with holes or by tunneling through one of the $Al_xGa_{1-x}As$ barriers.

For sufficiently thin barriers the latter process dominates, and a decay rate due to tunneling could be determined. It was found to agree with the calculated width of the resonant level Γ. Hence, there is no controversy at this point. The measured lifetimes depend on the system studied, but are about 100 ps. This is a long time compared with the time it would take an electron at the Fermi level in the conduction band of the doped GaAs to pass the scattering region, if the Al_xGa_{1-x} barriers had not been there. This time is around 10 fs and should give the approximate scale for the traversal time. The huge difference between the two numbers explains why the lifetime of these metastable states can be measured and why different approaches tend to give the same answer; the lifetime is dominated by the time spent in the quantum well; precisely how quickly they pass through a barrier has no numerical significance.

Another system for which *both* a traversal time and the lifetime of a metastable state have been measured is a Josephson junction.[15] Here it is a macroscopic state, rather than an electron that tunnels out of a long-lived state, as will be discussed in a later section. In this case the *traversal* time is about 100 ps, while the lifetime of the metastable state is approximately 10 μs. Both times were measured by Esteve *et al.*[15] Their landmark experiment has given the best information about a traversal time to date.

10.3. CLOCKS FOR MEASURING TRAVERSAL TIMES

We have found that the phase times and the dwell time derived from scattering theory are inadequate for measuring the traversal time for tunneling. Neither gives the answer to the question: How long did an electron, *which eventually penetrated through a barrier*, spend inside it? The reason in the first case is that phase times are asymptotic times that cannot distinguish between the time spent inside the barrier or any region of space, and the time spent in propagating to the asymptotic region. The dwell time on the other hand, provides a measure of the time spent in a specified region, but averages over all scattering channels. It does not isolate the contribution from electrons that are eventually transmitted through a barrier from that of electrons that are reflected from it.

We clearly have to find a different method for measuring the traversal time. It is useful to recall that times are measured by clocks which mark the passing of time at a well-defined frequency. Is it possible to find a suitable clock to measure traversal time? We need some dynamical interaction, characterized by a steady frequency, which influences the electron as it tunnels. It turns out that one can think of several such clocks. We shall discuss some of them: a time-modulated barrier, the Larmor clock as modified by Büttiker, polarization modes coupling to the tunneling electrons, and a Josephson junction shunted by a tunable transmission line (delay line).

10.3.1. Time-Modulated Barrier

In an influential 1982 paper,[9] Büttiker and Landauer (BL) proposed a thought experiment in which a time-modulated barrier was used to measure the traversal time. Their idea was that an electron passing through the barrier in a much shorter time than the modulation period sees a static barrier corresponding to a particular amplitude of the modulation, while a slowly tunneling electron interacts with several modulation periods of the oscillating barrier. If the modulation frequency is varied, a crossover between two distinct types of electron-barrier interactions should therefore emerge. The inverse of the crossover frequency gives the traversal time.

An electron tunneling through a time-modulated barrier may absorb, or emit, modulation quanta. Hence, in momentum (or energy) space the reflected and transmitted parts of a wave function used to represent the tunneling electrons will have a main feature at the initial energy E and sidebands at $E \pm n\hbar\omega$, as shown in Figure 10.1. As explained in more detail in Ref. 10, BL used the sideband intensities of the transmitted current to monitor the crossover. In this section we briefly review, and somewhat generalize, their discussion (see also Chapter 9).

Following BL we consider a time-modulated square barrier

$$V(x) = \begin{cases} V_0 + V_1 \sin(\omega t), & |x| < d/2 \\ 0 & \text{elsewhere} \end{cases} \qquad (10.3.1)$$

where d is the barrier thickness, $V_0 \equiv \hbar^2 k_0^2 / 2m$ is the average potential

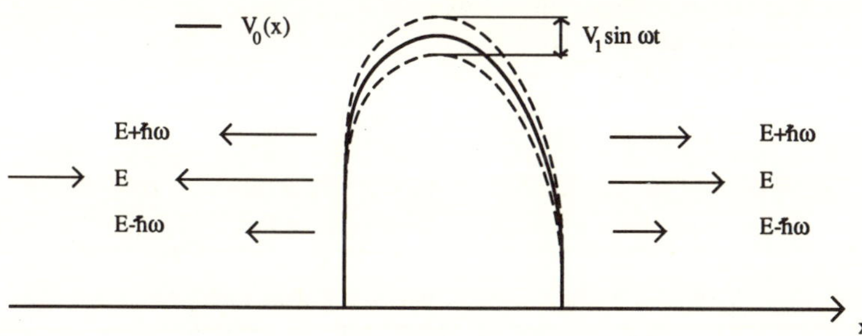

FIGURE 10.1. A time-modulated barrier. An electron interacting with this potential may emit or absorb modulation quanta $\hbar\omega$. In energy space the reflected and transmitted parts will therefore have a main feature at the initial energy E and sidebands at $E \pm n\hbar\omega$. The first sidebands are indicated.

height, V_1 is the modulation amplitude, and ω is the modulation frequency. For simplicity, let the barrier first be opaque so that $k_0 d \gg 1$. In this case Büttiker and Landauer found that the intensities of the sidebands $T_\pm[\equiv T(E \pm \hbar\omega), E = \hbar^2 k^2/2m$ being the energy of the incoming particle] are

$$T_\pm = (V_1/2\hbar\omega)^2 (e^{\pm\omega md/\hbar\kappa} - 1)^2 T(E) \tag{10.3.2}$$

where $\kappa = (k_0^2 - k^2)^{1/2}$ and $T(E) = |D(E)|^2$ are the transmission probability and the transmission amplitude of the static barrier, respectively. It is furthermore assumed that $V_1 \ll \hbar\omega$.

Next, BL define an asymmetry function

$$f(\omega) = \frac{T_+ - T_-}{T_+ + T_-} \tag{10.3.3}$$

which for the opaque barrier acquires the simple form

$$f_{\text{opaque}}(\omega) = \tanh(\omega md/\hbar\kappa) \tag{10.3.4}$$

Thus, the asymmetry function is characterized by a single quantity with the dimension of time, in this case

$$\tau^{\text{BL}}(d, k) = md/\hbar\kappa \tag{10.3.5}$$

which separates characteristic low- and high-frequency behaviors. Therefore BL identify τ^{BL} as the traversal time for tunneling. It follows that the corresponding speed of the tunneling electron is $d/\tau^{\text{BL}} = \hbar\kappa/m$. These estimates of the traversal time and the associated speed are the *same* as those Jonson found in his analysis of the dynamical image potential for a tunneling electron in the same opaque limit.[3] We shall return to this point.

Let us now consider general barriers. Büttiker and Landauer, in their original work,[9,10] discuss the crossover behavior of the asymmetry function *explicitly* only in the opaque limit, but their argument suggests the approach might have a more general validity. As pointed out, for instance, by Hauge *et al.*[16,35] and Jauho *et al.*,[36] the asymmetry function (10.3.3) can no longer be characterized by a single parameter of dimension time in the general case and is therefore of little use for defining a traversal time. Büttiker[17] has unambiguously stated that the crossover argument can only be applied to the case of opaque barriers (i.e., to cases where the WKB approximation applies).

Actually, in Ref. 10 a different approach was used to generalize the traversal time for opaque barriers (10.3.5) to general tunneling barriers. Büttiker and Landauer observed that in the static limit $\omega \to 0$ the sideband intensities (10.3.2) for opaque barriers can be written as

$$T_\pm = (V_1 \tau^{\text{BL}}/2\hbar)^2 T(E) \tag{10.3.6}$$

with τ^{BL} given by (10.3.5). For a general barrier the sideband intensities can be written in the same form as (10.3.6), provided τ_{BL} is replaced by

$$\tau_T^{\mathrm{BL}} \equiv \hbar \left| \frac{d \ln[D(E, \bar{V})]}{dV} \right| \qquad (10.3.7)$$

In (10.3.7) $D(E, \bar{V})$ is the complex transmission amplitude for an electron of energy E penetrating a barrier of average height $\bar{V}$. As shown by Jauho and Jonson,[36] with additional details provided in Appendix 1.2, this result is valid for an arbitrarily shaped potential $V_0(x)$ defined in the interval $-d/2 < x < d/2$, whose average is $\bar{V}$. In this sense it is a generalization of the discussion in Ref. 10 (an equivalent generalization using a different approach has been given by Støvneng and Hauge[35]). The generalized form (10.3.7) of the BL traversal time was first asserted by Leavens and Aers,[37,38] using an analogy with the BL result for a simple square barrier. In passing, we note that the detailed discussion in Appendix 1.2 also serves to refute some technical objections to Ref. 10 raised by Collins *et al.*[24] Finally, without going into details (see Appendix 1.2 though), we observe that a "reflection" time τ_R^{BL} can be defined[13] by an analogous study of the adiabatic limit of the reflection amplitudes $R_\pm$.

The traversal time defined by (10.3.7) emerges from an analysis of the time-modulated barrier in the adiabatic limit of vanishing modulation frequency. It is not supported by the physically appealing crossover argument, which refers to a finite crossover frequency and is valid only for opaque barriers. In fact, it is not very clear at this point how (10.3.7) can be justified as a traversal time for a general barrier (although additional support will come from Büttiker's modified analysis of the Larmor clock, reviewed in the next section). The main justification for (10.3.7), in our opinion, comes from the fact that it smoothly interpolates between the opaque barrier result—which is supported by the crossover argument—and the result for an electron whose energy is much larger than the barrier height. The latter result can be understood from classical physics using the correspondence principle.

It is interesting to observe that by adding the static part of the transmitted wave function, $D \exp(ikx - iEt/\hbar)$, and the one-phonon sideband terms, $D_\pm \exp(ik_\pm x - iE_\pm t/\hbar)$, the total transmitted wave function, in the limit of low frequency and amplitude of the modulation, can be written as a single term,[36]

$$\psi(x, t) = D(E, \bar{V}) \exp\{i[k(\tau_T^{\bar{V}})x - E(\tau_T^{\bar{V}})t/\hbar] + \eta\} \qquad (10.3.8)$$

(We have changed the notation from $\tau_D^{\bar{V}}$, used in Ref. 36 and in Chapter 9, to $\tau_T^{\bar{V}}$ in order to make the relation to Transmission clearer.) This result, which is valid for a general form of the potential barrier if $t \ll 1/\omega$ and $x \ll v(k)/\omega$, is obtained by relating the sideband amplitudes $D_\pm$ to the *static* transmission (D) and reflection (A) amplitudes. As reported in Ref. 36, and shown in detail in Appendix 1.2, a generalization of the discussion in Ref. 10 gives

$$D_{\pm} = -\frac{i}{2}\frac{V_1}{2\hbar\omega} D(E \pm \hbar\omega, \bar{V})$$

$$\times \left(\left(1 + \frac{k}{k_{\pm}}\right)\left[1 - \frac{D(E, \bar{V})}{D(E, \bar{V} \mp \hbar\omega)}\right] \right.$$

$$\left. + \left(1 - \frac{k}{k_{\pm}}\right) A(E, \bar{V})\left[1 - \frac{D(E, \bar{V})}{D(E, \bar{V} \mp \hbar\omega)} \frac{A(E, \bar{V} \mp \hbar\omega)}{A(E, \bar{V})}\right] \right)$$

$$(10.3.9)$$

where $\bar{V}$ is the average height of the barrier, $\bar{V}_{\mp} = \bar{V} \mp \hbar\omega$, $E_{\pm} = E \pm \hbar\omega$, and $k_{\pm} = \sqrt{2mE_{\pm}/\hbar^2}$. The sideband intensities are $T_{\pm} = |D_{\pm}|$.[2] Upon expanding (10.3.9) to lowest order in ω and summing the static and sideband terms as described above, one obtains (10.3.8), where

$$E(\tau_T^{\bar{V}}) = E + V_1 \omega \tau_T^{\bar{V}} \qquad (10.3.10)$$

and

$$k(\tau_T^{\bar{V}}) = k + \frac{V_1}{2E}(\omega\tau_T^{\bar{V}})k \qquad (10.3.11)$$

The complex quantity

$$\tau_T^{\bar{V}} = i\hbar \frac{d \ln D(E, \bar{V})}{d\bar{V}} \qquad (10.3.12)$$

has the dimension of time and is closely related to the complex times introduced earlier by Sokolovski and Baskin[14] and Leavens and Aers.[37,38,39] Note that this time is related to other times that have been discussed in the literature; for the symmetric barrier considered in Appendix 1.2, the real and imaginary parts of (3.3.12) are[16,17] the dwell time τ_{dwell} and the so-called Büttiker time[13] τ_z to be discussed further in the next section. We have $\tau_D^{\bar{V}} = \tau_{\text{dwell}} - i\tau_z$. The Büttiker–Landauer traversal time of (10.3.7) is obviously related to (10.3.12) as $\tau_T^{\text{BL}} = |\tau_T^{\bar{V}}|$.

In deriving (10.3.8)–(10.3.12), we took the modulating potential to be $V_1 \sin(\omega t)$; that is, $V_1\omega t$ for $t \ll 1/\omega$. From (10.3.10) it therefore is tempting to interpret the modified energy and momentum of the transmitted wave function to be the result of an adiabatic interaction between the tunneling electron and the rising barrier during a traversal time $\tau_T^{\bar{V}}$. Such an interpretation is not meaningful, however, as $\tau_T^{\bar{V}}$ is complex and any measurable traversal time must certainly be real. We do not agree with the proposition[14,39] that the traversal time can be a complex quantity.

A complex time equivalent to (10.3.12) appeared in a different context. Sokolovski and Baskin[14] proposed a formal generalization of the expression

$$\tau_{cl}^{\Omega}[x(t)] = \int_{t_1}^{t_2} dt \int_{\Omega} dx\, \delta(x - x(t)) \tag{10.3.13}$$

which measures the time spent by a classical particle in the arbitrary region Ω in the time interval $t_1 < t < t_2$. Here $x(t)$ is the one-dimensional classical path from $x(t_1)$ to $x(t_2)$. A natural generalization invokes Feynman's path integral formulation of quantum mechanics[31] and gives

$$\tau^{\Omega}(x(t_1), x(t_2)) = \langle \tau_{cl}^{\Omega}[x(\cdot)]\rangle_{\text{paths}} \tag{10.3.14}$$

Here $x(\cdot)$ is an arbitrary path between two given end points $x(t_1)$ and $x(t_2)$. In general, τ_{Ω} is complex. For the one-dimensional case discussed above and in Appendix 1.2, the time τ_T^{Ω} associated with a transmitted electron can be written as

$$\tau_T^{\Omega} = i\hbar \int_{\omega} dx\, \frac{\delta \ln D}{\delta V(x)} \tag{10.3.15}$$

where D is the complex transmission amplitude as above. An analogous time τ_R^{Ω} involving a functional derivative of the complex-reflection amplitude A can also be written. Equation (10.3.15) is equivalent to (10.3.12); the difference is in notation only.

The complex times τ^{Ω} (and therefore $\tau^{\bar{V}}$) can be defined for any volume Ω. From the definition (10.3.15) [or (10.3.12)] it is clear that these complex times are additive in the sense that (in one dimension)

$$\tau_T^{\Omega}(x_1, x_3; k) = \tau_T^{\Omega}(x_1, x_2; k) + \tau_T^{\Omega}(x_2, x_3; k) \tag{10.3.16}$$

Sokolovski and Baskin[14] proved another interesting relation connecting τ^{Ω} to the dwell time:

$$\tau_{\text{dwell}}(x_1, x_2; k) = T(k)\tau_T^{\Omega}(x_1, x_2; k) + R(k)\tau_R^{\Omega}(x_1, x_2; k) \tag{10.3.17}$$

This has exactly the form of the sum rule in (10.2.17) for mutually exclusive events. We maintain, however, that the traversal time cannot be a complex quantity.

The procedure of Sokolovski and Baskin is purely formal, and the interpretation of their complex time is not clear from its definition. In particular, the requirement that τ^{Ω} reduces to the classical result τ_{cl}^{Ω} can be achieved in a number of ways, as they themselves pointed out. For instance, by defining τ^{Ω} using the absolute value of the right-hand side of (10.3.14), it can be identified with τ^{BL} of (10.3.7). Because it was derived in a more physical context, some insight into the role of the equivalent complex time

$\tau_T^{\bar{V}}$ of (10.3.12) can be gained by calculating the transmitted electron and current densities. If x_0 and x are points beyond the barrier and $t_0, t \ll 1/\omega$, one finds to lowest order in ω (cf. Appendix 1.2)

$$\rho(x, t) = \rho(x_0, t_0)e^{-2V_1(\omega\tau_z)[(t-t_0)-(x-x_0)/v(k)]/\hbar} \tag{10.3.18}$$

and

$$J(x, t) = v(k)\rho(x, t) + \frac{V_1}{2E}(\omega\tau_{\text{dwell}})v(k)\rho(x_0, t_0) \tag{10.3.19}$$

One notes that the decrease in the transmitted density with time is related to the Büttiker time τ_z and to the imaginary part of $\tau_T^{\bar{V}}$. The result for the transmitted current density is a sum of two terms. The first simply reflects the change of the transmitted density with time, while the second comes about because the change in density inside the barrier is associated with a change in current (to conserve charge). The latter term can also be interpreted from (10.3.11) as arising from an adiabatic change of the momentum of the tunneling electron while it interacts with the barrier during the dwell time τ_{dwell}, given by the real part of $\tau_T^{\bar{V}}$. In light of (10.3.18) and (10.3.19), where the particle and current densities develop differently with time, we speculate that experiments probing different aspects of the transmitted wave function (10.3.8) may yield different, but complementary, information about the tunneling process. A related discussion of the nonuniqueness of the tunneling time appears in Ref. 40.

The result for the transmitted current discussed above is in a sense reminiscent of the results of Büttiker's analysis of the so-called Larmor clock.[13] This is the topic of the next section.

10.3.2. The Larmor Clock

In the mid-1960s, Baz'[8] proposed a thought experiment invoking the Larmor precession of a particle's spin in a magnetic field to measure the duration of a quantum mechanical collision event. His idea was to study a beam of spin-$\frac{1}{2}$ particles traveling in the y-direction and scattering from a potential $V(r)$ of finite range r_0. An infinitesimal homogeneous magnetic field $\mathbf{B} = B_0\hat{z}$ confined inside a sphere of radius $r_B > r_0$ would make the spins, initially polarized in the x-direction, precess with the Larmor frequency $\omega_L = g\mu_B B_0$ in the plane perpendicular to the magnetic field.[20] (g is the gyromagnetic ratio, μ_B is the Bohr magneton, and the infinitesimal strength of the magnetic field makes second-order effects of the field negligible.) Hence, the average spin of the transmitted particles would acquire a y-component proportional to the time spent inside the sphere $r < r_B$. The precession angle $\theta_{i,\perp}$ for particles emerging in a particular scattering channel

i could, in principle, be measured and

$$\tau_{yi}^{L}(r_B, k) \equiv \lim_{\omega_L \to 0} \frac{\theta_{i,\perp}}{\omega_L} = \lim_{\omega_L \to 0} \frac{\langle s_y \rangle_i}{-\frac{1}{2}\hbar\omega_L} \qquad (10.3.20)$$

is interpreted as the time spent inside the sphere. (We use the conventional subscript y rather than $\perp$ for this Larmor time.) Somewhat later, Rybachenko[41] applied this *Larmor clock* method to a one-dimensional tunneling problem to determine the time spent by a tunneling particle in the potential-barrier region $-d/2 < y < d/2$, or more generally in the region $x_1 < x < x_2$ including the barrier (cf. Figure 10.2).

More recently Büttiker modified Baz's analysis of the Larmor clock in the context of a one-dimensional barrier.[13] He pointed out that incoming electrons, even if they have no spin component in the z-direction, can be thought of as a superposition of a spin-up and a spin-down state. Because of the Zeeman interaction, these states have different energies and therefore the energy dependence of the barrier transmission probability leads to preferential transmittance of the spin-up component. As a result the average spin of the transmitted (or reflected) particles will have a nonzero z component. In analogy with (10.3.20) Büttiker used the corresponding tilt angle $\theta_{\parallel,i}$, in the plane parallel to the magnetic field, to define another Larmor time which we shall call the Büttiker time,

$$\tau_{zi}^{L}(r_B, k) \equiv \lim_{\omega_L \to 0} \frac{\theta_{i,\perp}}{\omega_L} = \lim_{\omega_L \to 0} \frac{\langle s_z \rangle_i}{\frac{1}{2}\hbar\omega_L} \qquad (10.3.21)$$

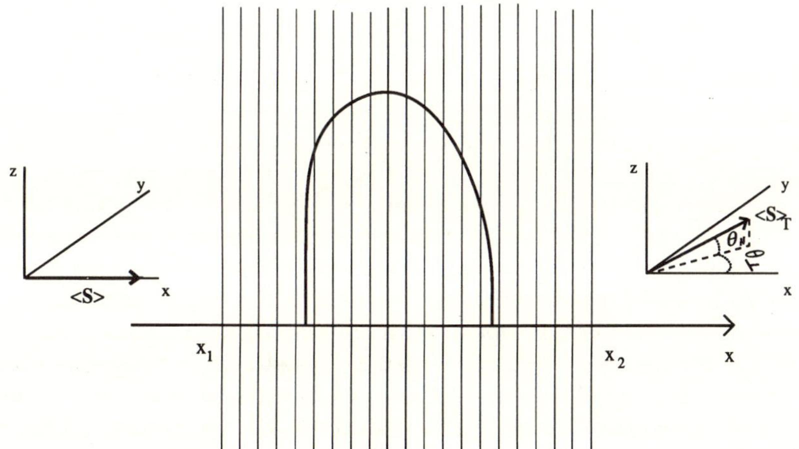

FIGURE 10.2. The Larmor clock in one dimension. A weak homogeneous magnetic field, $\mathbf{B} = B\hat{z}$, is confined to the interval $x_1 < x < x_2$, which here includes the barrier. The spin **S** of particles entering the field are polarized in the x-direction. Beyond $x = x_2$, the average spin of transmitted particles has been tilted a (small) angle $\theta_\perp$ in the plane perpendicular to the magnetic field and a (small) angle $\theta_\parallel$ in a plane parallel to it.

The interpretation of this quantity as a time is less obvious than for the case of (10.3.20). We shall discuss this point further.

The two tilt angles in planes parallel and perpendicular to the field combine to give a total tilt angle of the averaged spin away from its initial direction. This led Büttiker to define a time using this total tilt angle as

$$\tau_i^{\text{BL}} = \lim_{\omega_L \to 0} \frac{1}{\omega_L} \sqrt{\theta_{\perp,i}^2 + \theta_{\parallel,i}^2} = \sqrt{(\tau_{yi}^L)^2 + (\tau_{zi}^L)^2}, \qquad i = T, R \quad (10.3.22)$$

In the one-dimensional case there are, of course, only two "scattering channels," corresponding to transmission (T) and reflection (R), respectively.

For the simple square barrier Büttiker showed[13] that the traversal time (10.3.22) is identical to the BL traversal time τ_T^{BL} in (10.3.7), derived from the adiabatic limit of the sideband intensities. The generalization of this result to an arbitrary one-dimensional potential was provided by Leavens and Aers.[37] Hence, τ_T^{BL} is related to the sensitivity of the transmission *amplitude*—both of its phase and its absolute value—to a variation in the *barrier potential*. A comparison with the phase delay times (10.2.9), (10.2.18) shows that these are related to the sensitivity of (only) the phase to a variation of the particle energy. Pollak and Miller[42] have proposed a new interpretation of time in scattering theory which partly bridges this discrepancy; in addition to the phase delay time (10.2.18), they introduced a *quantal imaginary time*

$$\tau_{\text{Im}} \equiv \text{Im}\left[-i\hbar(S_{ij})^{-1} \frac{d}{dE} S_{ij} \right] = -\frac{\hbar}{2} \frac{1}{|S_{ij}|^2} \frac{d|S_{ij}|^2}{dE} \quad (10.3.23)$$

which is related to the Büttiker time τ_z. They showed that Smith's phase delay time (10.2.18) and their time (10.3.23) can be interpreted as the real and imaginary part of a flux–flux correlation function.[43] They furthermore suggested[43] that τ_{Im} is the relevant time scale for determining the interaction of the tunneling particle with additional ("perpendicular") degrees of freedom. At least in the opaque limit, this interpretation is consistent with our own analysis[3] of the dynamic image potential for a tunneling electron, to be reviewed in the next section.

Pollak and Miller do not propose any interpolation scheme such as (10.3.22) between the phase time and the quantal imaginary time, but they note that (10.3.23) dominates for opaque barriers, where (for the one-dimensional, square-barrier case) it reduces to the same limit (10.3.5) as the BL time defined in (10.3.12). In the opposite limit of almost transparent barriers, the phase delay time dominates.

The phase delay times and the "imaginary" time of Pollak and Miller are related to the sensitivity of the phase and the absolute value of a scattering matrix element to a variation of the particle energy respectively.

The BL time, on the other hand, is related to a sensitivity of essentially both these quantities to a variation in the potential-barrier height. Leavens and Aers[37] have investigated the numerical significance of this difference and found it to be rather unimportant except for low particle energies. The reason is that the dominant dependence on particle energy and potential enters in the combination $V(x) - E$.

Leavens and Aers[39] have introduced an interesting generalization of Baz's original thought experiment. By confining the magnetic field to a limited region—possibly inside the barrier—they define *local* Larmor times (see also Ref. 44 where the same authors proposed a different thought experiment to achieve the same generalization). The local Larmor times τ_{yi}^L and τ_{zi}^L are directly related, respectively to the real and complex parts of the complex time $\tau_i^{\bar{V}}$ (or τ_i^Ω). From (10.3.16) it therefore follows that they are additive:

$$\tau_{yi}^L(x_1, x_3; k) = \tau_{yi}^L(x_1, x_2; k) + \tau_{yi}^L(x_2, x_3; k)$$

$$\tau_{zi}^L(x_1, x_3; k) = \tau_{zi}^L(x_1, x_2; k) + \tau_{zi}^L(x_2, x_3; k), \qquad i = T, R \tag{10.3.24}$$

This in turn implies that the BL time (10.3.7) adds as the absolute values of complex numbers:

$$\tau_i^{\mathrm{BL}}(x_1, x_3; k)$$
$$= \sqrt{\tau_{yi}^L(x_1, x_3; k)^2 + \tau_{zi}^L(x_1, x_3; k)^2}$$
$$= \sqrt{(\tau_{yi}^L(x_1, x_2; k) + \tau_{yi}^L(x_2, x_3; k))^2 + (\tau_{zi}^L(x_1, x_2; k) + \tau_{zi}^L(x_2, x_3; k))}$$
$$\tag{10.3.25}$$

We can learn more from the relation with the complex time τ^Ω. Considering the real and imaginary parts of (10.3.18) separately, one finds

$$\tau_{\mathrm{dwell}}(x_1, x_2; k) = T(k)\tau_{y,T}^L(x_1, x_2; k) + R(k)\tau_{y,R}^L(x_1, x_2; k)$$

$$0 = T(k)\tau_{z,T}^L(x_1, x_2; k) + R(k)\tau_{z,R}^L(x_1, x_2; k) \tag{10.3.26}$$

From the definition of τ_z^L it is obvious that (10.3.26) expresses the fact that angular momentum (or momentum if we refer to the time-modulated barrier) is conserved.[13] The Larmor times τ_{yi}^L apparently obey the sum rule for mutually exclusive events. Could $\tau_{yT}^L(x_1, x_2; k)$ possibly measure the traversal time? No, when x_1 and x_2 are in the asymptotic region, Falck and Hauge[29] have shown that this Larmor time is equivalent to the phase time, which we know does not measure the traversal time. When the Larmor clock measures a "local" rather than an asymptotic time, both Leavens and Aers[45] and Hauge and Støvneng[16] have constructed a special example in which $\tau_{yT}^L(x_1, x_2; k)$ unambiguously does not measure the traversal time.

The concept of a local time, referring to a portion inside a potential barrier, say, is interesting in the context of inelastic tunneling. Kotler and

Nitzan[46] showed for an opaque one-dimensional barrier that the probability for a tunneling electron to make a transition between two states because of an interaction that couples them in the interval $x_1 < x < x_2$ *inside* the barrier is proportional to the time $m(x_2 - x_1)/\hbar\kappa$. This is a generalization of (10.3.6). We shall now go on to analyze the interaction between a tunneling electron and other degrees of freedom in some detail for the case of the dynamical image potential.

10.3.3. The Dynamical Image Potential for Tunneling Electrons

We know from classical electrostatics that a test charge q in vacuum at a distance x outside a metal surface feels an image potential

$$V_{\text{im}}(x) = -q^2/4x \tag{10.3.27}$$

The name image potential derives from the fact that the potential and electric field in the vacuum can be obtained by putting a fictitious "image" charge inside the metal. This is just a convenient computational trick, of course; the image potential is really due to polarization of the metal by the electric field of the test charge. Except very near the metal surface the dominant polarization modes are the surface plasmon modes.[47] These are collective charge oscillations at the surface, which in the long-wavelength limit have frequency ω_s. In this section we shall show how the surface plasmons can be used as a clock for measuring the speed of a tunneling electron, and how from the speed one can determine a traversal time. For this we need a quantum mechanical theory.

In a first attempt to go beyond the classical theory in describing the interaction of an electron with a metal surface, we consider an electron at the fixed position (x, ρ) in the vacuum $(x > 0)$. Assuming it interacts with dispersionless surface plasmons, the Hamiltonian for this model system is

$$H = \sum_{\mathbf{k}} \hbar\omega_s a_{\mathbf{k}}^\dagger a_{\mathbf{k}} + \sum_{\mathbf{k}} \Gamma_{\mathbf{k}}(a_{\mathbf{k}}^\dagger + a_{\mathbf{k}})e^{i\mathbf{k}\cdot\rho} \tag{10.3.28}$$

where[48]

$$\Gamma_{\mathbf{k}}(x) = \sqrt{\pi e^2 \hbar\omega_s/kA}\, e^{-kx} \tag{10.3.29}$$

and $\mathbf{k}$ is the two-dimensional wave vector of the surface plasmons. We solve (10.3.28) by simply completing the square:[49]

$$H = \sum_{\mathbf{k}} \hbar\omega_s \tilde{a}_{\mathbf{k}}^\dagger \tilde{a}_{\mathbf{k}} - \sum_{\mathbf{k}} \frac{\Gamma_{\mathbf{k}}^2}{\hbar\omega_s} \tag{10.3.30}$$

The last term is the self-energy of the electron. Its physical origin is that the electric field of the electron polarizes surface plasmon modes. These modes, when polarized, create electric fields which act back on the electron.

The self-energy term in (10.3.30), when evaluated using the interaction (10.3.29), gives precisely the classical image charge result of (10.3.27).

$$-\sum_{\mathbf{k}} \frac{\Gamma_{\mathbf{k}}^2}{\hbar\omega_s} = -\pi e^2 \int \frac{d^2k}{(2\pi)^2} \frac{e^{-2kx}}{k} = -\frac{e^2}{4x} \qquad (10.3.31)$$

Expressed differently, the surface plasmons are a collection of harmonic oscillators which are displaced to new equilibrium positions by the field of the electron. The energy gained is given by (10.3.31).

Now, consider the dynamics! Imagine that we switch on the interaction in (10.3.28) at a time $t = 0$. How long does it take for the surface plasmon "oscillators" to reach their new equilibrium positions? Clearly, the time scale for the required displacements of the oscillators is given by the inverse of the surface plasmon frequency. We want to include the dynamics of the surface plasmons in an improved model. We also need to incorporate the fact that in realistic cases the electron will have a finite velocity. If this velocity is large, there may not be enough time for the polarization modes to adjust to the changing position of the electron. Arguments like this led Mahan and others[47] in the early 1970s to study the *dynamical image potential.* They continued to treat the electron as a classical particle, but allowed it to move with a velocity v, which is a parameter in the theory. A straightforward solution[47] of the time-dependent Schrödinger equation for a modified Hamiltonian given by (10.3.28) and (10.3.29), where now $x = x(t) = x(0) - vt$, and $x(0) = \infty$, gives a semiclassical dynamical image potential

$$V_{\text{dyn im}} = -\frac{e^2}{4x} f\!\left(\frac{v}{2x\omega_s}\right), \qquad f(s) \equiv \int_0^\infty d\alpha \frac{e^{-\alpha}}{1 + \alpha^2 s^2} \qquad (10.3.32)$$

The interpretation of this result is obvious. The extent to which the image potential develops to its full classical value depends on the ratio between two time scales: one related to the electron velocity and the other to the inverse frequency of the surface plasmons. If the velocity of the electron is too large, there is not enough time for the polarization modes at the metal surface to adjust to the changing position of the electron. The limiting behavior of the function f in (10.3.32) is consistent with this interpretation

$$\lim_{v\to 0} f\!\left(\frac{v}{2x\omega_s}\right) = 1, \qquad \lim_{v\to\infty} f\!\left(\frac{v}{2x\omega_s}\right) = 0 \qquad (10.3.33)$$

Hence, for zero velocity the classical image potential is recovered, while the dynamical potential is reduced as the velocity increases and, in fact, disappears in the limit of infinite velocity.

In an experiment such as thermionic emission of a charged particle, say an electron from a metal surface, the outcome depends on the potential barrier it has to overcome. As the image potential is part of this barrier, its reduction from a full classical value can be measured in principle. At least in the simple theory leading to (10.3.32), the reduction depends on the speed of the particle, which in turn can be used to calculate the time spent by the particle in the region where the image potential is important. Surely, a more sophisticated theory would be needed, but we are content at this point to establish the principle that the dynamical image potential can be used as a clock. Can this clock be used for tunneling electrons also? Clearly the electron can no longer be treated classically. The theory has, however, been generalized to treat the electron quantum mechanically, first by Jonson[3] and later by others (see, for instance, Refs. 50 and 51). We shall briefly review the theory of Ref. 3, where a Schrödinger equation

$$\left\{ -\frac{\hbar^2}{2m} \nabla^2 + V(x) + V_{xc}(x, i) \right\} \Psi_i(\mathbf{r}) = E_i \Psi_i(\mathbf{r}) \qquad (10.3.34)$$

was solved for a many-body "exchange-correlation" potential:

$$V_{xc}(x, i) = \int d\mathbf{r}' \, \Sigma \, (\mathbf{r}, \mathbf{r}'; E_i) \frac{\Psi_i(\mathbf{r}')}{\Psi_i(\mathbf{r})} \qquad (10.3.35)$$

defined in terms of the eigenfunctions Ψ_i and a nonlocal self-energy Σ. For an electron moving in a one-dimensional single-particle potential $V(x)$, the eigenfunctions are separable with a simple plane wave in the "parallel" direction:

$$\Psi_i(\mathbf{r}) = \frac{1}{\sqrt{A}} e^{i \mathbf{p}_\| \cdot \boldsymbol{\rho}} \psi_i(x), \qquad E_i = \frac{\hbar^2 p_\|^2}{2m} + E_{i,x} \qquad (10.3.36)$$

In this case the xc-potential is conveniently expressed in a mixed real space and parallel momentum representation as

$$V_{xc}(x, \mathbf{p}_\|, E_i) = \int dx' \sum_{\mathbf{k}} \Gamma_{\mathbf{k}}(x) \Gamma_{-\mathbf{k}}(x') G_{\mathbf{k}+\mathbf{p}_\|}(x, x'; E_i - \hbar\omega_s) \frac{\psi_i(x')}{\psi_i(x)} \qquad (10.3.37)$$

Here $\Gamma_{\mathbf{k}}(x)$ is the interaction of (10.3.29), and the partially Fourier-transformed Green's function obeys an effective one-dimensional Schrödinger equation

$$\{\varepsilon - H(x')\} G_{\mathbf{p}}(x, x'; \varepsilon) = \delta(x - x'),$$

$$H(x') = \frac{\hbar^2 p^2}{2m} - \frac{\hbar^2}{2m} \frac{\partial^2}{\partial x'^2} + V(x') + V_{xc}(x', \varepsilon + \hbar\omega_s) \qquad (10.3.38)$$

Equations (10.3.34) and (10.3.36)–(10.3.38) should be solved self-consistently. This has been done by Rudberg and Jonson[52] for an n-GaAs/Al$_x$Ga$_{1-x}$As/n-GaAs structure, and we shall review some of their results shortly. To get a feeling for the physics, however, it suffices to study the limit of large surface plasmon frequency, where it is possible to extract some analytical results. In this limit the Green's function in (10.3.37) is evaluated at a large negative energy and is therefore sharply peaked around $x' \sim x$. For an electron traveling in the x direction ($\mathbf{p}_\parallel = 0$), we find to lowest order in the interaction

$$G_{\mathbf{k}}(x, x'; E_i - \hbar\omega_c) \approx -\frac{m}{\hbar^2 q}\, e^{-q|x-x'|},$$

$$q = \sqrt{\frac{2m}{\hbar^2}\left(\hbar\omega_c + V(x) - E_{i,x}\right) + k^2} \tag{10.3.39}$$

Because of this sharply localized Green's function, we can expand the rest of the integrand in (10.3.37) around $x' = x$. Hence, we make the approximation

$$\frac{\psi_i(x')}{\psi_i(x)} \approx e^{ik_\psi(x'-x)}, \qquad k_\psi \equiv -i\frac{d}{dx}\ln\psi_i(x) \tag{10.3.40}$$

With the simplifications furnished by (10.3.39) and (10.3.40) we can evaluate the xc-potential analytically. The result for its real part is

$$V_{xc} = -\frac{e^2}{4x}\, h\!\left(\frac{\hbar k_\psi/m}{2x\omega_s}\right), \qquad h(z) \equiv \mathrm{Re}\int_0^\infty d\alpha\,\frac{e^{-\alpha}}{1 + i\alpha z} \tag{10.3.41}$$

In the present discussion we neglect the imaginary part of the xc-potential. If we are concerned with a scattering solution Ψ_i to the Schrödinger equation (10.3.34), the imaginary part represents scattering out of the channel i. We assume this is compensated by "scattering in" terms from other channels and put the imaginary part of the xc-potential to zero in the present scheme. This approximation ensures that the current in the ith channel is conserved.

In order to analyze the xc-potential (10.3.41), we first assume that an electron approaching a simple barrier $V(x)$ from the right as in Figure 10.3 can be described by an incoming plane wave $\exp(-ipx)$. We have in mind an n-GaAs/Al$_x$Ga$_{1-x}$/As-GaAs heterostructure, where the Al$_x$Ga$_{1-x}$ provides a barrier for the electrons moving in the conduction band, and where the leftmost GaAs part is heavily doped and "metallic." If the energy of the incoming electron is much larger than the barrier height, the probability for reflection is small and the wave function in the barrier region and the quantity k_ψ are, respectively,

$$\psi(x) \sim e^{-ip'x}, \quad p' = \sqrt{p^2 - 2mV(x)/\hbar^2}, \quad k_\psi = -p' \tag{10.3.42}$$

In this limit we recover the dynamical image potential result (10.3.32) if we, as usual, interpret $\hbar p'/m$ as the speed of the electron (speed being the absolute value of the velocity).

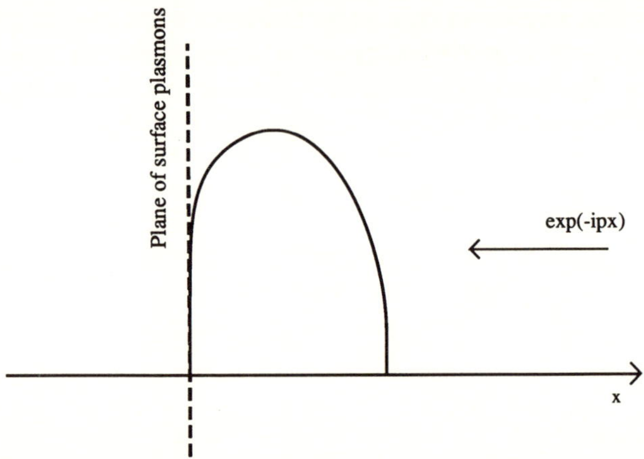

FIGURE 10.3. The dynamical image potential. An electron represented by a plane wave approaches a model static potential from the right. Due to dynamic interactions with polarization waves (surface plasmons) at the interface (indicated by the dashed line), the electron will see a different, effective potential. The degree to which the effective potential differs from the simple one-electron potential shown depends on the relation between the time scale associated with the surface plasmons and the velocity of the tunneling electron.

In the opposite limit where the energy of the incoming electron is much less than the barrier height, the electron will most probably be reflected, but the wave function has a finite amplitude in the barrier region which is exponentially decreasing for smaller x (as the electron is moving from right to left). We have

$$\psi(x) \sim e^{\kappa x}, \quad \kappa = \sqrt{\frac{2mV(x)}{\hbar^2} - p^2}, \quad k_\psi = -i\kappa \qquad (10.3.43)$$

It follows from (10.3.41), that for such an opaque barrier the classical image potential will be decreased by an amount which depends on the dimensionless quantity $\hbar\kappa/2m\omega_s$. By analogy with the case where the electron energy is much larger than the barrier, it is natural to interpret the quantity $\hbar\kappa/m$ as the *speed* of the tunneling electron.[3] If this speed is large compared to $x\omega_s$, the timescale for the surface plasmons, the xc-potential (10.3.41) will be reduced from the classical image potential result in much (but not identically) the same way as when the electron moves with the same speed above the barrier, in the classically accessible region. Actually, as k_ψ in (10.3.43) is imaginary, the velocity of the tunneling electron, if defined as $\hbar k_\psi/m = -i\hbar\kappa/m$, is also imaginary.

The same imaginary velocity appears in the instanton approach to tunneling out of a metastable state.[2] In that case a certain action integral along a tunneling trajectory is minimized by the *classical* path in an *inverted* potential. Consequently, the tunneling instanton has an imaginary velocity. Although this might look like a mere mathematical trick, Leggett[53] has asserted that the speed and associated "bounce time" of the instanton has physical significance as the velocity and traversal time for the tunneling particle (or, more generally, the tunneling state). This is the same interpretation we have made here and earlier[3] in the context of the dynamical image potential.

We now propose that the expression

$$v_\psi(x) \equiv \frac{\hbar|k_\psi|}{m} \equiv \frac{\hbar}{m}\left|\frac{d}{dx}\ln\psi_i(x)\right| = \left|\frac{j}{\rho(x)} - i\frac{\hbar}{2m}\frac{d\ln\rho(x)}{dx}\right| \quad (10.3.44)$$

gives the approximate speed of an electron of arbitrary energy, smaller or larger than the barrier. In the second part of (10.3.44) we have written $v_\psi(x)$ in terms of the (constant) current density j and number density $\rho(x)$, which are more robust quantities than the wave function. The limiting results discussed above, when the electron energy is much larger ($v_\psi = \hbar p'/m$) or much smaller ($v_\psi = \hbar\kappa/m$) than the barrier height, is obviously recovered in the appropriate limits. Equation (10.3.44) should be understood as an interpolation between these limits and does not necessarily have any fundamental significance. After all, it was derived by a somewhat less than straightforward "deconvolution" of the xc-potential in (10.3.41). In the same spirit we can, for the one-dimensional case, define a traversal time as

$$\tau_\psi(x_1, z_2) = \int_{x_1}^{x_2} \frac{dx}{v_\psi(x)} \quad (10.3.45)$$

We emphasize again that (10.3.45) is not to be understood as a rigorous result but as a reasonable interpolation between two better-understood limits. We proposed earlier that the Büttiker–Landauer time (10.3.8) should be interpreted as an interpolation formula between the same limits. The main objective here is to demonstrate that such an interpolation formula may not be unique.

In a number of experiments on different n-GaAs/Al$_x$Ga$_{1-x}$As/n-GaAs structures, covering a wide range of barrier heights and thicknesses, Guéret et al.[54,55] have recently established that tunneling electrons indeed see a dynamical image potential. The two n-doped GaAs regions on both sides of the Al$_x$Ga$_{1-x}$ barrier were doped to levels corresponding to a Fermi energy of about 12 meV above the bottom of the conduction band (using an effective mass of 0.067 m). The corresponding Fermi velocity, which sets the scale for the velocity of the tunneling electrons, is $v_F = 2.5 \times 10^5$ m/s, and the inverse surface plasmon frequency is $\omega_s^{-1} = 7 \times 10^{-14}$ s (using the GaAs dielectric constant of 10.9 for the entire structure). By varying the aluminum content x, the Al$_x$Ga$_{1-x}$-barrier height could be varied from 8 to 90 meV above the Fermi level, and the barrier thicknesses could be varied from 10 to 100 nm. The tunnel conductance was hence varied over seven orders of magnitude, and the dimensionless measure

$$\frac{v_\psi}{2x\omega_s} \sim \frac{v_\psi}{v_F}\frac{10\ \text{nm}}{x} \tag{10.3.46}$$

of the importance of dynamical corrections can, with x equal to half the barrier width, be expected to vary from about 0.1 to perhaps 2 or 3, covering small and large corrections to the classical image potential.

Guéret *et al.* measured the conductance G_{meas} through the barrier at low bias and compared their result to a calculated value G_{calc} with no image potential correction to the "bare" single-particle square-box potential barrier of the Al_xGa_{1-x}. All measured structures had opaque barriers *without image potential correction,* so that the tunneling current can be expected to vary exponentially with barrier thickness and height. In general, $\log(G_{\text{meas}}/G_{\text{calc}})$ was found to deviate from unity. When plotted as a function of $\omega_s\tau^{\text{BL}}$, with τ^{BL} given by (10.3.5), it started to deviate appreciably from unity at about $\omega_s\tau^{\text{BL}} \sim 2$ (or $\omega_p\tau^{\text{BL}} \sim 3$, as they used $\omega_p = \sqrt{2}\,\omega_s$, the bulk plasmon frequency).

Although we support the conclusion in Refs. 54 and 55 that dynamical image potential effects are seen, and that these vary with the traversal time, we do not agree with the theoretical analysis in detail. In particular, we disagree with the model potential used to describe the dynamical effects. Rudberg and Jonson[52] have analyzed the same experiments by calculating the transmission probability for the electrons at the Fermi surface with momentum perpendicular to the barrier. These are the most likely to be transmitted through the barrier and therefore give the dominant contribution to the current. Equations (10.3.34) and (10.3.36)–(10.3.38) were solved self-consistently with the trivial extension that the tunneling electron couples to surface plasmons at both interfaces. Figure 10.4a–c gives results for the self-consistent potential in three typical cases, with potential parameters giving $\tau_\psi = 0.4\omega_s^{-1}$, $2.4\omega_s^{-1}$, and $8.2\omega_s^{-1}$ respectively. [If we use the BL traversal time (10.3.7), we find the almost identical traversal times $\omega_s\tau_T^{\text{BL}} = 0.4, 2.5,$ and 8.2.]

For the longest traversal time, $\omega_s\tau_\psi = 8.2$, Figure 10.4c shows that the dynamical image potential is very well approximated by the classical image potential; the surface plasmon modes have time to respond almost adiabatically to the motion of the tunneling electron. In the case of the shortest traversal time $\omega\tau_\psi = 0.4$, on the other hand, Figure 10.4a reveals that the dynamical image potential is quite insignificant; the electron tunnels so fast the image potential does not have time to develop fully.

In Figure 10.5 we have plotted $\log(T/T_0)$ as a function of $\omega_s\tau_\psi$ (T is the self-consistent transmission probability and T_0 is the transmission probability with *no* image potential at all) for some 30 different potentials. This is a relevant and convenient measure of the dynamical image potential effects on *opaque* barriers where the transmission probability depends exponentially on the barrier height. Note that the transmission probability varies *smoothly* with traversal time, and that the results fall within a rather narrow band. Potentials characterized by the same traversal time can have widely different transmission probabilities T_0 (and T); hence, the traversal time is the relevant parameter for characterizing the dynamical image potential effects. Although the dependence on traversal time is smooth and monotonic in Figure 10.5, if one had to pick a traversal time large enough that the image potential is appreciable, $\omega_s \tau_\psi \sim 1$–2 seems a reasonable value.

We have shown in this section how the surface plasmons can be used as a clock for measuring the speed of a tunneling electron. Based on an *approximate* deconvolution of the xc-potential (10.3.41), we proposed that

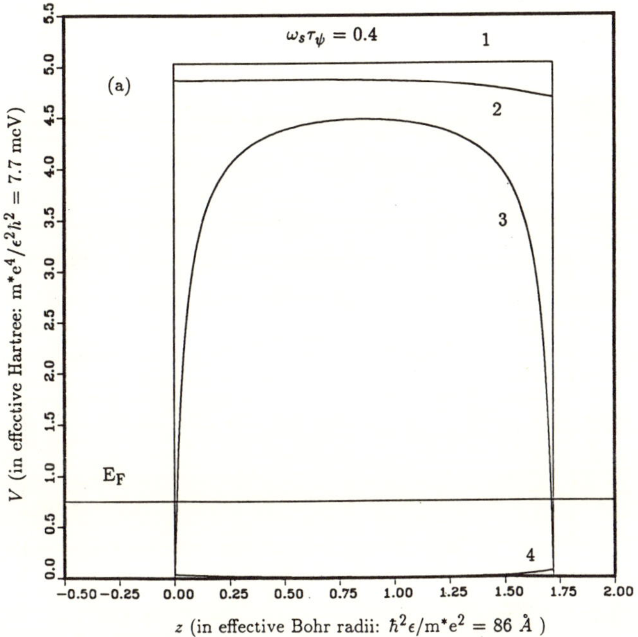

FIGURE 10.4a–c. Dynamical image potentials calculated for an electron of (the same) energy E_F tunneling through potential barriers characterized by three different traversal times τ_ψ: (a) $\omega_s \tau_\psi = 0.4$, (b) $\omega_s \tau_\psi = 2.4$, (c) $\omega_s \tau_\psi = 8.2$. Four potentials are shown in each panel: (1) a bare single-electron square-box potential of a certain height and width; (2) an effective potential, including the self-consistent dynamical image potential; (3) an effective potential, including the classical image potential; and (4) the imaginary part of the self-consistent image potential (ignored in the self-consistent calculation scheme; see text). The larger the value of $\omega_s \tau_\psi$, the closer the dynamical image potential is to its classical value. Note the changes in scale between panels.

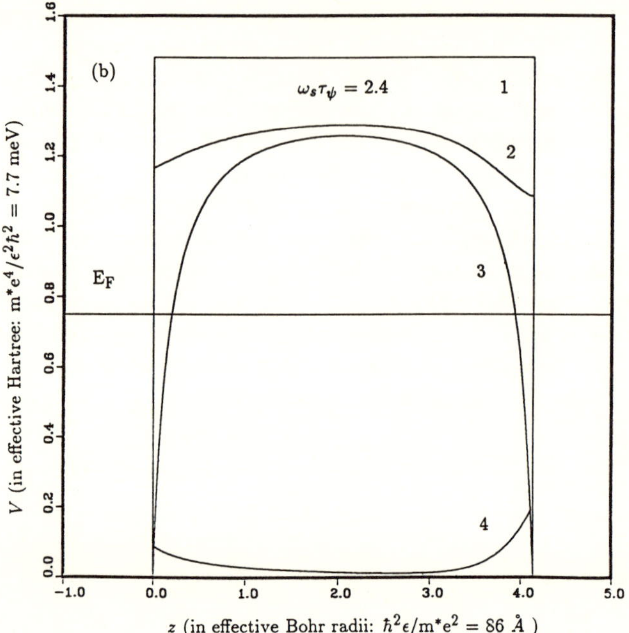

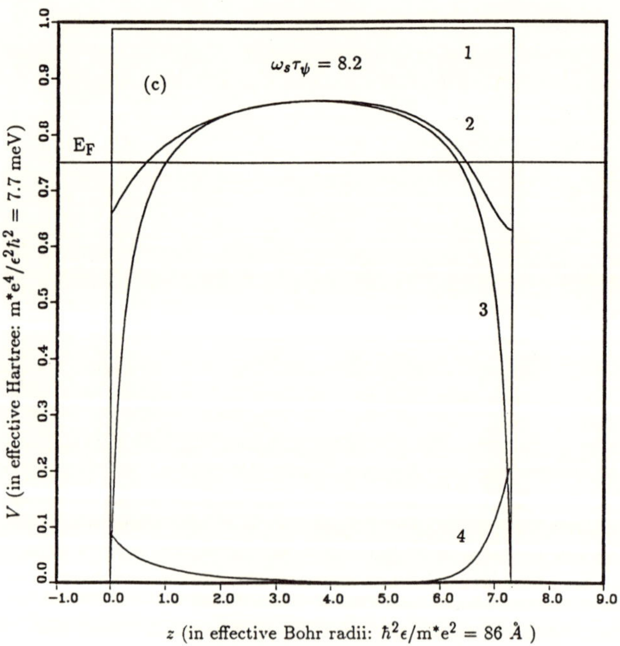

FIGURE 10.4. Continued.

FIGURE 10.5. Relative transmission probability T/T_0 for an electron at the Fermi energy tunneling through an effective potential (including the self-consistent image potential) as a function of the traversal time τ_ψ. T_0 is the transmission probability in the absence of any image potential, and ω_s is the surface-plasmon frequency. The dynamical image potential effects scale quite well with the traversal time. The only point outside the band of results corresponds to a case where the barrier is almost transparent and serves to remind us that in such a case the reduction in barrier height does not translate into an exponential increase of the transmission probability.

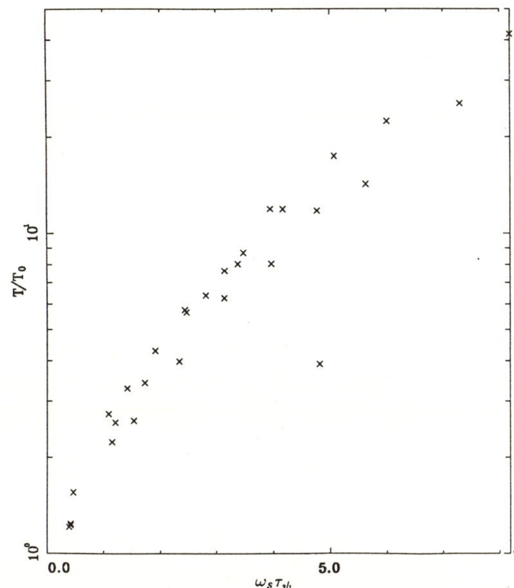

(10.3.44) gives the speed and (10.3.45) the traversal time. It is interesting to note that with only the first term, $j/\rho(x)$, in the expression for v_ψ one recovers the dwell time of (10.2.15) from (10.3.45). The other term contributing to v_ψ is related to the change in density of the tunneling state and, in this sense, is similar to the Büttiker time τ_z. Whether we plot $\log(T/T_0)$ as a function of τ_T^{BL} or τ_ψ in Figure 10.5, the result is essentially identical, whereas if we use the opaque limit (10.3.5) of the BL traversal time the points are considerably more scattered. Viewed as two interpolation schemes between the same limits, τ_T^{BL} and τ_ψ are therefore numerically very similar in this case.

10.3.4. The Shunted Josephson Junction

We have so far discussed the coupling between a tunneling particle and other degrees of freedom. In this sense we have been concerned with microscopic tunneling, which in the absence of coupling is what is normally discussed in introductory quantum mechanics texts. For the quantum tunneling of a macroscopic variable, coupling to the external environment is unavoidable, and tunneling in the presence of the resulting dissipation has indeed been extensively studied in what has grown to become the field of macroscopic quantum tunneling.[2] A particularily well-studied system is the current-biased Josephson junction.[56] This device is made of two superconducting layers separated by an thin insulator, and can be characterized by its capacitance C and critical current I_0 (for current bias $I > I_0$ the superconducting state is destroyed). The current across the junction is due to pairs of electrons that tunnel between the superconducting electrodes on either side of the insulating barrier. The state of the Josephson junction is characterized by the difference in phase θ between the macroscopic wave functions of the two weakly coupled superconductors. This phase difference is connected to the voltage V across the junction, and for an ideal current biased junction one has

$$V = \frac{\hbar}{2e}\frac{\partial \theta}{\partial t}, \qquad I = I_0 \sin\theta + \frac{1}{C}\frac{\partial V}{\partial t} \qquad (10.3.47)$$

Although the underlying physics of the Josephson junction is certainly quantum mechanical in origin, it is an entirely separate question whether the macroscopic variable θ can show quantum behavior. It goes without saying that the classical equations (10.3.47) for the phase difference do not describe any quantum mechanical effects. It turns out, however, that coupling to the environment can induce quantum fluctuations in θ and its conjugate variable Q (the difference in charge between the two superconductors). We then have to abandon (10.3.47) and consider the relevant junction Hamiltonian which can be "derived" from it:[57]

$$H = -\frac{(2e)^2}{2C}\frac{\partial^2}{\partial \theta^2} + E_J f(\theta)$$

$$(10.3.48)$$

$$E_J = \frac{\hbar}{2e} I_0, \qquad f(\theta) = -(s\theta + \cos\theta), \quad s = \frac{I}{I_0}$$

Here we have used the commutation relation $[Q, \theta] = i2e$ to replace Q by $-i2e(\partial/\partial\theta)$.

From (10.3.48) it is clear that the Hamiltonian for the current-biased Josephson junction is analogous to that of a particle in a tilted washboard potential. To make this analogy more transparent, one can introduce an arbitrary length scale x_0 and define the particle coordinate and mass as

$$x = x_0\theta, \qquad m = \frac{C}{x_0^2}\left(\frac{\hbar}{2e}\right)^2 \qquad (10.3.49)$$

Under appropriate conditions the ground state of the function corresponds to a local minimum in the washboard potential. The phase is thus constant in time, and from (10.3.47) we conclude that the junction is in a zero-voltage state. This state is only metastable, however; in the mechanical analog, the particle can escape out of the local potential minimum either by thermal excitation above the barrier or by quantum mechanical tunneling through it. At low enough temperature the latter mechanism dominates. The velocity of the escaping particle corresponds to a voltage over the Josephson junction, which can be detected. After detection the circuit can be reset to a zero-voltage state by reducing the current bias (hence eliminating the tilt of the washboard potential). When the bias current is increased again, it is possible to measure the time delay until a voltage pulse is detected. Repeated measurements give a value for the average lifetime of the metastable state.

In order to determine the voltage, the Josephson junction has to be connected to a measuring circuit, which then constitutes its environment. As far as the junction is concerned, this environment is completely described by an ideal current source in parallel with a frequency-dependent impedance $Z(\omega)$, as shown in Figure 10.6. Coupling to the additional electromagnetic degrees of freedom of the impedance $Z(\omega)$, assumed to be in thermal equilibrium at a temperature T, will induce a viscous damping force on the junction variable θ. This in turn will affect the probability for tunneling out of the metastable zero-voltage state and consequently change its lifetime. One way to view this is that in the tunneling process, electromagnetic modes of energy $\hbar\omega$ are excited with some probability $P(\omega)$. To conserve energy, the tunneling variable θ then has to find a final state at the energy $E_0 - \hbar\omega$ lower than the initial state energy E_0. In effect, it therefore sees a higher potential barrier; tunneling is less likely, and the lifetime of the metastable state is increased. To be important, the characteristic energy $\hbar\omega$ associated with dissipation has to be larger than the uncertainty in the energy of the tunneling electron, h/τ_T (cf. the discussion of the Coulomb blockade in Ref. 58).

In a beautiful experiment Esteve *et al.*[15] were able to control the external impedance $Z(\omega)$ and even tune it in a controlled manner *in situ*. The adjustable impedance was achieved by connecting the Josephson junction to a transmission line partially covered by a microwave absorbing block. The portion of the transmission line left uncovered defines a delay line, of length l, characteristic impedance Z_0, and propagation velocity c. In the frequency range of interest, the covered portion behaves as a terminating resistor Z_t for the delay line. The impedance seen by the junction is therefore[15]

$$Z(\omega) = Z_0 \frac{1 + a\,\exp(2i\omega t_d)}{1 - a\,\exp(2i\omega t_d)}, \qquad a = \frac{Z_t - Z_0}{Z_t + Z_0}, \quad t_d = \frac{l}{c} \quad (10.3.50)$$

If a voltage step is applied to a circuit described by this impedance, the time development of the current response is obtained from the Fourier

FIGURE 10.6. An equivalent circuit for the experiment in Ref. 15. The Josephson junction is characterized by its capacitance C and critical current I_0. The function is biased by a current $I < I_0$ and is shunted by an imped-ance $Z(\omega)$, representing its electromagnetic environment. The important contribution to the impedance comes from a variable-length transmission line which connects the junction to a "terminating" resistor.

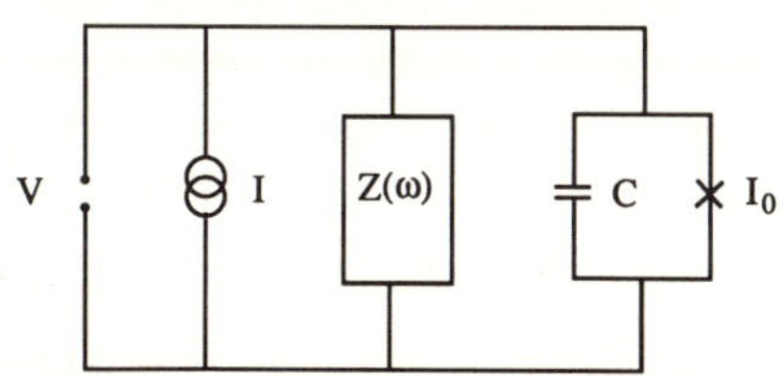

transform of $[i\omega Z(\omega)]^{-1}$ and consists of a series of steps,[15] each separated by a delay time $2t_d$. The appropriate picture is that the voltage induces an electromagnetic wave, which is reflected at the terminating resistance and returns to the junction after a delay time. By changing *in situ* the length l, Esteve *et al.* were able to change the rise time of the current response to the voltage step associated with the tunneling of the junction out of its zero-voltage state. To the extent that there is a current response, it will interfere with the tunneling process itself and increase the lifetime of the metastable state as qualitatively described above. However, if the response time associated with dissipation is longer than the traversal time, no interference effects are expected. In a complementary picture, we may argue that for a long transmission line there is an abundance of low-energy transmission line modes.[58] This means thaat the probability of exciting an energetic mode is negligible, and the final-state energy $E_0 - \hbar\omega \approx E_0$ on the average. Hence, the tunneling will be unaffected by excitation of the transmission line modes. For a short uncovered transmission line, however, there is a minimum excitation energy of $\hbar(2\pi/l)c = h/t_d$ due to spatial quantization of the energy levels. Hence, the final-state energy is at least this much below the initial-state energy E_0, and tunneling will become less probable if $h/\tau_d > h/\tau_T$.

In the experiment, it was indeed found that the measured lifetime decreased as a function of the length of the uncovered portion of the transmission line (i.e., as a function of the delay time t_d) until it reached a constant value for a sufficiently long line (see Figure 5 of Ref. 15). The crossover value of the delay time was then identified with the traversal time. The measured value agrees quite well with the opaque barrier limit (10.3.5) of the BL traversal time. The theoretical analysis of the experiment is based on work of Leggett.[59] For illustrative purposes, we shall take a simpler approach, which nevertheless gives rather accurate results for both the lifetime of the metastable state (without coupling to the transmission line modes) and the traversal time.

For the tunneling probability to be significant, the local minimum in the washboard potential shown in Figure 10.7 has to be quite shallow. In this regime, for a good approximation we can expand the potential energy around the minimum of the potential well at $\theta = \theta_{\min}$ and keep only quadratic terms. In the same manner we can expand the potential around the maximum of the potential barrier at $\theta = \theta_{\max}$ and keep only quadratic terms. Using the notation introduced in (10.3.49) for the equivalent, and perhaps more familiar, mechanical problem, it is a simple exercise to show that

$$V(x) \approx \tfrac{1}{2}m\omega_0^2(x - x_{\min})^2 \qquad (10.3.51)$$

close to the potential minimum, and

$$V(x) \approx \Delta U - \tfrac{1}{2}m\omega_0^2(x - x_{\max})^2 \qquad (10.3.52)$$

FIGURE 10.7. Part of the "washboard" potential well (here scaled by the Josephson energy E_J) for the Josephson junction as a function of the phase variable θ. Also shown are the approximate, parabolic potentials used in the text for crude estimates of the traversal time through the barrier and the lifetime of the metastable state, whose energy is E_0. Due to interactions with its environment, the tunneling electron dissipates an energy $\hbar\omega$ with some probability $P(\omega)$ and therefore sees a higher potential. The dissipated energy, absorbed by the additional degrees of freedom of the system, is related to the characteristic time scale of the environment. The amount of dissipation depends on how this time scale compares with the traversal time.

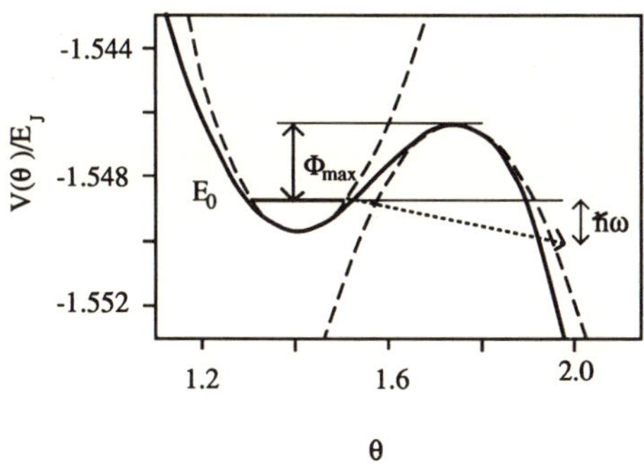

in the vicinity of the potential maximum. Note that the curvature at the minimum is the same as at the maximum and is characterized by the frequency ω_0. One finds

$$\omega_0 = \left(\frac{2eI_0}{\hbar C}\right)^{1/2}(1 - s^2)^{1/4},$$

$$\tag{10.3.53}$$

$$\Delta U = E_J[f(\theta_{max}) - f(\theta_{min})] = \frac{4\sqrt{2}}{3}E_J(1 - s)^{3/2}$$

Using experimental parameters from Ref. 15, $s = 0.9855$, $C = 2.7 \times 10^{-12}$ F, $I_0 = 7.0 \times 10^{-6}$ A, we can determine the numerical values of the potential parameters:

$$E_J = 14.5 \text{ meV}$$

$$\Delta U = 0.048 \text{ meV}$$

$$\tag{10.3.54}$$

$$E_0 = \tfrac{1}{2}\hbar\omega_0 = 0.012 \text{ meV}$$

$$\Phi_{max} = \Delta U - E_0 = 0.036 \text{ meV}$$

Here ΔU is the difference in potential energy between the local minimum and the local maximum, E_0 is the energy of the metastable state metastable state measured from the local potential minimum, and Φ_{max} is the effective barrier height measured from E_0. As Φ_{max}/k_B is only 280 mK, the experiment has to be performed at low temperature to suppress thermal excitations over the barrier (temperatures down to 18 mK were used in Ref. 15).

We determine the lifetime τ of the metastable state (in the absence of coupling to the transmission line) from the width of the quasibound energy level,[60]

$$\tau = \frac{2d}{v}\frac{2}{T} \tag{10.3.55}$$

where (v is the velocity of the electron in the potential well of effective width d) $v/2d = \omega_0/2\pi$ is an attempt frequency, while

$$T = \frac{1}{1 + e^{-2\pi\varepsilon}}, \qquad \varepsilon = \frac{\Phi_{\max}}{\hbar\omega_0} \tag{10.3.56}$$

is the transmission probability through the inverted parabolic potential.[20] Inserting numbers we find the lifetime of the metastable state to be $\tau \sim 4\ \mu$s, which is reasonably close to the measured value of $10\ \mu$s, considering the simplified assumptions we made about the potential.

The transmission probability T of (10.3.56) is very small. We are therefore in the opaque-barrier limit and the BL traversal time reduces to[9]

$$\tau^{\mathrm{BL}} = \frac{m}{\hbar} \int_{x_1}^{x_2} \frac{dx}{\sqrt{(2m/\hbar^2)(V(x) - E_0)}} \tag{10.3.57}$$

where x_1 and x_2 are the classical turning points and $V(x_{1,2}) = E_0$. It is interesting that for an inverted parabola, the traversal time of (10.3.57) is independent of the turning points (or, equivalently, independent of the particle energy E_0) and simply given by

$$\tau^{\mathrm{BL}} = \pi/\omega_0 \tag{10.3.58}$$

In other words, the traversal time is only related to the curvature of the potential barrier. Inserting numbers, one finds $\tau^{\mathrm{BL}} \sim 85$ ps, in good agreement with the value of 78 ps cited in Ref. 15.

The experiment discussed in this section is very important in that a time scale associated with the barrier traversal time has clearly been measured. We shall discuss its significance further in the concluding section.

10.4. CONCLUSIONS

We have reviewed a (large) number of different theoretical expressions for the traversal time of tunneling in search of an answer to the old question: "How long a time did a particle *that eventually penetrated* a barrier spend inside it?" For a long time this question was pondered with very little support from experimental information. Recently, however, such information is emerging from experiments where a *dynamical* interaction characterized by a well-defined frequency provides a clock for measuring the traversal time. Experiments do not yet provide convincing proof that one theoretical formula or another is the correct one in *all* cases. However, both the experiments by Guéret *et al.*[54,55] probing the dynamical image potential for a tunneling electron and the experiment by Esteve *et al.*[15] involving dissipative tunneling of a macroscopic state in a Josephson junction seem to show that in the opaque-barrier limit the traversal time is consistent with (10.3.57). This formula for the traversal time was explicitly proposed by Büttiker and Landauer,[9] somewhat more implicitly assumed in our analysis of the dynamical image potential,[3] and appears in other contexts as well. The range of different barrier types studied, however, is not wide enough to support a statement that the traversal time is given by that formula. The Josephson junction experiment, for example, gives results for one potential barrier only, and a rather peculiar one with a high degree of symmetry between the local potential minimum and the barrier. It is interesting, though, that two such different experiments as those of Ref. 15, involving tunneling of a macroscopic state of a Josephson junction, and those of those of Refs. 54 and 55, involving electrons tunneling through an AlGaAs structure, can be analyzed in terms of the same traversal time (in the opaque limit).

There is hope that further evidence can be provided by future experiments, perhaps using the method proposed by Lucas *et al.*,[61] where a laser frequency provides the clock as well as induces tunneling between an STM tip and a substrate. Until more results are available, we conclude with Landauer[62] (in a comment on Ref. 15) that "... more important than the exact result and its relation to theoretical controversies is the fact that a time scale associated with the barrier traversal time can be measured, and is a real (not imaginary) quantity ...."

Although recent experimental developments do not imply that the traversal time problem has been solved, they are consistent with the opaque-barrier limit of the BL traversal time. Since we know the quantum mechanical traversal time for a particle tunneling "over" a barrier from classical physics and the correspondence principle, it is natural to construct an interpolation formula between these limits. We propose that the general form of the Büttiker–Landauer formula (10.3.7) is approximately valid in at least this sense. However, by way of giving an example of a different—but in the cases studied by Rudberg and Jonson[52] numerically almost equivalent—interpolation formula (10.3.45), we suggest that such an interpolation formula may not be unique.

With decisive experimental information about the traversal time through a general potential barrier lacking, we turn to theoretical evidence. We have reviewed recent work that makes it quite clear that neither the phase times of Bohm,[4] Wigner,[5] and Eisenbud,[6] nor the dwell time of Smith[7] measure the traversal time of tunneling. Now an energy ΔE and a time τ can always be related by Planck's constant, $\tau = \Delta E / \hbar$. One may therefore ask whether a measured traversal time is really the "intrinsic" time the electron spends in the tunneling barrier, or is it instead fundamentally related to some energy scale. In their review Hauge and Støvneng[16] point out that the BL traversal time (10.3.7) does not obey the sum rule (10.2.17) for mutually exclusive events, which in their view shows that it cannot be the proper measure of the traversal (or reflection) time in general. In fact, adding the criterion that the measure of a traversal time must be a real quantity, they show that *no* proposed theoretical expression for the traversal time can be accepted. Although they recognize that the time given by, say, (10.3.57) may be the most relevant time scale for a particular experiment, they associate it with an energy sensitivity of the tunneling process and not to an intrinsic traversal time. We find the interpretation of the results of Esteve *et al.* in terms of a traversal time and a delay time satisfactory and natural. We also prefer to interpret the experiments of Guéret *et al.*[54,55] in terms of characteristic time scales determining the strength of the dynamical image potential. (However, other experiments claiming to measure the traversal time from a reduction of the tunneling current due to a deflecting static magnetic field[63] can equally well be interpreted without invoking a traversal time.[62,64]) With this said, it must be emphasized again that the sum rule for mutually exclusive events *is not satisfied* by the BL traversal time. If interpreted as an interpolation formula, this may not be a problem as the limiting expressions do obey the sum rule, but this point needs to be studied further.

We conclude by observing that although most technical controversies in this field have been resolved, different opinions about interpretation remain. In particular, the more philosophical questions as to whether an intrinsic tunneling time can be defined without reference to a measurement or clock, and if different clocks (interactions) could in principle give different traversal times, remain largely open.

ACKNOWLEDGMENTS

This work was supported by DOE under contract DE-AC05-84OR21400 with Martin Marietta Energy Systems, Inc. I am grateful to G. D. Mahan for encouraging me to write this chapter and to his colleagues at the Solid State Division of ORNL and the University of Tennessee for their hospitality during the academic year 1989–90. I am also grateful to A. P. Jauho and B. Rudberg, with whom parts of the work described in this chapter were carried out, for many discussions. Thanks are also due to M. Büttiker and R. Landauer for enlightening correspondence, and to E. Hauge and J. A. Støvneng for discussions and for sending their papers prior to publication.

APPENDIXES

Appendix 1.1

In this appendix we derive the equivalence (10.2.16) between the dwell time $\tau_{\text{dwell}}(R, k)$ and the average phase time defined in (10.2.14) for the specific case of a plane wave scattered by a central force $U(r)$. Starting from the definition (10.2.15) of the dwell time and using the wave packet (10.2.11), we have

$$
\begin{aligned}
\tau_{\text{dwell}}(r, k) &\equiv \int_{-\infty}^{\infty} dt \int_{r<R} d\mathbf{r}\, |\Psi(\mathbf{r}, t)|^2 \\
&= \sum_{k''k'} A^*(k'' - k)A(k' - k) \\
&\quad \times \int_{r<R} d\mathbf{r}\, \chi_{k''}^*(r, \theta)\chi_{k'}(r, \theta) \int_{-\infty}^{\infty} dt\, e^{i(E_{k''} - E_{k'})t/\hbar}
\end{aligned} \qquad \text{(A1.1.1)}
$$

Here the wave packet has been normalized to contain one particle in the volume $\Omega = LA$. Note that this differs from the normalization used in (10.2.11), which is the conventional one in scattering problems. For the purpose of this discussion the choice of normalization is unimportant.

The integral over time in (A.1.1.1) gives a delta function that can be used to perform the sum over k'':

$$
\int_{-\infty}^{\infty} dt\, e^{i(E_{k''} - E_{k'})t/\hbar} = 2\pi\hbar\delta(E_{k''} - E_{k'}) = \frac{2\pi}{v_k}\delta(k'' - k') = \frac{L}{v_k}\kappa_{k'',k'} \qquad \text{(A1.1.2)}
$$

Summing over k'' in (A1.1.1) and integrating over angles [cf. (10.2.12)], one gets

$$
\tau_{\text{dwell}}(R, k) = \sum_{k'} |A(k' - k)|^2 \frac{L}{v_{k'}} \sum_{l=0}^{\infty} \frac{\pi(2l+1)}{k'^2} \frac{1}{\Omega} \int_0^R r^2\, dr |R_{k'l}(r)|^2 \qquad \text{(A1.1.3)}
$$

To proceed, we write the integrand of the remaining r-integral as a derivative.[7] It is convenient to define a new function $\phi_l(r) \equiv rR_l(r)$ which obeys the one-dimensional effective Schrödinger equation:

$$
\frac{d^2\phi_l(r)}{dr^2} + (k^2 - \tilde{u}_l(r))\phi_l(r) = 0, \qquad \tilde{u}_l(r) = \frac{2m}{\hbar^2}U(r) + \frac{l(l+1)}{r^2} \qquad \text{(A1.1.4)}
$$

Using this equation and a relation obtained by operating on it with $d/d(k^2)$,

$$
\frac{d^3\phi_l(r)}{dr^2\, d(k^2)} + (k^2 - \tilde{u}_l(r))\frac{d\phi_l(r)}{d(k^2)} = -\phi_l(r) \qquad \text{(A1.1.5)}
$$

we find that

$$
\begin{aligned}
\frac{d\phi_l(r)}{d(k^2)}\frac{d^2\phi_l^*(r)}{dr^2} - \phi_l^*(r)\frac{d^3\phi_l(r)}{d(k^2)\,dr^2} &= \frac{d}{dr}\left(\frac{d\phi_l(r)}{d(k^2)}\frac{d\phi_l^*(r)}{dr} - \phi_{(r)}^{*l}\frac{d^2\phi_l(r)}{dr\,d(k^2)}\right) \\
&= |\phi_l(r)|^2 = r^2|R_l(r)|^2
\end{aligned} \qquad \text{(A1.1.6)}
$$

The integral over r in (A1.1.3) is now trivial. When R is in the asymptotic region where (10.2.4) holds, we have

$$\left[\frac{d\phi_l(r)}{d(k'^2)}\frac{d\phi_l^*(r)}{dr} - \phi_l^*(r)\frac{d^2\phi_l(r)}{dr\,d(k'^2)}\right]_{r=R}$$

$$= v_{k'}\left(\frac{2R}{v_{k'}} - i\hbar S_l^{-1}\frac{d}{dE_{k'}}S_l + (-1)^{l+1}\frac{2\hbar}{E_{k'}}\sin[2(k'R + \delta_l)]\right) \quad \text{(A1.1.7)}$$

This expression should be substituted for the r-integral in (A1.1.3). We are assuming that the normalized envelope function $A(k' - k)$ is sharply peaked around $k' = k$ with a small width σ. We can nevertheless neglect the sine function in (A1.1.7) because it will be averaged out in the summation over k' *for sufficiently large* $R \gg \sigma^{-1}$. Because of the sharpness of the envelope function, we can simply replace k' by k in the other two terms and integrate out the envelope function. The result is

$$\tau_{\text{dwell}}(R, k) = \sum_{l=0}^{\infty}\frac{\pi(2l + 1)}{k^2}\frac{1}{A}\left(\frac{2R}{v_k} - i\hbar S_l^{-1}\frac{d}{dE}S_l\right)$$

$$= \sum_{l=0}^{\infty}p_l\tau_l^{\phi}(R, k) \equiv \langle\tau^{\phi}(R, k)\rangle_{\text{av}} \quad \text{(A1.1.8)}$$

where p_l is given by (10.2.14) and $\tau_l^{\phi}(R, k)$ by (10.2.8) and (10.2.9). This proves (10.2.16).

Appendix 1.2

Consider the problem of an electron tunneling through a general, one-dimensional potential. By adding an infinitesimal time-dependent part, the total potential is

$$V(x) = V_0(x) + V_1\sin(\omega t) \quad \text{(A1.2.1)}$$

Particles that interact with this time-dependent potential will emit or absorb energy in quanta of $\hbar\omega$. In this appendix we generalize the derivation of the BL traversal time in Ref. 10 by considering the adiabatic limit, $\omega \to 0$, of this inelastic scattering process.

We envisage that the spatial extent of $V(x)$ is limited to the interval $-d/2 \le x \le d/2$. This means that for $|x| > d/2$, we have a free-particle Hamiltonian whose eigenfunctions are of the form $e^{\pm ikx}$, where $k = (2mE/\hbar^2)^{1/2}$. In the scattering region the Hamiltonian is

$$H = -\frac{\hbar^2}{2m}\frac{d^2}{dx^2} + V_0(x) + V_1\sin(\omega t) \quad \text{(A1.2.2)}$$

and the two independent eigensolutions to the corresponding time-dependent Schrödinger equation can be written as

$$\Phi_n(x, t; E, \bar{V}) = \phi_n(x, E, \bar{V}) \exp\left(-i\frac{Et}{\hbar}\right) \exp\left(i\frac{V_1}{\hbar\omega}\cos(\omega t)\right),$$

$$n = 1, 2 \quad (A1.2.3)$$

Here we have, in addition to x, t and the eigenenergy E, labeled the eigenfunctions by $\bar{V}$, the *average* potential in the scattering region. This will be useful below. For an infinitesimal amplitude of the time-dependent potential, $V_1 \ll \hbar\omega$, we may expand the eigensolutions of (A1.2.3) to lowest order in V_1 as

$$\Phi_n(x, t; E, \bar{V}) = \phi_n(x, E, \bar{V})e^{-iEt/\hbar}\left[1 + i\frac{V_1}{2\hbar\omega}e^{i\omega t} + i\frac{V_1}{2\hbar\omega}e^{-i\omega t}\right],$$

$$n = 1, 2 \quad (A1.2.4)$$

The sidebands at $E \pm \hbar\omega$ generated by the scattering potential will have to be matched to solutions of corresponding energy outside the scattering region. Hence, for an electron of energy E impinging on the scattering region, there will be reflected and transmitted intensity at energies $E, E + \hbar\omega$, and $E - \hbar\omega$. For a vanishingly small amplitude it suffices to consider these one-phonon sidebands. In a general case one would have to consider how the potential couples the one-phonon bands to two-phonon bands, and so on. Although such multiphonon solutions can be constructed, we shall follow the analysis in Ref. 10—with a straightforward generalization to allow for a space-dependent potential—and keep terms to lowest order in V_1 only. Hence, the wave function in the region to the left of the barrier will be of the form

$$\Psi_{\text{inc+refl}} = (e^{ik(x+d/2)} + Ae^{-ik(x+d/2)})e^{-iEt/\hbar}$$

$$+ A_+e^{-ik_+(x+d/2)}e^{-i(E+\hbar\omega)t/\hbar}$$

$$+ A_-e^{-ik_-(x+d/2)}e^{-i(E-\hbar\omega)t/\hbar} \quad (A1.2.5)$$

where $k_\pm = (2m/\hbar^2)^{1/2}(E \pm \hbar\omega)^{1/2}$. Equation (A1.2.5) represents an incident plane wave of unit amplitude and three reflected waves, one of amplitude A at the incident energy and two of amplitudes $A_\pm$ at energies $E \pm \hbar\omega$. To the right of the barrier we have

$$\Psi_{\text{transm}} = De^{ik(x-d/2)}e^{-iEt/\hbar}$$

$$+ D_+e^{ik_+(x-d/2)}e^{-i(E+\hbar\omega)t/\hbar} + D_-e^{ik_-(x-d/2)}e^{-i(E-\hbar\omega)t/\hbar} \quad (A1.2.6)$$

where D is the transmission amplitude at the energy of the incident wave and $D_\pm$ are the transmission amplitudes of the sidebands. In the scattering region, finally, we find

$$\Psi_V = [B\phi_1(x, E, \bar{V}) + C\phi_2(x, E, \bar{V})]e^{-iEt/\hbar}\left[1 + i\frac{V_1}{2\hbar\omega}e^{i\omega t} + i\frac{V_1}{2\hbar\omega}e^{i\omega t}\right]$$

$$+ [B_+\phi_1(x, E + \hbar\omega, \bar{V}) + C_+\phi_2(x, E + \hbar\omega, \bar{V})]e^{-i(E+\hbar\omega)t/\hbar}$$

$$+ [B_-\phi_1(x, E - \hbar\omega, \bar{V}) + C_-\phi_2(x, E - \hbar\omega, \bar{V})]e^{-i(E-\hbar\omega)t/\hbar} \quad \text{(A1.2.7)}$$

where $\phi_{1,2}(x, E, \bar{V})$ are the eigenfunctions of (A1.2.4). The coefficients A, $A_\pm$, B, $B_\pm$, C, $C_\pm$, D, and $D_\pm$ are determined by matching the wave functions and their derivatives at $x = \pm d/2$ in the usual manner. The extra phase factors $e^{\pm ikd/2}$ in (A1.2.5) and (A1.2.6) are introduced for convenience, as they make the matching relations look simpler. The matching conditions must hold for all times; therefore, we have to match each time Fourier component separately. Hence, we readily identify Ae^{-ikd} and De^{-ikd} with the static reflection and transmission amplitudes, respectively. Note that $A = A(k, E, \bar{V})$ and $D = D(k, E, \bar{V})$. Furthermore, the following relations obtained from the matching of the static terms at the boundaries of the scattering region hold:

$$1 + A(k, E, \bar{V}) = B(k, E, \bar{V})\phi_1(-d/2, E, \bar{V})$$
$$+ C(k, E, \bar{V})\phi_2(-d/2, E, \bar{V})$$
$$ik[1 - A(k, E, \bar{V})] = B(k, E, \bar{V})\phi_1'(-d/2, E, \bar{V})$$
$$+ C(k, E, \bar{V})\phi_2'(-d/2, E, \bar{V}) \quad \text{(A1.2.8)}$$

$$B(k, E, \bar{V})\phi_1(d/2, E, \bar{V}) + C(k, E, \bar{V})\phi_2(d/2, E, \bar{V})$$
$$= D(k, E, \bar{V})$$
$$B(k, E, \bar{V})\phi_1'(d/2, E, \bar{V}) + C(k, E, \bar{V})\phi_2'(d/2, E, \bar{V})$$
$$= ikD(k, E, \bar{V}) \quad \text{(A1.2.9)}$$

where ' indicates derivates with respect to x. For a general potential, $V_0(x)$, we of course do not know the explicit form of the reflection and transmission amplitudes. Nevertheless, it is possible, for small modulation frequencies, to express the sideband coefficients in terms of $A(k, E, \bar{V})$, $D(k, E, \bar{V})$, and their derivatives with respect to E and $\bar{V}$. To see this, consider the matching of the sideband terms at $x = -d/2$. From (A1.2.5) and (A1.2.7) one finds

$$A_\pm = i\frac{V_1}{2\hbar\omega}\left[B(k, E, \bar{V})\phi_1\left(-\frac{d}{2}, E, \bar{V}\right) + C(k, E, \bar{V})\phi_2\left(-\frac{d}{2}, E, \bar{V}\right)\right]$$

$$+ \left[B_\pm\phi_1\left(-\frac{d}{2}, E \pm \hbar\omega, \bar{V}\right) + C_\pm\phi_2\left(-\frac{d}{2}, E \pm \hbar\omega, \bar{V}\right)\right] \quad \text{(A1.2.10)}$$

and

$$-ik_{\pm}A_{\pm} = i\frac{V_1}{2\hbar\omega}\left[B(k, E, \bar{V})\phi_1'\left(-\frac{d}{2}, E, \bar{V}\right) + C(k, E, \bar{V})\phi_2'\left(-\frac{d}{2}, E, \bar{V}\right)\right]$$

$$+ \left[B_{\pm}\phi_1'\left(-\frac{d}{2}, E \pm \hbar\omega, \bar{V}\right) + C_{\pm}\phi_2'\left(-\frac{d}{2}, E \pm \hbar\omega, \bar{V}\right)\right]$$

$$(A1.2.11)$$

Now, from matching the static parts we know from (A1.2.8) and (A1.2.9) that the first portions of (A1.2.10) and (A1.2.11) equal

$$i(V_1/2\hbar\omega)(1 + A(k, E, \bar{V}))$$

and

$$(ik)i(V_1/2\hbar\omega)(1 - A(k, E, \bar{V}))$$

respectively. Rather than using $B_{\pm}$ and $C_{\pm}$, we combine them in a particular way to get two other variables, $\alpha_{\pm}$ and $\beta_{\pm}$, where

$$B_{\pm} = -i\frac{V_1}{2\hbar\omega}[\alpha_{\pm}B(k, E, \bar{V} \mp \hbar\omega) + \beta_{\pm}B(k_{\pm}, E \pm \hbar\omega, \bar{V})]$$

$$(A1.2.12)$$

$$C_{\pm} = -i\frac{V_1}{2\hbar\omega}[\alpha_{\pm}C(k, E, \bar{V} \mp \hbar\omega) + \beta_{\pm}C(k_{\pm}, E \pm \hbar\omega, \bar{V})]$$

The peculiar form of the coefficients of $\alpha_{\pm}$ and $\beta_{\pm}$ are chosen in order to make maximum use of the static parts of the wave functions. To do so, we observe that for small modulation frequency, $\hbar\omega \ll E, \bar{V}$ (but keeping $\hbar\omega \gg V_1$), the values of B, C, ϕ_1, and ϕ_2 are unchanged if we replace the arguments $(E, \bar{V})$ by $(E \pm \hbar\omega, \bar{V} \pm \hbar\omega)$. This is an exact relation within the WKB approximation,[20] where $V(x)$ and E always appear in the combination $V(x) - E$. We can now manipulate the matching relation (A1.2.10) to take the simple form

$$A_{\pm}\left(i\frac{V_1}{2\hbar\omega}\right)^{-1} = (1 + A(k, E, \bar{V})) - \alpha_{\pm}(1 + A(k, E, \bar{V} \mp \hbar\omega))$$

$$- \beta_{\pm}(1 + A(k\pm, E \pm \hbar\omega, \bar{V})) \qquad (A1.2.13)$$

while (A1.2.11) becomes

$$-ik_{\pm}A_{\pm}\left(i\frac{V_1}{2\hbar\omega}\right)^{-1} = ik(1 - A(k, E, \bar{V})) - ik\alpha_{\pm}(1 - A(k, E, \bar{V} \mp \hbar\omega))$$

$$-ik_{\pm}\beta_{\pm}(1 - A(k\pm, E \pm \hbar\omega, \bar{V})) \qquad (A1.2.14)$$

We now have four equations to solve for the six unknowns $A_{\pm}$, $\alpha_{\pm}$, and $\beta_{\pm}$. This is not enough, but one also has to consider the matching of the sidebands at $x = d/2$. One finds

$$i\frac{V_1}{2\hbar\omega}\left[B(k, E, \bar{V})\phi_1\left(\frac{d}{2}, E, \bar{V}\right) + C(k, E, \bar{V})\phi_2\left(\frac{d}{2}, E, \bar{V}\right)\right]$$

$$+ \left[B_\pm\phi_1\left(\frac{d}{2}, E \pm \hbar\omega, \bar{V}\right) + C_\pm\phi_2\left(\frac{d}{2}, E \pm \hbar\omega, \bar{V}\right)\right] = D_\pm \quad \text{(A1.2.15)}$$

and

$$i\frac{V_1}{2\hbar\omega}\left[B(k, E, \bar{V})\phi_1'\left(\frac{d}{2}, E, \bar{V}\right) + C(k, E, \bar{V})\phi_2'\left(\frac{d}{2}, E, \bar{V}\right)\right]$$

$$+ \left[B_\pm\phi_1'\left(\frac{d}{2}, E \pm \hbar\omega, \bar{V}\right) + C_\pm\phi_2'\left(\frac{d}{2}, E \pm \hbar\omega, \bar{V}\right)\right] = ik_\pm D_\pm$$

$$\text{(A1.2.16)}$$

Using the static matching relations (A1.2.9) and (A1.2.12) we can manipulate (A1.2.15) and (A1.2.16) to read

$$D(k, E, \bar{V}) - \alpha_\pm D(k\pm, E \pm \hbar\omega, \bar{V}) - \beta_\pm D(k, E, \bar{V} \mp \hbar\omega) = D_\pm\left(i\frac{V_1}{2\hbar\omega}\right)^{-1}$$

$$\text{(A1.2.17)}$$

and

$$ikD(k, E, \bar{V}) - ik_\pm\alpha_\pm D(k\pm, E \pm \hbar\omega, \bar{V})$$

$$- ik\beta_\pm D(k, E, \bar{V} \mp \hbar\omega) = ik_\pm D_\pm\left(i\frac{V_1}{2\hbar\omega}\right)^{-1} \quad \text{(A1.2.18)}$$

We have now four more equations but only two more unknowns, $D_\pm$, and we can therefore solve the eight equations (A1.2.13), (A1.2.14), (A1.2.17), and (A1.2.18) for the eight unknowns $A_\pm$, $D_\pm$, $\alpha_\pm$, and $\beta_\pm$. This is straightforward, but before we proceed we stress that we have been careful in using labels $A(k, E, \bar{V})$. But in (A1.2.13), (A1.2.14), (A1.2.17), and (A1.2.18) the first two labels always appear as k, E or $k_\pm, E \pm \hbar\omega$; hence, we can drop the separate label k in A and D in what follows, and let these amplitudes depend on k through E. Solving for $\alpha_\pm$ and $\beta_\pm$, we find

$$\alpha_\pm = \frac{D(E, \bar{V})}{D(E, \bar{V} \mp \hbar\omega)} \quad \text{(A1.2.19)}$$

and

$$\beta_\pm = \frac{1}{2}\left(1 + \frac{k}{k_\pm}\right)\left[1 - \frac{D(E, \bar{V})}{D(E, \bar{V} \mp \hbar\omega)}\right]$$

$$+ \frac{1}{2}\left(1 - \frac{k}{k_\pm}\right)\left[A(E, \bar{V}) - \frac{A(E, \bar{V} \mp \hbar\omega)D(E, \bar{V})}{D(E, \bar{V} \mp \hbar\omega)}\right] \quad \text{(A1.2.20)}$$

For the amplitudes of the transmitted and reflected sidebands we get

$$D_\pm = -\frac{i}{2}\frac{V_1}{2\hbar\omega} D(E \pm \hbar\omega, \bar{V})$$

$$\times \left(\left(1 + \frac{k}{k_\pm}\right)\left[1 - \frac{D(E, \bar{V})}{D(E, \bar{V} \mp \hbar\omega)}\right] + \left(1 - \frac{k}{k_\pm}\right) A(E, \bar{V})\left[1 - \frac{D(E, \bar{V})}{D(E, \bar{V} \mp \hbar\omega)} \frac{A(E, \bar{V} \mp \hbar\omega)}{A(E, \bar{V})}\right] \right)$$

$$(A1.2.21)$$

and

$$A_\pm = \tfrac{1}{2} A(E, \bar{V})\left(1 + \frac{k}{k_\pm}\right)\left(i\frac{V_1}{2\hbar\omega}\right)\left[1 - \frac{D(E, \bar{V})}{D(E, \bar{V} \mp \hbar\omega)} \frac{A(E, \bar{V} \mp \hbar\omega)}{A(E, \bar{V})}\right]$$

$$- \tfrac{1}{2} A(E \pm \hbar\omega, \bar{V})\left(1 + \frac{k}{k_\pm}\right)\left(i\frac{V_1}{2\hbar\omega}\right)\left[1 - \frac{D(E, \bar{V})}{D(E, \bar{V} \mp \hbar\omega)}\right]$$

$$- \tfrac{1}{2} A(E \pm \hbar\omega, \bar{V})\left(1 - \frac{k}{k_\pm}\right) A(E, \bar{V})\left(i\frac{V_1}{2\hbar\omega}\right)$$

$$\times \left[1 - \frac{D(E, \bar{V})}{D(E, \bar{V} \mp \hbar\omega)} \frac{A(E, \bar{V} \mp \hbar\omega)}{A(E, \bar{V})}\right]$$

$$+ \frac{1}{2}\left(1 - \frac{k}{k_\pm}\right)\left(i\frac{V_1}{\hbar\omega}\right)\left[1 - \frac{D(E, \bar{V})}{D(E, \bar{V} \mp \hbar\omega)}\right] \qquad (A1.2.22)$$

If we first turn to the transmitted wave, we should consider the expression

$$e^{ik(x-d/2)-iEt/\hbar}\left(D(E, \bar{V}) + D_+ e^{i(k_+-k)(x-d/2)-i\omega t} + D_- e^{i(k_--k)(x-d/2)+i\omega t}\right)$$

$$(A1.2.23)$$

We want to keep terms to lowest order in ω. With $k_\pm = k \pm \omega/v_k$, where $\hbar v_k = dE/dk$, one has

$$e^{i(k_\pm - k)x \pm i\omega t} = 1 \pm i\omega(x/v_k - t) + O(\omega^2)$$

Expanding $D_\pm$ (and later $A_\pm$), we need

$$1 - \frac{D(E, \bar{V})}{D(E, \bar{V} \mp \hbar\omega)} \frac{A(E, \bar{V} \mp \hbar\omega)}{A(E, \bar{V})}$$

$$= \pm i\omega(\tau_T^{\bar{V}} - \tau_R^{\bar{V}}) + \tfrac{1}{2}\omega^2(\tau_T^{\bar{V}} - \tau_R^{\bar{V}})^2$$

$$- \frac{i}{2}\hbar\omega^2 \frac{d}{d\bar{V}}(\tau_T^{\bar{V}} - \tau_R^{\bar{V}}) + O(\omega^3) \qquad (A1.2.24)$$

as well as

$$1 - \frac{D(E, \bar{V})}{D(E, \bar{V} \mp \hbar\omega)} = \pm\omega\tau_T^{\bar{V}} + \tfrac{1}{2}(\omega\tau_T^{\bar{V}})^2 - \frac{i}{2}\,\hbar\omega^2\,\frac{d}{d\bar{V}}\,\tau_T^{\bar{V}} + O(\omega^3) \quad \text{(A1.2.25)}$$

and

$$D(E \pm \hbar\omega, \bar{V}) = D(E, \bar{V})(1 \mp i\omega\tau_T^E) + O(\omega^2)$$
$$A(E \pm \hbar\omega, \bar{V}) = A(E, \bar{V})(1 \mp i\omega\tau_R^E) + O(\omega^2) \quad \text{(A1.2.26)}$$

We have defined four complex quantities [cf. equation (6) of Ref. 39] which have the dimension of time:

$$\tau_T^{\bar{V}} = i\hbar\,\frac{d \ln D(E, \bar{V})}{d\bar{V}}$$

$$\tau_R^{\bar{V}} = i\hbar\,\frac{d \ln A(E, \bar{V})}{d\bar{V}}$$

$$\tau_T^E = i\hbar\,\frac{d \ln D(E, \bar{V})}{dE}$$

$$\tau_R^E = i\hbar\,\frac{d \ln D(E, \bar{V})}{dE} \qquad \text{(A1.2.27)}$$

At this stage it is not clear that these times have any meaningful interpretation. There are no clocks that measure complex times! These complex times are, on the other hand, related to real quantities with dimension of time that have been discussed in the literature. In the adiabatic limit we find

$$|D_{\pm}|^2 \to \left(\frac{V_1|\tau_T^{\bar{V}}|}{2\hbar}\right)^2 |D(E, \bar{V})|^2 \qquad \text{(A1.2.28)}$$

With $|D_{\pm}|^2 = T_{\pm}$ and $|D(E, \bar{V})|^2 = T(E)$, this is (10.3.6) and (10.3.7), used by Büttiker and Landauer to define a traversal time for a general potential barrier. The corresponding definition of a reflection time appeals to the adiabatic limit of the reflected sidebands, $R_{\pm} = |A_{\pm}|^2$, which gives

$$|A_{\pm}|^2 \to \left(\frac{V_1|\tau_R^{\bar{V}}|}{2\hbar}\right)^2 |A(E, \bar{V})|^2 \qquad \text{(A1.2.29)}$$

where $|A(E, \bar{V})|^2 = R(E)$ is the static reflection probability. We further note that from (A1.2.28) it follows directly that

$$T \operatorname{Im} \tau_T^{\bar{V}} + R \operatorname{Im} \tau_R^{\bar{V}} = 0, \qquad T \operatorname{Im} \tau_T^E + R \operatorname{Im} \tau_R^E = 0 \qquad \text{(A1.2.30)}$$

These relations merely express the fact that momentum is conserved in the scattering process. It can be shown that[14,15]

$$\tau_{\text{dwell}} = T \operatorname{Re} \tau_T^{\bar{V}} + R \operatorname{Re} \tau_R^{\bar{V}} \qquad \text{(A1.2.31)}$$

For a symmetric barrier, where $\operatorname{Re} \tau_{\text{dwell}}^{\bar{V}} = \operatorname{Re} \tau_R^{\bar{V}}$, one has therefore

$$\tau_T^{\bar{V}} = \tau_{\text{dwell}} = i\tau_z, \qquad \tau_R^{\bar{V}} = \tau_{\text{dwell}} + i\frac{T}{R}\,\tau_z \qquad \text{(A1.2.32)}$$

For a simple square barrier, explicit expressions for Büttiker's time τ_z and the dwell time τ_{dwell} can be found in, for instance, Refs. 16 and 24.

Using the expansions above, the transmitted wave function to lowest order in ω can be written as in (10.3.8). The expressions for the charge and current densities in (10.3.18) and (10.3.19) follow from the usual definitions

$$\rho(x, t) = |\psi(x, t)|^2, \qquad J(x, t) = \frac{1}{2m}\left[\psi^*(-i\hbar)\frac{d\psi}{dx} + \psi(i\hbar)\frac{d\psi^*}{dx}\right]$$

REFERENCES

1. M. Büttiker and R. Landauer, *J. Phys. C* **21**, 6207 (1988).
2. For a review see A. J. Leggett, S. Chakravarty, A. T. Dorsey, M. P. A. Fisher, A. Garg, and W. Zwerger, *Rev. Mod. Phys.* **59**, 1 (1987).
3. M. Jonson, *Solid State Commun.* **33**, 743 (1980).
4. D. Bohm, *Quantum Theory*, pp. 257–263 (Prentice-Hall, Englewood Cliffs, NJ (1951).
5. E. P. Wigner, *Phys. Rev.* **98**, 145 (1955).
6. L. Eisenbud, Princeton thesis (unpublished) (1948).
7. F. T. Smith, *Phys. Rev.* **118**, 349 (1960).
8. A. I. Baz', *Sov. J. Nucl. Phys.*, **4**, 182 (1967).
9. M. Büttiker and R. Landauer, *Phys. Rev. Lett.* **49**, 1739 (1982).
10. M. Büttiker and R. Landauer, *Phys. Scripta* **32**, 429 (1985).
11. See, for example, F. Capasso, K. Mohammed, and A. Y. Cho, *IEEE J. Quantum Electron.* **QE 22**, 1853 (1986).
12. For an example of recent work see, for instance, B. N. J. Persson, *Phys. Scripta* **38**, 282 (1988).
13. M. Büttiker, *Phys. Rev. B* **27**, 6178 (1983).
14. D. Sokolovski and L. M. Baskin, *Phys. Rev. A* **36**, 4604 (1987).
15. D. Esteve, J. M. Martinis, C. Urbina, E. Turlot, and M. H. Devoret, *Phys. Scripta* **T29**, 121 (1989).
16. E. H. Hauge and J. A. Støvneng, *Rev. Mod. Phys.* **1**, 1 (1990).
17. M. Büttiker, in: *Electronic Properties of Multilayers and Low Dimensional Semiconductor Structures* (J. M. Chamberlain, L. Eaves, and J. C. Portal, eds.), Plenum, New York (1989).
18. D. Lowe and S. Collins, *J. Phys. C* **21**, 6210 (1988).
19. C. R. Leavens and G. C. Aers, *Phys. Rev. B* **39**, 1202 (1989).
20. L. D. Landau and E. M. Lifshitz, *Quantum Mechanics*, Pergamon Press, Oxford (1977).
21. This expression differs from a similar one giving the partial cross section, i.e., the probability for the scattered wave to have angular momentum l. The latter quantity has[20] an extra factor of $4\sin^2\delta_l$.
22. T. E. Hartmann, *J. Appl. Phys.* **33**, 3427 (1962).
23. A. P. Jauho and M. M. Nieto, *Superlatt. Microstruct.* **2**, 407 (1986).
24. S. Collins, D. Lowe, and J. R. Barker, *J. Phys. C* **20**, 6213 (1987).
25. E. H. Hauge, J. P. Falck, and T. A. Fjeldly, *Phys. Rev. B* **36**, 4203 (1987).
26. N. Teranishi, A. M. Kriman, and D. K. Ferry, *Superlatt. Microstruct.* **3**, 509 (1987).
27. C. R. Leavens and G. C. Aers, in: Basic concepts and applications of *Scanning Tunneling Microscopy and Related Techniques* (R. J. Behm, ed.), Kluwer Academic, Dordrecht (1989).
28. M. Büttiker and R. Landauer, *IBM J. Res. Develop.* **30**, 451 (1986).
29. J. P. Falck and E. H. Hauge, *Phys. Rev. B* **38**, 3287 (1988).

30. W. JAWORSKI and D. M. WARDLAW, *Phys. Rev. A* **37**, 2843 (1988).

31. R. P. FEYMAN and A. R. HIBBS, *Quantum Mechanics and Path Integrals*, McGraw-Hill, New York (1965).

32. M. L. GOLDBERG and K. M. WATSON, *Collision Theory*, Wiley, New York (1964).

33. See, for instance, W. O. AMREIN and M. B. CIBILS, *Helv. Phys. Acta* **60**, 481 (1987).

34. M. TSUCHIYA, T. MATSUSUE, and H. SAKAKI, *Phys. Rev. Lett.* **59**, 2356 (1987); M. K. JACKSON, M. B. JOHNSON, D. H. CHOW, T. C. MCGILL, and C. W. NIEH, *Appl. Phys. Lett.* **54**, 522 (1989); N. SAWAKI, R. A. HÖPFEL, E. GORNIK, and H. KANO, *Appl. Phys. Lett.* **55**, 1996 (1989).

35. J. A. STØVNENG and E. H. HAUGE, to appear in *J. Stat. Phys.*

36. A. P. JAUHO and M. JONSON, *J. Phys.: Condens. Matter* **1**, 9027 (1989).

37. C. R. LEAVENS and G. C. AERS, *Solid State Commun.* **63**, 1101 (1987).

38. C. R. LEAVENS and G. C. AERS, *J. Vac. Sci. Technol.* **A6**, 305 (1988).

39. C. R. LEAVENS, *Solid State Commun.* **67**, 1135 (1988).

40. D. SOKOLOWSKI and P. HÄNGGI, *Euroophys. Lett.* **7**, 7 (1988).

41. V. F. RYBACHENKO, *Sov. J. Nucl. Phys.* **5**, 635 (1967).

42. E. POLLAK and W. H. MILLER, *Phys. Rev. Lett.* **53**, 115 (1984).

43. E. POLLAK, *J. Chem. Phys.* **83**, 1111 (1985).

44. C. R. LEAVENS and G. C. AERS, *Solid State Commun.* **63**, 1107 (1987).

45. C. R. LEAVENS and G. C. AERS, *Phys. Rev. B* **40**, 5387 (1989).

46. Z. KOTLER and A. NITZAN, *J. Chem. Phys.* **88**, 3871 (1988).

47. G. D. MAHAN, in: *Collective Properties of Physical Systems*, Nobel Symposium XXIV (B. I. LUNDQVIST and S. LUNDQVIST, eds.) Academic, New York (1974); M. SUNJIC, G. TOULOUSE, and A. A. LUCAS, *Solid State Commun.* **11**, 1629 (1972); J. Heinrichs, *Phys. Rev. B* **8**, 1346 (1973).

48. A. A. LUCAS, E. KARTHEUSER, and R. G. BRADO, *Phys. Rev. B* **2**, 2488 (1970).

49. G. D. MAHAN, *Many Particle Physics*, Ch. 4, Plenum, New York (1981).

50. P. M. ECHENIQUE, A. GRAS-MARTI, J. R. MANSON, and R. H. RITCHIE, *Phys. Rev. B* **35**, 7357 (1987).

51. B. N. J. PERSSON and A. BARATOFF, *Phys. Rev. B* **38**, 9616 (1988).

52. B. RUDBERG and M. JONSON, *Phys. Rev. B* (1991).

53. A. J. LEGGETT, *Progr. Theor. Phys.*, Suppl. **69**, 80 (1980).

54. P. GUÉRET, E. MARCLAY, and H. MEIER, *Appl. Phys. Lett.* **53**, 1617 (1988).

55. P. GUÉRET, E. MARCLAY, and H. MEIER, *Solid State Commun.* **68**, 977 (1988).

56. See, for example, M. H. DEVORET, J. M. MARTINIS, and D. ESTEVE, *Helv. Phys. Acta* **61**, 622 (1988); E. TURLOT, D. ESTEVE, C. URBINA, J. M. MARTINIS, and M. H. DEVORET, *Phys. Rev. Lett.* **62**, 1788 (1989).

57. P. W. ANDERSON, *Lectures on the Many Body Problem*, p. 113, Academic, New York (1963).

58. S. M. GIRVIN, L. I. GLAZMAN, M. JONSON, D. R. PENN, and M. D. STILES (unpublished, submitted to *Phys. Rev. Lett.*); M. H. DEVORET, D. ESTEVE, H. GRABERT, G.-L. INGOLD, H. POTTIER, and C. URBINA, *Phys. Rev. Lett.* **64**, 1824 (1990).

59. A. J. LEGGETT, *Phys. Rev. B* **30**, 1208 (1984).

60. M. JONSON and A. GRINCWAJG, *Appl. Phys. Lett.* **51**, 1729 (1987).

61. A. A. LUCAS, P. H. CUTLER, T. E. FEUCHTWANG, T. T. TSONG, T. E. SULLIVAN, Y. YUK, H. NGUYEN, and P. J. SILVERMAN, *J. Vac. Sci. Technol.* **A6**, 461 (1988).

62. R. LANDAUER, *Nature* **341**, 567 (1989).

63. P. GUÉRET, A. BARATOFF, and E. MARCLAY, *Europhys. Lett.* **3**, 367 (1987).

64. L. EAVES, K. W. H. STEVENS, and F. W. SHEARD, in: *The Physics and Fabrication of Microstructures and Microdevices*, (M. J. Kelly and C. Weisbuch, eds.) Springer, Berlin (1986).

Chapter 11

Wigner Function Modeling of the Resonant Tunneling Diode

A. M. Kriman, N. C. Kluksdahl, David K. Ferry, and C. Ringhofer

11.1. INTRODUCTION

In recent years a number of techniques have been developed which have allowed semiconductor devices to be fabricated with ever-smaller feature sizes. Using MBE and MOCVD epitaxial growth techniques,[1] the composition of semiconductor wafers can be controlled on atomic length scales. This leads to potential variations along one (growth) direction that vary on scales as small as one atomic layer. Using electron-beam and x-ray lithographies, it has also become possible to pattern gates with characteristic features on the scale of a few tens of nanometers. These small scales are comparable to the wavelengths of the electrons and holes within the devices. In this regime, device modeling cannot be based on a classical picture augmented by quantum corrections and parameters; it must incorporate quantum mechanics at a fundamental level.

At present, the electronic devices which exhibit intrinsically quantum mechanical transport effects fall into two main categories: quasi-one-dimensional devices, in which transport takes place along the direction

A. M. Kriman and N. C. Kluksdahl • Department of Electrical Engineering, Arizona State University, Tempe, Arizona 85287, USA.   D. K. Ferry • Center for Solid State Electronic Research, College of Engineering and Applied Science, Arizona State University, Tempe, Arizona 85287, USA.   C. Ringhofer • Department of Mathematics, Arizona State University, Tempe, Arizona 85287, USA. The present address of A. M. Kriman is Department of Electrical and Computer Engineering, State University of New York at Buffalo, Buffalo, New York 14260, USA. The present address of N. C. Kluksdahl is Ford Aerospace, Houston, Texas 77058, USA.

Quantum Transport in Semiconductors, edited by David K. Ferry and Carlo Jacoboni. Plenum Press, New York, 1991.

defined by epitaxial growth;[2-4] and surface, or quasi-two-dimensional, devices, in which transport takes place primarily within a single epitaxially grown layer.[5-7] This chapter is concerned with the former class of device. We describe the modeling of quasi-one-dimensional devices by means of the Wigner distribution function.

The paradigmatic quasi-one-dimensional device is the resonant tunneling diode (RTD).[2-4] This is a finite superlattice: a sequence of barriers and wells formed by epitaxial growth of alternating layers of semiconductor material. In the typical double-barrier resonant tunneling diode, a thin GaAs quantum well separates two AlGaAs barriers. The AlGaAs barrier layers have a wider band gap than GaAs and a higher conduction band edge. The band edge can be modeled as a potential, and the AlGaAs layers serve as potential barriers, whose height is controlled by adjusting the composition at the time of growth. RTDs can be grown in other material systems, subject primarily to the requirement that lattice constants of the two materials be close enough to permit epitaxial growth. The GaAs/AlGaAs system has the advantage that it is lattice-matched at all aluminum (alloy) compositions. Its main drawback is that the satellite valleys in AlGaAs are close in energy to the central conduction band minimum. (Indeed, at aluminum concentrations above $x \approx 0.3$, the alloy gap goes from direct to indirect.) This complicates the conduction process, particularly in long-barrier RTDs. In this chapter we will discuss only a single-valley model.

The resonant tunneling diode generally has barriers that are much higher than the typical electron energies. As a result, low-bias conduction is dominated by tunneling, a quantum mechanical process. In such a device, there can be no question of using a classical simulation: in the classical limit, the current is essentially zero. The double-barrier structure of the RTD gives rise to a current-voltage (I-V) characteristic with a prominent negative differential conductance (NDC) region (Figure 11.1). This feature of the transport behavior of the RTD can be understood in terms of a quasibound localized within the central quantum well region, which is the lowest-energy resonance of the transmission probability. As the barrier height is increased and the confinement of the electron to the well region becomes more exact, the lifetime of the resonance increases and the resonance becomes sharper. That is, the imaginary part of its energy becomes small and the resonance becomes an exact bound state.

When an electron, incident on the double-barrier structure from one side, has an energy in resonance with the quasibound state, the transmission probability has a maximum. In practice, this resonance condition is satisfied by biasing the RTD so that the potential drop across the barrier on the cathode side equals the energy needed to bring the cathode electrons into resonance with the quasibound state. This resonance appears as a peak in the I-V characteristic. At higher bias, the quasibound state has an energy which lies below the incident electron energy, giving rise to an NDC.

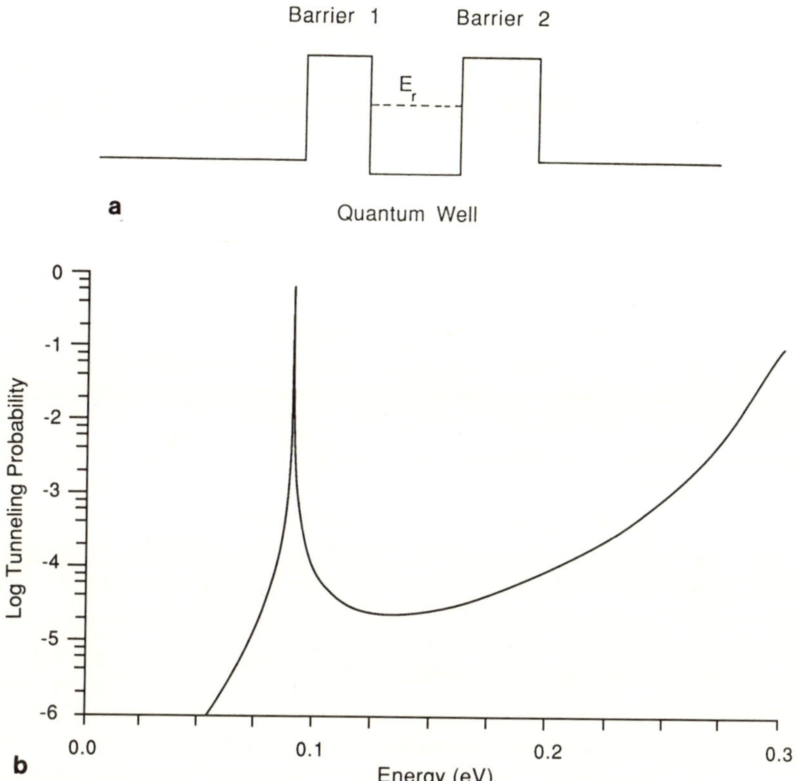

FIGURE 11.1. (a) A generic double barrier system. (b) The tunneling probability for the resonant tunneling system, as a function of longitudinal kinetic energy.

Detailed theoretical treatment of these quantum devices has lagged the progress made in their fabrication and electrical performance. The first attempts at modeling the tunneling structures made simple assumptions about the underlying potential, specifically ignoring the part of the self-consistent potential arising from the Coulomb interaction between electrons. From that point, modeling can be relatively straightforward.[8-10] For a given potential $V(x)$, a tunneling probability $T(E)$ can be calculated as a function of energy E using a variety of methods, both exact and approximate (most commonly, WKB, which is simple, but not strictly applicable to heterostructure potentials, as these do not vary slowly on scales of a de Broglie wavelength.[11] Among the exact methods are the phase integral approach[12] as well as a number of formulations of the transfer matrix approach.[8,10,13,14] The current through the device is then calculated by integrating the charge density times, the tunneling probability, and the carrier velocity $v(E)$. The current is a function of the differences in the distributions functions and Fermi energies in the two sides.[15,16] For a one-dimensional system, the current from left to right is

$$J_{lr} = q \int dE \, T_{lr}(E)n(E)f(E)v(E) \tag{11.1.1}$$

where $f(E)$ is the Fermi–Dirac distribution function, $-q$ is the electron charge, and $n(E)$ is the density of states. A second equation is obtained for J_{rl}, and the total current is then the net $J_{lr} - J_{rl}$.

A major criticism of this approach is that it depends on a knowledge of the distribution of the electrons at each side of the tunneling interface, rather than the bulklike distribution far from the tunneling interface, although the latter is often used. The distribution at the interface is completely different from the bulk distribution due to quantum repulsion from the barrier,[17,18] and the application of a small bias can affect the high-momentum tails of this distribution dramatically. This difference can profoundly affect the low-bias behavior.

As noted, many treatments ignore self-consistent potential effects. It is known that band bending—the self-consistent modification of potentials due to redistribution of free charge—may lead to the formation of bound states or other types of quantization in the interface region.[19] Accumulation or depletion layers, and their effects upon the potential and, consequently, the tunneling probability, can drastically affect the local distributions and the tunneling current.

The primary evidence that one must go beyond the simple tunneling models, however, lies in the poor fit of predicted *I-V* curves to those measured experimentally. Figure 11.2 shows a comparison of an *I-V* curve calculated by the tunneling approach and the experimentally observed *I-V* curve for a device with the same structural parameters. Typically the current predicted for the valley of the curve is considerably smaller than observed. Part of the reason for low theoretical predictions of the valley current is that the simple models do not take account of barrier lowering or ballistic injection over the barriers, both of which can be important experimentally.

In this chapter, we present a fully self-consistent model of the RTD based on the quantum mechanical Wigner function. Though the Wigner function has been around for quite some time, and its general formal properties have been studied extensively,[20-23] it has only recently begun to be applied to electronic transport.

The Wigner formalism offers many advantages for quantum modeling. First, it is a phase-space description, similar to classical Boltzmann distributions and therefore accessible to classical intuition. In the Wigner formalism, scattering is a local phenomenon.[23] In particular, it is conceptually possible to use the correspondence principle to determine where quantum corrections enter a problem. At the boundaries, the phase-space description permits separation of incoming and outgoing components of the distribution, which

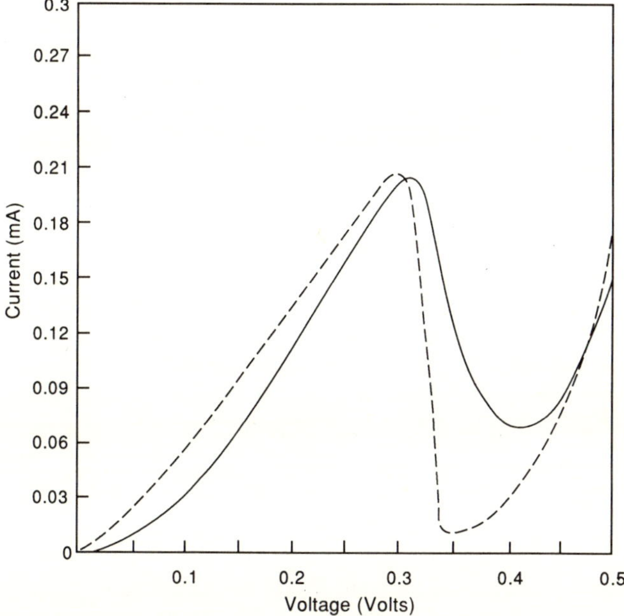

FIGURE 11.2. Experimental (solid) and theoretical (dashed) *I-V* curves for a resonant tunneling diode. The theoretical curve is calculated by a simple tunneling model.

thus permits one to model an ideal contact, and hence an open system. The clean separation of inflow and outflow boundaries is also associated with the possibility of defining "nonreflecting" boundary conditions which are important in modeling a system with nonlocal operators. Still another advantage is that the Wigner function is purely real, which simplifies calculation and interpretation of results. By coupling the Wigner function equation of motion to the Poisson equation, we obtain a fully self-consistent model of the RTD. This then allows us to examine charge redistribution effects and many properties of the devices.

In the field of quantum optics, closely related Wigner function formalisms have been used widely. The quantum phase-space distribution has been used to describe coherence of optical fields and polarization and transient superposition effects.[24] The description has been applied to finding solutions for a laser master equation (the Fokker–Planck equation[25]) and for describing quantum noise in lasers.[26] More recently, Wigner functions have been applied to optical systems and signals, where they provide a link between Fourier optics and geometric optics.[27,28] Two-dimensional Wigner optical distributions have been generated,[29] as well as slices of four-dimensional Wigner optical distributions.[30] The Wigner function description of optical signals has been investigated for use in elementary pattern recognition.[31] It is clear that the Wigner function is useful for exhibiting phase interference and quantum resonance.

A number of articles have described the utility of the Wigner function formalism for quantum electronic transport, with a view toward the advantages offered by a phase-space representation.[21–23,32,33] Some early attempts to model transport in the RTD used the Wigner formalism,[18,34] while other models advocated use of the density matrix.[35] The density matrix approach is quantum mechanically correct and indeed completely equivalent to the Wigner function approach. However, in comparison with the latter it exhibits some serious drawbacks. The density matrix is complex, and except for the diagonal terms which yield the particle density $n(x)$, the density matrix does not correspond directly to any classical function. Additionally, scattering is nonlocal in the density matrix formalism, which complicates the inclusion of scattering in the transport model. For these reasons the density matrix description of electronic transport has made slower progress than the Wigner function description. Until now, in fact, it has still not been used to treat self-consistent potentials.

In the next section, we will review the development of the Wigner function and its equation of motion. The roles of nonlocality and correlation in the Wigner function will be discussed. The critical nature of a quantum mechanically correct initial distribution, and its subsequent temporal development, will be examined. Major concerns in modeling are the stability and the convergence of the numerical techniques used in the model, which will be discussed. Because an open system is modeled, the nature of the boundaries will be treated in some detail. Poisson's equation will be coupled with the Wigner function equation of motion to include fully self-consistent potentials, and simple scattering will be added to the model.

Finally, the self-consistent Wigner function model will be applied to RTDs of various dimensions, and the steady-state and transient behavior will be examined. Bistability, a controversial topic in RTD studies, is found to occur under a range of conditions that will be explored within the model. A zero-bias anomaly, previously observed in tunneling structures, is found to occur primarily as a result of the distribution functions near the barriers. Finally, the effect of variations in doping of the RTD structure will be explored.

11.2. THE WIGNER DISTRIBUTION FUNCTION

11.2.1. Description of a Statistical System Using the Wigner Function

The devices we model, though small enough to exhibit atomiclike resonance effects, are large enough to be describable in statistical terms.

The general statistical state is described by the density operator $\hat{\rho}$. In terms of this, the observable corresponding to a Hermitian operator $\hat{A}$ can be evaluated as

$$\langle A \rangle = \mathrm{Tr}(\hat{A}\hat{\rho}) \tag{11.2.1}$$

where Tr indicates a trace taken over the full Hilbert space of quantum states. For a single-particle system (i.e., on in which many-body effects such as exchange may be ignored), the information in this operator is equivalently described by the density matrix, which is its coordinate representation

$$\rho(x_1, x_2) = \langle x_1 | \hat{\rho} | x_2 \rangle \tag{11.2.2}$$

This density matrix contains essentially the same information for a quantum statistical system that the single-particle distribution function does for a classical system. However, the latter is a function of one position and one momentum, whereas (11.2.2) is a function of two positions. Clearly, a quantum analog of the classical distribution function must be a linear functional of the density matrix, but determining which one is not immediately obvious. A number of functions, as, for example, $\langle p | \hat{\rho} | x \rangle$, have the appropriate classical limit, but have some undesirable mathematical property. Typically, the function is not real in the quantum regime.

In this study, we use a quantum analog of the classical function which was introduced by Wigner.[20] To define it, it is useful to rewrite the density matrix in terms of the separation vector x between two points x_1 and x_2,[36,37]

$$x \equiv x_1 - x_2 \tag{11.2.3}$$

and the center-of-mass vector X (the mean of the positions x_1 and x_2),

$$X \equiv \tfrac{1}{2}(x_1 + x_2) \tag{11.2.4}$$

Then the density matrix may be described in terms of these two new coordinates as

$$\rho(x_1, x_2) = \rho(X + \tfrac{1}{2}x, X - \tfrac{1}{2}x)$$

Since the electron number density is given by

$$n(x) = \rho(x, x) \tag{11.2.5}$$

we expect the center-of-mass X to be the appropriate coordinate for the distribution function analog. (Note, however, that a distribution defined in terms of the center-of-mass coordinates may be nonzero at positions X where the wave functions, and probability density, are zero. This is almost unavoidable if the electron lies in a concave region, as illustrated in Figure 11.3.) Moreover, the density in momentum space can be obtained from an integral of ρ over x alone, so the momentum coordinate should arise from eliminating this separation vector.

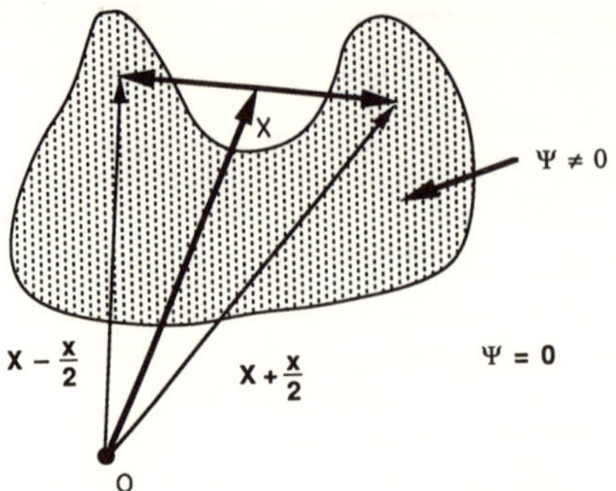

FIGURE 11.3. Illustration of the nonlocal nature of the Wigner function:[49] For an electron confined to a nonconvex region (shaded region), there are exterior points with nonzero Wigner function. [Some points in the unoccupied region that are centers of mass for pairs of points within the allowed (shaded) region.]

The Wigner function is defined as

$$f_w(X, p) \equiv \frac{1}{2\pi\hbar} \int dx\, e^{ipx/\hbar} \rho(X + \tfrac{1}{2}x, X - \tfrac{1}{2}x) \qquad (11.2.6)$$

It is easily demonstrated that this function is real, in contrast to other distribution function analogs. However, it may become negative, severely restricting any attempt to interpret it as a probability density.[38] The possibility of negative values can be reconciled with a limited probability interpretation, and the negative regions can be understood in terms of the Heisenberg uncertainty relations. We will return to this issue at various points. Other issues, concerning the "locality" of the coordinates, are examined in Refs. 23 and 32.

The transformation (11.2.6) used to obtain the Wigner function is a special case of the Wigner–Weyl transformation.[21,22,36] This transformation can be used to obtain a time-evolution equation (the quantum Liouville equation) for the Wigner function from the corresponding (von Neumann) equation for the density matrix. For a general noninteracting Hamiltonian, with a parabolic energy function,

$$H = \frac{\hat{p}^2}{2m} + V(x) \qquad (11.2.7)$$

this is[20–23]

$$\left[\frac{\partial}{\partial t} - \frac{p}{m}\frac{\partial}{\partial x}\right] f_w(x, p, t) = \frac{1}{2\pi\hbar} \int dP\, M(x, P) f_w(x, p + P, t) \qquad (11.2.8)$$

where

$$M(x, P) = \int dy\, e^{iPy/\hbar}[V(x + \tfrac{1}{2}y) - V(x - \tfrac{1}{2}y)] \qquad (11.2.9)$$

The kinetic term is formally identical to the kinetic term of the Boltzmann equation. The potential term in (11.2.8), however, is nonlocal. That is,

$f_w(x, p, t + \delta t)$ $(\delta t \to 0^+)$ depends on $f_w(x', p', t)$ for *all* x' and p'. This nonlocality is strictly a quantum correction to the equation of motion: as $\hbar \to 0$, the classical Liouville equation is recovered.[32]

For a general Hamiltonian H, the equation of motion is[21]

$$\frac{\partial}{\partial t} f_w(x, p, t) = \int d\eta \int d\xi \int d\theta \int d\tau \, f_w(\eta, \xi, \tau) e^{i[\tau(\eta - p) + \theta(\xi - x)]}$$

$$\times [H(\xi - \tfrac{1}{2}\hbar\tau, \eta + \tfrac{1}{2}\hbar\theta) - H(\xi + \tfrac{1}{2}\hbar\tau, \eta - \tfrac{1}{2}\hbar\theta)] \qquad (11.2.10)$$

The Wigner function is closely related to the correlation function $G^<$ in the quantum Boltzmann equation of the Kadanoff–Baym–Keldysh formalism.[37–39] While the Wigner function is a function of two positions and one time, the function $G^<$ is a two-position, *two-time* function, written $G^<(x_1, t_1, x_2, t_2)$, or $G^<(X, T, x, \tau)$ in terms of the center-of-mass coordinates. The relative coordinates x and τ may be Fourier-transformed (in the homogeneous, steady state) into momentum and frequency variables, giving $G^<(X, T, p, \omega)$. The Wigner function is the integral of $G^<(X, T, p, \omega)$ over the frequency ω, or equivalently the single-time $(\tau \to 0)$ limit of $G^<(X, T, p, \tau)$. Since $G^<$ is explicitly a correlation function, the single-time limit of $G^<$ is also a correlation function; the Wigner function thus automatically incorporates spatial correlations. Earlier, it was noted that the Wigner function may be nonzero at phase-space points with position coordinates in regions of vanishing density. These regions are thus seen to exhibit a component of the Wigner function which arises from two-point correlations.

11.2.2. Evaluation of Observables

In order to avoid having to transform back into a coordinate representation in order to obtain physical quantities, we want a version of (11.2.1) within the Wigner–Weyl description. We can obtain this by starting from the coordinate representation of (11.2.1) in the single-particle sector,

$$\langle A \rangle = \int dX \int dx \, A(X + \tfrac{1}{2}x, X - \tfrac{1}{2}x)\rho(X + \tfrac{1}{2}x, X - \tfrac{1}{2}x) \qquad (11.2.11)$$

with $A(X + \tfrac{1}{2}x, X - \tfrac{1}{2}x) = \langle X + \tfrac{1}{2}x|\hat{A}|X - \tfrac{1}{2}x\rangle$. Inserting an integral representation of the delta function,

$$\delta(x) = \frac{1}{2\pi} \int dp \, e^{ipx} \qquad (11.2.12)$$

we find

$$\langle A \rangle = \iint dp \, dx \, A_w(x, p) f_w(x, p) \qquad (11.2.13)$$

where $A_w(x, p)$ is the Wigner–Weyl transform of $A(x_1, x_2)$. For our purposes, $A_w(x, p)$ can be found by expressing the quantum mechanical operator $\hat{A}$ as a polynomial in $\hat{x}$ and $\hat{p}$, and substituting x and p for the corresponding operators. This procedure is valid whenever the noncommutation of $\hat{x}$ and $\hat{p}$ does not affect $\hat{A}$. In particular, this includes all functions of the form $A_1(x) + A_2(p)$.

The observable quantity of central importance in device modeling is the current. The current is an integral of the current density, which is given by the operator

$$\hat{j} = e\hat{p}/m \tag{11.2.14}$$

In a classical system, a local current density $J(x)$ may be exactly defined, since the distribution is a true probability distribution, and both position and momentum are exactly known. According to the Heisenberg uncertainty principle, however, exact knowledge of both position and momentum is impossible in a quantum system, so the current density at any given point in space cannot be known. In fact, the current density at a point may be written as $(e\hat{p}/m)\delta(x - \hat{x})$, so the expression (11.2.13) is not applicable. Indeed, the total current in any given direction is usually evaluated at a particular point along that direction by an integration over the remaining two dimensions. In the quasi-one-dimensional devices which we treat, this procedure is justified by stationarity: The current is independent of coordinate. In general, however, caution must be exercised.

11.3. INITIAL CONDITIONS

11.3.1. Relationship to the Equation of Motion

The problem we treat is essentially the time evolution of a statistical system, starting from an initial condition of thermal equilibrium. The initial condition requires quantum mechanics in its computation much as does the time evolution. Further observations may be made, however, regarding the relationship of the Wigner function equation of motion (11.2.8) to the specification of the initial condition.

The first point concerns the use of the time-evolution equation itself as a means to find the initial condition. In general, there is an infinite set of solutions to the full time-evolution equations. This includes at least the one-parameter family of solutions generated by varying the temperature. In the classical case, boundary conditions—which may represent contact with a thermal reservoir—may be used to find a unique stationary solution which can serve as the equilibrium initial condition. The distribution function is subject only to the condition of positivity; any nonnegative, normalized function is physically allowable as a probability distribution on the classical phase-space. In the quantum analog of this stationary-solution approach, the Wigner function is no longer required to be positive. Nevertheless, the Wigner function *is* required to arise from a mixed-state superposition of quantum mechanically allowed solutions of the Schrödinger equation. This more complicated requirement is known to impose a number of interesting necessary conditions, such as that f_w be bounded above by a number of order h^{-D}. It is possible to find an adjoint equation, supplementing the Wigner function time-evolution equation, whose satisfaction is a necessary and sufficient condition for the function to be an admissible Wigner function.[22]

Another point regards the energy of quantum mechanical effects through the nonlocal potential term. It is straightforward to show that a quadratic potential in (11.2.7) leads to precisely the same force term

$$\nabla V \cdot \nabla f_w(x, p, t) \tag{11.3.1}$$

which arises in the classical Boltzmann equation.[40,41] Thus, in that particular problem, all of the difference between the classical and the quantum mechanical systems can be traced back to the initial condition. In certain kinds of hydrodynamic equation approaches, this is reflected in the fact that quantum correction terms do not enter as corrections through the potential (forcing) term.[42]

In the case just considered, there is a lowest-energy bound state with energy above the potential minimum. It is then immediately clear that a classical and quantum thermal distributions are not equivalent. The question arises whether, in a problem *without* bound states, one can use the classical distribution to find an initial condition. The answer is that to a certain limited extent, one can.

We are modeling here a device which is embedded in an infinite system, with the potential approaching a constant at infinity. Within a single-particle description, the classical distribution function and the quantum (Wigner) distribution function are equal for a constant potential. For a system in thermal equilibrium, the distribution is determined by the boundary conditions: the equilibrium distribution is the stationary solution determined by the thermal baths with which the system is in contact. In principle, then, one may determine a thermal initial condition from the boundary conditions. (It is necessary to include inelastic processes in order to thermalize the occupancy of the bound states.)

In practice, as we shall see, boundary conditions for a nonlocal equation like (11.2.8) are so tricky that this approach is more of a problem than a solution. Furthermore, application of this approach requires the simulation region to be extended as much as possible away from the device, since the distribution function approaches the constant potential one only far from potential deviations from a constant. "Far" here means further away than a healing distance. For a nondegenerate system, this is comparable to the thermal wavelength λ_{th} given by[17]

$$\lambda_{th} = \hbar\sqrt{\beta/2m} \qquad (11.3.2)$$

In a GaAs device at 300 K, this length is nearly 100 nm.

In view of these difficulties, we use a different approach to determine the initial condition: we "simply" evaluate the equilibrium distribution by taking a thermal sum over all states. The Wigner function equation of motion enters this approach only in an auxiliary role: after finding the initial condition we time-evolve the Wigner function that it gives rise to, in order to confirm that it is accurately stationary.

11.3.2. Computation of the Initial State

The only certain way to compute an initial condition is to start with the system in thermal equilibrium (the zero-bias state), There we have available well-known statistical mechanical expressions for the density matrix, from which we can compute the Wigner distribution function. In evaluating those expressions it is important to keep in mind that the system being modeled is considerably larger than the region which we can comfortably simulate. Thus, the boundaries of our simulation region do not correspond to physical walls where we might apply simple boundary conditions on the allowed wave functions.

If we impose hard-wall boundary conditions at the ends of the simulation region, we generate nonphysical dips in the density; if we impose periodic or Neumann boundary conditions, we still obtain spurious features in momentum space associated with Fourier transformation of the density matrix. One solution is to use a much larger box in computing the initial conditions than in performing the time evolution. This nevertheless gives rise to fine structure in the energy distribution, associated with quantization in the normalization box. Also, the energy of a state, which goes into the Gibbs factor, has a contribution from the microstructure potential that is fractionally of the same order as the size of the double barrier relative to the normalization box. This goes to zero, as it should for extended states, but it does so only as the inverse of the normalization box size.

To avoid these difficulties, we compute the initial conditions using states normalized in the continuum.[17,43] In the continuum, one complete basis of states for the problem is composed of all the bound states plus all of the scattering states. (These correspond to the "outgoing" states of ordinary scattering theory, associated with the retarded Green's functions.[11]) Far from the scattering region, each scattering state is composed of a single incident plane wave plus a number of outgoing plane waves, and can be labeled uniquely by its incident wave vector. We consider potentials which become constant at a finite distance from the origin: $V(x) = V^-$ for $x < x_-$ and $V(x) = V^+$ for $x > x_+$, so the scattering states attain their asymptotic form in $x < x_-$ and $x > x_+$. In the equilibrium case we need to consider, $V^\infty \equiv V^- = V^+$ ("zero bias"), and therefore the left-incident states (states with positive incident wave vector k) take the form

$$\psi_k(x) = \frac{1}{\sqrt{2\pi}} [e^{ikx} + r(k)e^{-ikx}] \tag{11.3.3a}$$

for $x < x^-$, and

$$\psi_k(x) = \frac{1}{\sqrt{2\pi}} t(k)e^{ikx} \tag{11.3.3b}$$

for $x > x_+$, where $t(k)$ is the transmission amplitude and $r(k)$ the reflection amplitude for wave vector k. States with negative k are defined in a similar manner, with the forms of the wave function on the left and right sides interchanged: For $k < 0$, (11.3.3b) describes a single wave traveling to the left, and is used to describe the transmitted wave in $x < x_-$, while (11.3.a) applies to $x > x_+$. The reflection and transmission amplitudes are essentially the analog for this problem of the T-matrix, or the deviation of the S-matrix from unity, that is defined in ordinary (three-dimensional) potential scattering. In analogy with the latter, they satisfy a number of properties arising from conservation laws (Levinson's theorem; optical and generalized optical theorems).[44] Also, reflecting the principle of causality, one has Wigner inequalities as in the ordinary case.[45] These are essentially quantum versions of the statement that a scattered electron cannot emerge from a bounded scatterer until after an incident electron has reached the scatterer.

The scattering states are computed by numerically integrating the Schrödinger equation from a point on the transmitted side described by (11.3.3b) ($x < x_-$ for $k < 0$, $x > x_+$ for $k > 0$), starting from an "initial" condition $\psi = 1$, $\psi' = ik$. The ordinary scattering states are in one-to-one correspondence with these ("Jost") states, differing only by a normalization factor and an irrelevant phase. Once the states have been propagated (the Schrödinger equation integrated) into the incident region where (11.3.3a) holds, the normalization factor is computed by decomposing the numerically determined wave function and imposing the constraint that the incident part has amplitude given by (11.3.3a). As a practical matter, a number of equivalent techniques can be used to integrate the Schrödinger equation. We use a transfer matrix method to integrate between the grid points on which the potential is defined.

The scattering states labeled by wave vector satisfy the same orthonormality relations satisfied by plane waves. (This can be shown formally, using a property of the Lippmann–Schwinger equation,[17,43] or by explicit calculation, on the basis of identities that express the conservation of probability.[43]) The states defined through (11.3.3) therefore have the appropriate normalization, and the density matrix is defined by

$$\rho(x, x') = \frac{1}{Z}\left[\sum_n \psi_n(x)\psi_n^*(x')f(E_n) + \int dk\, \psi_k(x)\psi_k^*(x')f(E_k)\right] \quad (11.3.4)$$

where Z is the partition function and $f(E)$ is the distribution function (Boltzmann or Fermi–Dirac). The discrete sum is over (nondegenerate, because one-dimensional) bound states with energies E_n. In the continuum sum, the energy $E(k)$ of a scattering state with incident wave vector k is given by the dispersion relation in the asymptotic region, by construction.

The integral is evaluated by a random sampling of k-values (we have typically used 10^4). Taking a random sample eliminates the problem of spurious structures in momentum space. A further small advantage concerns the fact that, when the energy equals the potential over any finite region, the usual (trigonometric expressions for the) transfer matrices becomes indeterminate. In this case degenerate forms must be used. Random sampling in energy automatically eliminates the need for this kind of exception handling, since the degeneracy then occurs with probability essentially zero (actually with probability $\approx$ floating-point precision $\div$ largest floating-point number).

The bound state sum that appears in (11.3.4) may or may not be empty. A well-known theorem of elementary quantum mechanics says that in one dimension, a potential with a negative region ($V(x) < V^\infty$) has at least one bound state. It is important to remember, however, that a condition on that theorem is that the potential be nowhere positive (i.e., $\sup(V) \le V^\infty$). In the RTD without a spacer layer (see below) we have found that bound states do occur. As a practical matter, in looking for bound states, one should expect them to come in pairs of nearly degenerate symmetric and antisymmetric states. This occurs because the barriers are generally so high, relative to the energies of the bound states, that the regions to the left and right of the left and right barriers, respectively, are each nearly independent half-spaces.

The partition function is essentially the exponential of the chemical potential and so contains information about the density. We therefore evaluate it by fixing the density at a point far from the microstructure, in the bulk where it is determined by the doping density. In the Boltzmann limit, it is given by

$$Z^{-1} = 2\sqrt{\pi}\, \lambda_{\text{th}} e^{\beta V^\infty} \rho_\infty \quad (11.3.5)$$

For a semi-infinite barrier, in the Boltzmann limit, it is possible to perform the scattering state sums analytically. In Figure 11.4, we display the Wigner distribution function for this potential, neglecting the self-consistent electrostatic potential. Oscillations can be seen clearly. It should

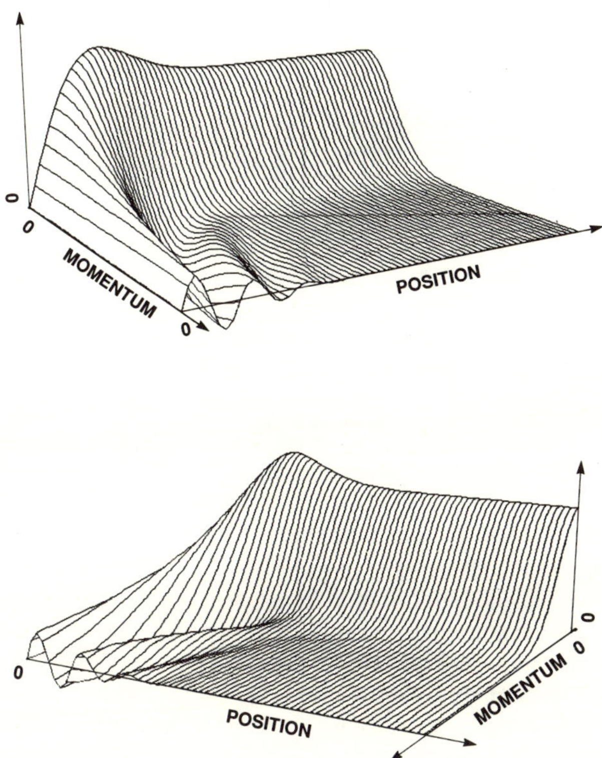

FIGURE 11.4. Wigner distribution near an infinite barrier. Two perspectives are shown to illustrate both quantum repulsion by the barrier and the related high-momentum tails in the distribution.

be understood that these are *not* Friedel oscillations, in the sense that they are not a consequence of the sharpness of the momentum distribution. Indeed, Figure 11.4 is computed from Boltzmann statistics, implying a smooth high-momentum cutoff. Instead, the oscillations arise from the sharpness of the barrier in space. Because the Wigner distribution graphically separates the components at different momenta, the oscillations in Figure 11.4 essentially represent standing-wave patterns caused by reflection from the barrier. These occur closer to the barrier at higher momentum (smaller wavelength).

In this view, the smoothing-out of the oscillations with increasing distance from the barrier is unexpected; the oscillations die out because the Schrödinger equation couples different momenta. This quantum repulsion causes the density of carriers to deplete, although this depletion is not a depletion layer in the classical sense. The depletion is caused by quantum

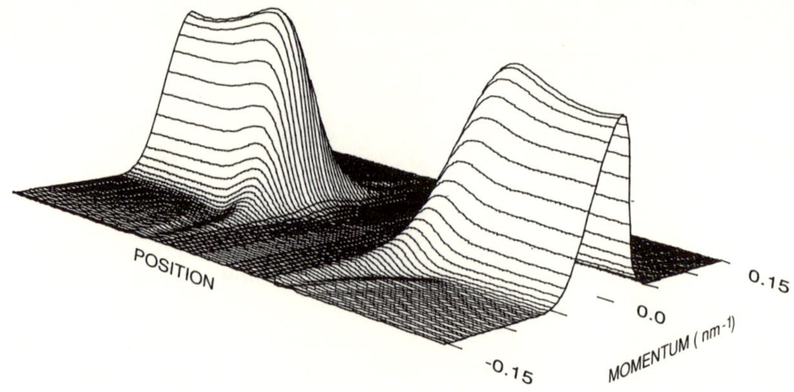

FIGURE 11.5. The self-consistent equilibrium Wigner function for a resonant tunneling diode, calculated with a scattering-state basis. Depletion at the barrier interface stems from quantum repulsion.

repulsion rather than by band bending. This quantum repulsion is, in a sense, complementary to barrier penetration; just as a nonzero density penetrates a finite distance into a classically forbidden region, a density deficit extends a finite distance into a classically allowed region.

In Figure 11.5, we show the results of a similar calculation, performed on a resonant tunneling diode potential.[41] Significantly, although this computation involves electrons that are only somewhat degenerate, the oscillations are less prominent than in Figure 11.4, where the electrons are completely nondegenerate. This confirms that the oscillations are not a Friedel-type effect. Instead, in the second calculation it is the potential which is less sharp, first because the potential barriers are finite, and second because the self-consistent corrections act in such a way as to smooth out the total potential. About the self-consistent effects we may observe in particular that the potential barriers, by excluding electrons, raise the electrostatic potential, which partially lowers the overall potential. We return to this issue when we consider the zero-bias anomaly.

11.4. NUMERICAL TECHNIQUES

Formally, the problem we are interested in involves continuous functions (Wigner function, potential, etc.) of continuous variables (position, momentum, time). As a practical matter, we must use a finite procedure to approximate the infinite information represented by such a problem. Concretely, all of the arguments become discrete, and the partial differential equations turn into a system of coupled ordinary difference equations. The possibility of solving even this finite, but still large, system of equations depends upon the choice of numerical approximations, discretization, and boundary conditions. We now consider these issues.

11.4.1. Discretization

The Wigner function equation of motion is evaluated on a two-dimensional discretized grid for the position and momentum variables x and p. The modeled region is a domain from $x = 0$ to $x = L$ and is divided into a spatial mesh, with a mesh size Δx chosen so that features of interest, such as potential barriers, are adequately represented. Since the potential in a RTD varies over a distance of a few atomic monolayers, an appropriate spatial mesh Δx is of the order of a unit cell of the GaAs lattice, and we approximate that with $\Delta x = 0.25$ nm. The mesh size for the momentum variable is found by considering the Fourier transform in (11.2.3) which defines the Wigner function. The discrete Wigner function is periodic in momentum with period $\hbar$. (For convenience, and especially in graphs, the momentum is expressed in terms of the wave vector $k = p/\hbar$.) This period in momentum is discretized into a convenient number of meshes and ranges from $-\pi/2\,\Delta x$ to $\pi/2\,\Delta x$. The discretized grid may conveniently be split into two regions: one half with positive momentum and the other half with negative momentum.[34,46] This will become important in the following discussion.

11.4.2. Stability and Convergence

A common numerical technique for solving discretized temporal equations is explicit differencing. Stability of an explicit scheme requires that error in the discretized equation remain bounded. When self-consistency is ignored, the system is linear and a Fourier analysis of the growth of error is possible. It leads to the Courant–Friedrichs–Lewy stability criterion[47] (CFL condition), which takes a standard form for explicit finite difference approximation schemes: For a mesh size Δx and a time step Δt, a necessary and sufficient condition for stability is that

$$\Delta x/\Delta t \le v \qquad\qquad (11.4.1)$$

where v is the velocity of the fastest component of the solution. For a time step Δt which satisfies (11.4.1) for the maximum velocity, convergence and stability of the solution are ensured for linear problems. For nonlinear systems, much less is known exactly, and the CFL condition may be regarded as a likely necessary, but probably not sufficient, condition.

Equation (11.2.8) is discretized using the Lax–Wendroff explicit time differencing.[48] The Lax–Wendroff method retains the second-order terms in the Taylor expansion of $f(t + \Delta t)$. This introduces a second-order spatial difference term into the equation of motion. This second-order term represents an artificial diffusion, which counteracts spurious numerical diffusion that always arises from the first-order terms. The Lax–Wendroff scheme has proven to be the technique of choice in numerical solutions of the Schrödinger equation.[18,49]

Each point in a discretized equation has a characteristic direction, which is to say that each point has a direction of propagation. For a phase-space representation, the velocity which defines propagation is directly proportional to the momentum. For positive momentum, propagation is in the positive x-direction, while negative momentum causes propagation in the negative x-direction. For any given point, the characteristic direction is defined by the local momentum. In the discretized Wigner function, there is a "slice" of the mesh over which the momentum is constant. These slices of the mesh can be viewed as systems of equations which are coupled through the potential term of (11.2.8). Each slice of the mesh is characterized by the same characteristic velocity; for positive momentum, information flows into the domain from the boundary at $x = 0$ and moves toward the boundary at $x = L$ where it leaves the domain. For negative momentum, the characteristic direction and the roles of the incoming and outgoing boundaries are reversed. Recognition of these characteristic directions solves an inherent problem.

A difficulty with the Lax–Wendroff discretization involves the boundaries. A second-order finite difference term at a point x_i involves the points x_{i-1} and x_{i+1}. In the interior of the device, this creates no problem, nor does it cause difficulty with the incoming boundary, where the distribution is specified. The problem occurs on the outgoing boundaries. The function is not known beyond the boundary, making the definition of a second-order difference at the boundary impossible. The solution to this dilemma is to use first-order upwind differencing[48] to propagate the function to the outgoing boundary along the characteristic direction, which is

$$\frac{\partial f}{\partial x} = \begin{cases} \dfrac{f(x_i) - f(x_{i-1})}{\Delta x} & \text{if } p(x_i) > 0 \\[2mm] \dfrac{f(x_{i+1}) - f(x_i)}{\Delta x} & \text{if } p(x_i) < 0 \end{cases} \tag{11.4.2}$$

Stability of the first-order terms also depends on satisfaction of the CFL condition. The discretization is illustrated in Figure 11.6. The characteristic directions on the mesh are indicated, as are the incoming and outgoing boundaries on the edges of the mesh and the type of discretization used for each mesh point.

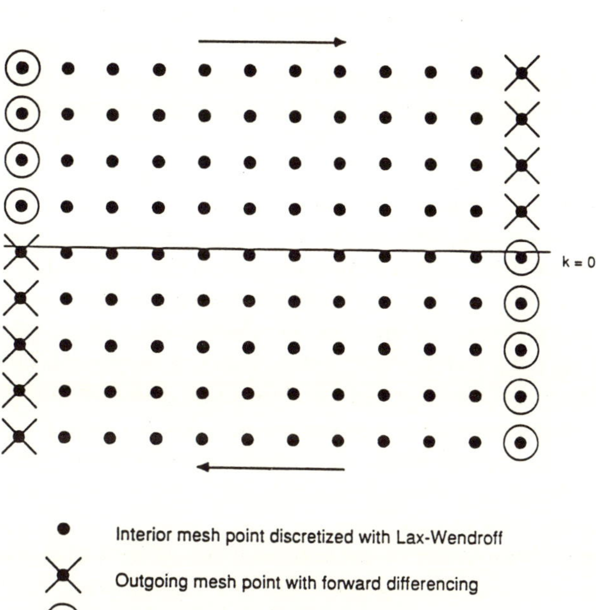

FIGURE 11.6. Phase-space discretization scheme for the Wigner function. Each row of the grid has a uniform velocity and can be thought of as a subsystem, or slice, of the distribution.

11.4.3. Boundaries and Contacts

Simulation of a real device includes some model of the interface between the device and an external circuit. At the very minimum, the external circuit consists of a "battery," which fixes a potential across the device, and "wires" which carry current from the "battery" to the device. These circuit parameters are usually included in device simulations as boundary conditions. In many models, the contact serves as an infinite reservoir of thermally distributed carriers.[16,50–55] This reservoir maintains a fixed distribution at the contact where the particles enter the simulation domain.

Conditions at the contact must be consistent with physical reality. If a current is flowing through the device, current continuity requires that an identical current be flowing through the external circuit. This implies that the external circuit, and thus the contact, must be characterized by some distribution which reflects current flow, such as a shifted thermal distribution.

In (11.2.8), we have left and right boundaries at $x = 0$ and $x = L$, respectively. Mathematical constraints permit us to specify only one boundary, however. As discussed earlier, the model may be thought of as a coupled set of systems, each of which is a slice of constant momentum (see Figure 11.6). Each of these momentum slices has a boundary, or contact, from which electrons enter the slice, and they leave the slice from the opposite boundary.

The "entering" contact is constrained by the current. Just inside the device, a current exists. Because an identical current must be present in the contact, and the distribution within the contact is assumed to be a shifted thermal distribution, the amount of shift and therefore the distribution within the contact are known. The distribution within the contact is then matched to the corresponding momentum slice of the model. This procedure must be self-consistently carried out, since the current within the device is a function of the boundary conditions, and the boundary conditions are functions of the current within the device. Because each momentum slice has only one boundary prescribed, the boundary conditions are mathematically consistent with (11.2.8).

A second requirement on the contact is that it remove all carriers which are leaving the device. This defines the "leaving" contact. In the so-called "ideal" contact, the carriers are perfectly extracted from the device without reflection. Studies of (11.2.8) with Gaussian wave packets have shown that the momentum nonlocality in the equation couples the incoming and outgoing boundaries at the contacts and serves as a source of artificial numerical reflections.[34] Such phenomena have been observed in other well-posed numerical systems[52] and arise strictly from the coupling at the boundaries.

One solution in principle is to put the boundary of the simulation so far from the region of interest that the artificially induced wave components do not appreciably affect the region of interest during the simulation time. As usual, this brute-force repair does not fix the problem because there is not enough such "brute force" in the computer. A more sophisticated approach would decompose the time-evolving function at the boundary into ingoing and outgoing waves and simply set the incoming part equal to the specified contact distribution. Unfortunately, the Wigner function equation of motion is nonlocal in momentum, so the separation into ingoing and outgoing components must also be nonlocal. In order to perform this separation exactly, we would in principle have to know exact solutions throughout the simulation region. Since determining these is the goal of the calculation, that would clearly be an inappropriate approach. Instead, we use an approximate numerical algorithm, which only relies on the asymptotic form of high-momentum solutions. The high-momentum components are the most important source of error, and the low-order treatment given here effectively decouples ingoing and outgoing waves.

Engquist and Majda treat shallow-water waves,[56] and the algorithm we use to tame artificial reflections is adapted[57] from their method. To illustrate the method, we first consider the formal problem of wave motion described by an equation of the general form

$$\frac{\partial \mathbf{u}}{\partial t} = \mathbf{A}\frac{\partial \mathbf{u}}{\partial x} + \mathbf{B}(x)\mathbf{u} \tag{11.4.3}$$

Here $\mathbf{u}$ is a vector in an arbitrary number of components. If momentum is regarded as a discrete variable, then (11.2.8) takes this form with momentum as the vector index and with $\mathbf{A}$ a diagonal matrix of velocities. If $\mathbf{B}(x)$ is zero, then left-moving (right-moving) waves can be identified trivially as those associated with negative (positive) diagonal elements of $\mathbf{A}$. The nonlocality of the forcing term in (11.2.8) corresponds to off-diagonal terms in $\mathbf{B}(x)$. This nonlocality mixes positive and negative velocity components, and our task is to unmix them.

We begin by rewriting (11.4.3) in the form

$$\frac{\partial \mathbf{u}}{\partial x} = \Lambda\frac{\partial \mathbf{u}}{\partial t} + \mathbf{C}(x)\mathbf{u} \tag{11.4.4}$$

with $\Lambda = \mathbf{A}^{-1}$, $\mathbf{C} = -\mathbf{A}^{-1}\mathbf{B}$. Using a pseudodifferential operator[56,58] $\mathbf{U}$, we can define a new vector $\mathbf{v}$ by

$$\mathbf{v} = \mathbf{U}\left(x, \frac{\partial}{\partial t}\right)\mathbf{u} \tag{11.4.5}$$

which satisfies

$$\frac{\partial \mathbf{v}}{\partial x} = \left\{ \mathbf{U}\left[\Lambda\frac{\partial}{\partial t} + \mathbf{C}\right]\mathbf{U}^{-1} - \frac{\partial \mathbf{U}}{\partial x}\mathbf{U}^{-1}\right\}\mathbf{u} \tag{11.4.6}$$

The object in making this transformation is to obtain an equation in the form

$$\frac{\partial \mathbf{v}}{\partial x} = \begin{bmatrix} \Lambda^i\,\partial/\partial t & 0 \\ \Omega_{21} & \Omega_{22} \end{bmatrix}\mathbf{v} \tag{11.4.7}$$

Here we have separated the components into blocks by the sign of the velocity. The earlier indices ["earlier" in the vector component ordering, indexing the upper and left halves of the matrix in (11.4.7)] correspond to incoming waves; Λ^i is the matrix of all the positive or all the negative inverse velocities. The higher-value indices describe the opposite sign of velocity—the outgoing waves. We define $\mathbf{P}$ to be the projection operator into the space spanned by the incoming components (thus, for example, $\Lambda^i = \mathbf{P}\Lambda$). While (11.4.7) is not a complete diagonalization, it does ensure that the outgoing component $\mathbf{v} - \mathbf{P}\mathbf{v}$ does not produce an incoming wave $\mathbf{P}\mathbf{v}$, and it allows us to specify separately the distribution of incoming waves.

As noted earlier, an exact solution of this problem is very difficult. Thus, we seek a **U** such that (11.4.7) is satisfied asymptotically for the highest velocities. To do this we make the formal identification

$$\frac{\partial}{\partial t} \sim i\xi \tag{11.4.8}$$

and make a Laurent expansion of the pseudodifferential operator **U**:

$$\mathbf{U}(x, i\xi) = \mathbf{U}_0(x) + \frac{1}{\xi}\mathbf{U}_1(x) + \frac{1}{\xi^2}\mathbf{U}_2(x) + \cdots \tag{11.4.9}$$

If the expansion is terminated at order ξ^{-1}, the nonreflection condition $\mathbf{Pv} = 0$ can be written

$$\mathbf{PU}_0(x_{\mathrm{bdry}})\frac{\partial \mathbf{u}}{\partial t} + i\mathbf{PU}_1(x_{\mathrm{bdry}})\mathbf{u}(x_{\mathrm{bdry}}, t) = 0 \tag{11.4.10}$$

When the above procedure is performed on the Wigner function equation of motion (11.2.8), the matrix multiplications become integrations over momentum. After some essentially straightforward algebra, one obtains explicit nonreflecting boundary conditions for a simulation in the region $[x_-, x_+]$ in the form (11.4.10). At the left boundary $x = x_-$, the incoming waves (those with momentum $p > 0$) satisfy

$$\frac{\partial}{\partial t} f_w(x_-, p, t) + \frac{1}{2\pi\hbar}\int_{-\infty}^{0} dP\, M(x_-, P - p)\frac{P}{P - p} f_w(x_-, P) = 0 \tag{11.4.11a}$$

while at the right boundary $x = x_+$, the incoming waves (with momentum $p < 0$) satisfy

$$\frac{\partial}{\partial t} f_w(x_+, p, t) + \frac{1}{2\pi\hbar}\int_{0}^{\infty} dP\, M(x_+, P - p)\frac{P}{P - p} f_w(x_+, P) = 0 \tag{11.4.11b}$$

11.4.4. Tests of the Numerical Algorithms

To test the accuracy of the numerical time evolution and of the boundary conditions, we study the evolution of a Gaussian wave packet (GWP).[41,59] The Wigner function for a Gaussian wave packet is

$$f_w(x, k) = \frac{1}{\pi\hbar} e^{-x^2/4a^2} e^{-4(p-p_0)^2 a^2/\hbar^2} \tag{11.4.12}$$

which is a Gaussian distribution in both position and momentum. From

(11.4.12), we find that the root mean square uncertainty in position (the standard deviation) is $\Delta x = \sqrt{2}a$, and the similarly defined momentum uncertainty is $\Delta p = \hbar/2\sqrt{2}a$. From this we can confirm the well-known fact that Gaussian wave packets are the "most classical" possible wave packets, in the sense of having the smallest position-momentum uncertainty product allowed by the Heisenberg uncertainty principle:[11]

$$\Delta x\, \Delta p = \hbar/2 \qquad (11.4.13)$$

In the absence of a potential, the Wigner function time-evolution equation (11.2.8) becomes

$$\frac{\partial}{\partial t} f_w(x, p, t) - \frac{p}{m}\frac{\partial}{\partial x} f(x, p, t) = 0 \qquad (11.4.14)$$

Thus, f_w may be decomposed into separate momentum components which evolve independently. On average, the packet maintains a constant velocity, but over time the various components' different velocities lead to a spread of the packet. The Gaussian wave packet has the property that as it spreads it maintains its Gaussian character, with the only changes being a time-dependent shift of the centroid (at the group velocity p_0/m) and an increase in the wave packet width.

We have tested our numerical programs by starting with (11.4.12) as an initial condition and integrating (11.4.14) numerically in time. The integral of (11.4.12) over momentum and position is normalized to unity. The formal equation of motion conserves total probability, so any variation in the value of this integral is attributable solely to numerical errors in the algorithm. After 200 time steps, the Lax–Wendroff scheme shows less than 1% numerical error. In comparison, simple first-order finite differencing of (11.2.8) produces a greater than 10% error after the same number of time steps.[60] We conclude that the hybrid discretization is stable and convergent.

Next, the GWP is allowed to interact with a pair of quantum barriers separated by a quantum well.[41] The barriers are each 3 nm thick and 0.3 eV in height, while the well is 5 nm in width. Figure 11.7a shows the initial wave packet as it begins to interact with the barriers, with the positions of the barriers indicated by the dark lines. In Figure 11.7b, the majority of the wave packet has reflected from the barrier, while a small portion has tunneled through. Finally, in Figure 11.7c, most of the wave packet has reflected from the barrier. Some of the wave packet has tunneled through the barriers and is leaving the region of simulation.

A particularly interesting feature of Figure 11.7c is the rapidly oscillating structure along the $k = 0$ axis, particularly prominent on the incident side. Like the larger regions of "negative probability," this structure is rendered unobservable by the Heisenberg uncertainty principle: any

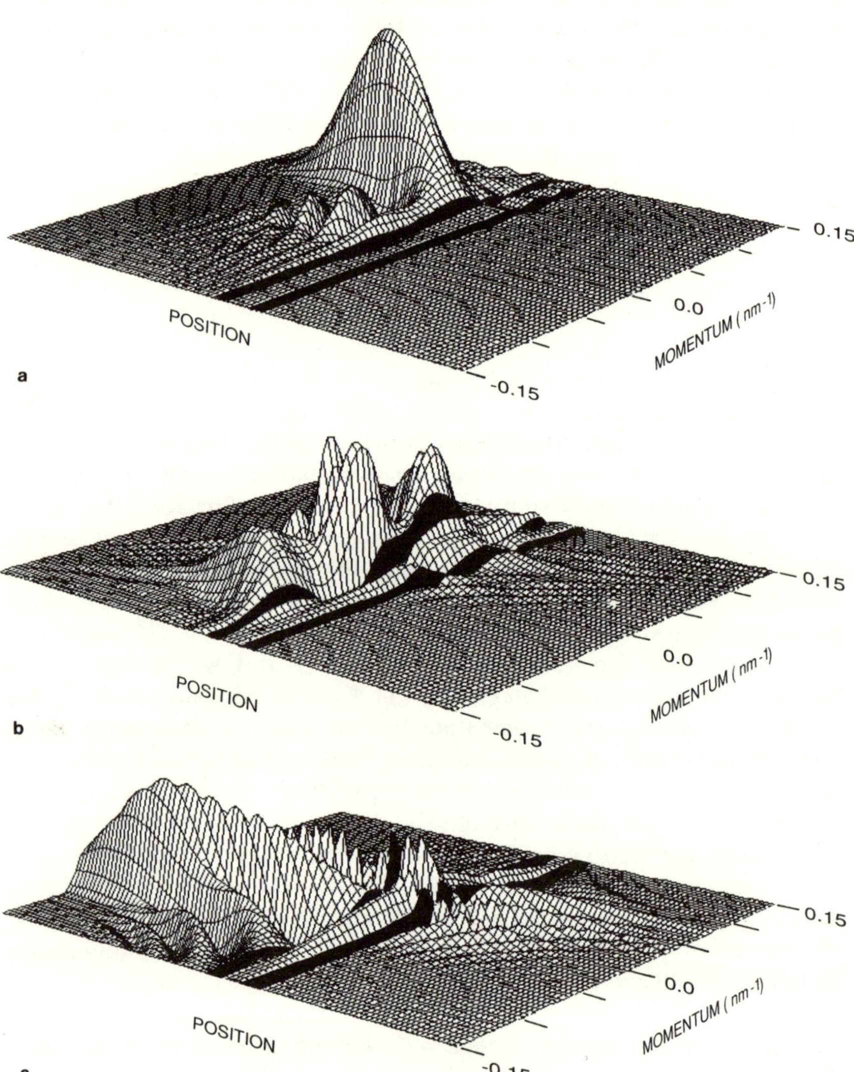

FIGURE 11.7. Gaussian wave packet interacting with resonant quantum potential barriers. The barriers are indicated by the dark bands. (a) The incident wave packet, approaching the barriers from the left, is just beginning to interact. ($t = 20$ fs.) (b) Gaussian wave packet during interaction. The incident and reflected components are visible, as is the correlation centered around $k = 0$. Part of the packet is tunneling through the barrier. ($t = 50$ fs.) (c) Gaussian wave packet packet after reflection. Most of the wave packet has beeen reflected. Tunneling packet is visible to the right of the barriers. ($t = 95$ fs.)

measurement designed to detect the oscillations directly must be so large as to encompass many oscillations. Thus, for example, integration over momentum yields no density near the barrier after the main pulses have passed, although the $k = 0$ oscillations persist. Nevertheless, the presence of these oscillations is associated with measurable consequences.

To understand the oscillations, it is useful to recall that the time-evolution equation for the Wigner distribution, like Schrödinger's equation for the wave function, is time-reversal-invariant. The Wigner function, however, unlike the wave function, is able to describe mixed states as well as pure states. If we take a film of the wave function for the situation of Figure 11.7 and run it backwards, we must have an equally valid physical situation: two synchronized wave packets approaching the barriers and interfering so that only one packet leaves the barrier. Described in terms of wave functions, this system has only packets and no ancillary features: Initially there are just two packets approaching each other; after a long time there is just one packet moving away. We are not surprised, because any system described by a single wave function has perfect phase coherence and is capable of interfering.

If the same general situation is described by a Wigner function, there are simultaneously more possibilities and apparently less freedom to describe those possibilities. On one hand, the Wigner function must describe the relative phase, if there is any, or the degree of coherence within and between the packets. On the other hand, this must be accomplished using a function which is restricted to be real. In fact, the information about phase coherence of the state is distributed nonlocally. In Figure 11.7c, the oscillations along $k = 0$ are the most obvious sign that the two outward-moving packets are correlated and not the time-reversed image of two independent packets approaching simultaneously from infinity. As long as this information is retained, the system has time-reversal symmetry. When the correlation is removed, the original wave packet can no longer be recovered.[41,49] Any dissipative processes will act to remove this correlation and thus destroy reversibility.

The Gaussian packets can also be used to illustrate the need for, and the success of, the nonreflecting boundary conditions (11.4.11). The simulation results in Figure 11.7 were obtained using those conditions. In Figure 11.8 we compare the results of simulating without (a) and with (b) such numerically absorbing boundary conditions. It is clear that (11.4.11), though approximate, represents a very effective abatement algorithm for the spurious reflections, eliminating waves of amplitude comparable to the wave packet peak height. It is also clear that the dominant component of the spurious reflection is found at high momentum, justifying the approach taken in the previous section of numerically eliminating them by an asymptotic analysis valid for high velocity.

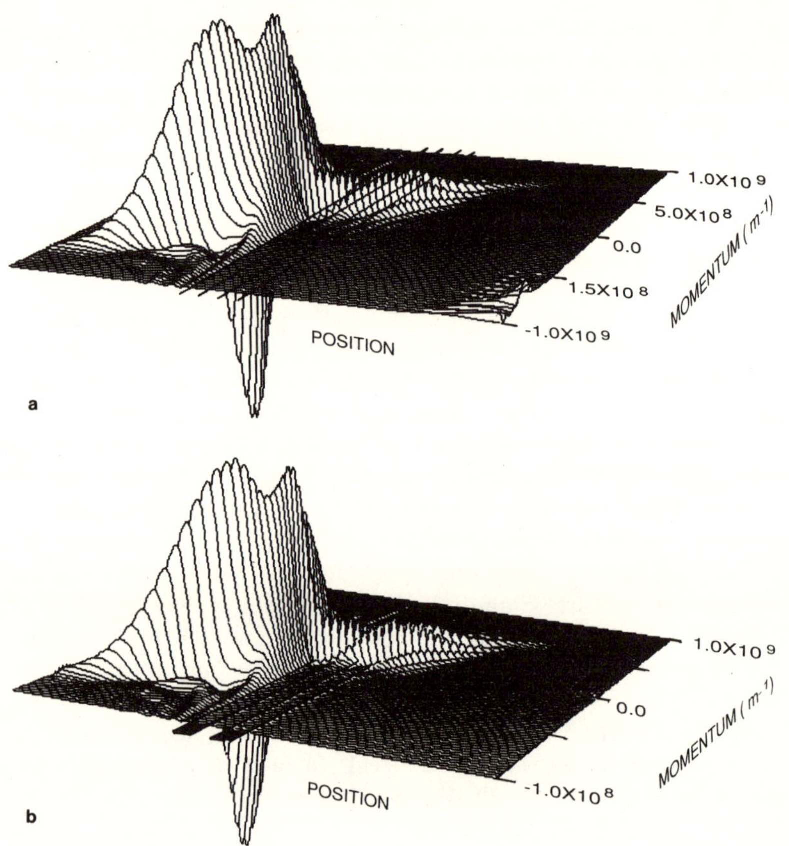

FIGURE 11.8. A comparison of Gaussian wave packet transmission through a double-barrier ($t = 20$ fs) absorbing boundary structure, without (a) and (b) nonreflecting boundary conditions ($P_0 = 3.0 \ F \ 8 \ \mathrm{m}^{-1}$).

Equation (11.2.8) may readily be used to check the correctness of an initial distribution. The initial potential and distribution are inserted into (11.2.8), and the system is stepped forward in time. The correct equilibrium distribution will not vary under the equation of motion. Because the numerical scheme is stable and convergent, no errors will be introduced through the equation. The only source of error, therefore, is from an incorrect initial distribution. We check this by using the initial distribution calculated in the previous section. The system is stepped 500 fs in time, or 5000 time steps, with zero applied bias, after which the state of the system is compared to the initial state. For an initial Wigner function normalized to have a maximum value of 1.0, the maximum error in the time-evolved Wigner function was 2.18×10^{-6}, and the L_2-norm of the error was less than 10^{-10}.

As a further check on the temporal stability of the initial Wigner function, the current at the contacts was calculated. With zero applied bias, the current calculated at the left contact was $0.038\ \text{pA/cm}^2$, while the current at the right contact was $-0.013\ \text{pA/cm}^2$, in comparison to the normal operating current densities of mA/cm^2. Thus, the integrated error in current is about 10^{-5} of the actual current.

11.4.5. Self-Consistent Potentials

Fully self-consistent potentials are introduced by adding Poisson's equation to the system of equations. A self-consistent initial Wigner distribution is calculated which includes the background charge from the ionized donors. The temperature and doping level determine the Fermi level within the bulk GaAs,[61] and this specifies the ionized donor concentration and the electron concentration far from the barriers. Since the electron density is a function of the potential and the ionized donor charge and potentials are interdependent, an iterative procedure is required to find the correct initial distribution. A guess is made for the potential, which is used to calculate the density $n(x)$, which is the trace of the density matrix found from the scattering states. The potential is also used to calculate the ionized charge density $N_d^+(x)$. The total charge is then used to solve Poisson's equation to get a new potential. Because the ionized charge density depends exponentially on the potential, the two quantities must be made consistent by iteration. The new potential is then used to calculate the density. The entire process is iterated until full consistency between $n(x)$, $V(x)$, and $N_d^+(x)$ is achieved. The density matrix is then transformed into the Wigner function, which together with the background charge $N_d^+(x)$ and the self-consistent potential $V(x)$ is input to (11.2.8) and Poisson's equation. The solution proceeds by stepping (11.2.8) forward, then adjusting the potential via Poisson's equation. In addition to an unchanging Wigner distribution, steady-state conditions may be tested by examining the time-variant fluctuations in the internal potentials. When both the Wigner function and the potentials become invariant, steady-state conditions have been reached.

11.4.6. Scattering

Scattering events represent the failure of a transport picture based on electrons moving in a perfect lattice, and so represent a complication relative to the simplest electron description. Inelastic events, also, represent times when the evolution of the electron system cannot be considered alone and must be time-evolved, at some level of approximation, simultaneously with the phonon system. Thus, a proper inclusion of dissipation is difficult, since it involves solving not merely for a more complicated motion but for a motion that takes place in a larger space. For the densities considered, the relevant scattering mechanism is scattering by phonons. Without introducing something to function in some sense as a phonon coordinate, the effect of collisions must then be expressed by some linear functional of the density, or the Wigner distribution function.

We use a single relaxation-time approximation:[62]

$$\left(\frac{\partial f_w}{\partial t}\right)_{\text{collision}} = \frac{f_w - f_w^{(0)}}{\tau} \qquad (11.4.15)$$

where $f_w^{(0)}$ is the equilibrium distribution and τ is the relaxation time. This very simple model has simple virtues. Physically, it reproduces two important qualitative features of the exact scattering term: First, it destroys correlation, which would be detectable in the Wigner function as described in section 11.4.4. Second, under stationary boundary conditions it returns the system to the correct (equilibrium) distribution. In principle, the scattering time, or a scattering time distribution, might be calculated using Fermi's golden rule.[62] This requires microscopic parameters which are not known accurately. Instead, τ is adjusted to reproduce the correct mobility, establishing effective contact with a known experimental quantity. Numerically, (11.4.15) has the considerable advantage of being fast and easy to compute.

Other approaches are available to treat scattering within a one-electron formalism, among them Levinson's formalism,[22] in which the interaction terms of the Hamiltonian are Wigner–Weyl-transformed and incorporated in the new equation of motion. The quantum Boltzmann equation[37-39] also includes scattering terms in its formulation, and the time integral of one (correlation) Green's function is the Wigner function, but this formalism requires simultaneous treatment of three independent functions, each of which depends on one more variable than the WDF.

11.5. APPLICATION TO THE RESONANT TUNNELING DIODE

11.5.1. Structure To Be Simulated

Here we will describe the results of using the self-consistently determined Wigner function to model resonant tunneling diodes (RTDs). The structures we study all consist of two identical rectangular barriers of AlGaAs. The barrier height is 0.3 eV, which is approximately the highest conduction band discontinuity, relative to GaAs, that is possible without making the alloy band gap indirect. Outside the barriers, the device is GaAs with a donor density of 10^{18} cm^{-3}. The barriers and the quantum well are undoped. We have investigated structures over a range of barrier and quantum well widths, all generally close to our fiducial structure of 5-nm barriers and 5-nm well region.

11.5.2. *I-V* Characteristics

The steady-state *I-V* curve of the device is calculated by applying an incremental bias potential to the cathode contact, then stepping (11.2.8) and Poisson's equation until steady-state conditions are reached. Earlier work with the Wigner function model of the RTD showed that large-signal transients decay exponentially and that steady-state conditions are achieved in a few hundred femtoseconds.[41,63] We step the system to a time of 1.5 ps to ensure that steady state has been achieved while checking for invariance of the distribution function and the potential. A transient analysis (discussed below) with the self-consistent model shows that transients have essentially vanished within 700 fs.[64] When steady-state conditions are achieved, the current in the device is calculated and the bias potential is again incremented.

Figures 11.9 and 11.10 show the results of this simulation, for a fixed bulk mobility and for a range of barrier and well sizes, respectively. It is clear from Figure 11.9 that the voltage of the resonant peak is essentially independent of the barrier thickness. This is consistent with the fact that the resonance wave function essentially does not penetrate the barriers, which consequently do not affect its energy. On the other hand, the current

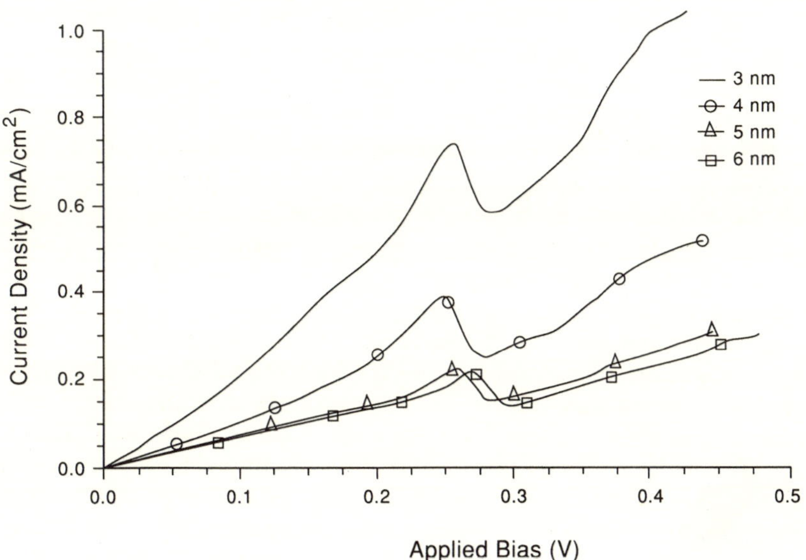

FIGURE 11.9. *I-V* curves for an RTD with 5-nm well and barriers ranging from 3 to 6 nm.

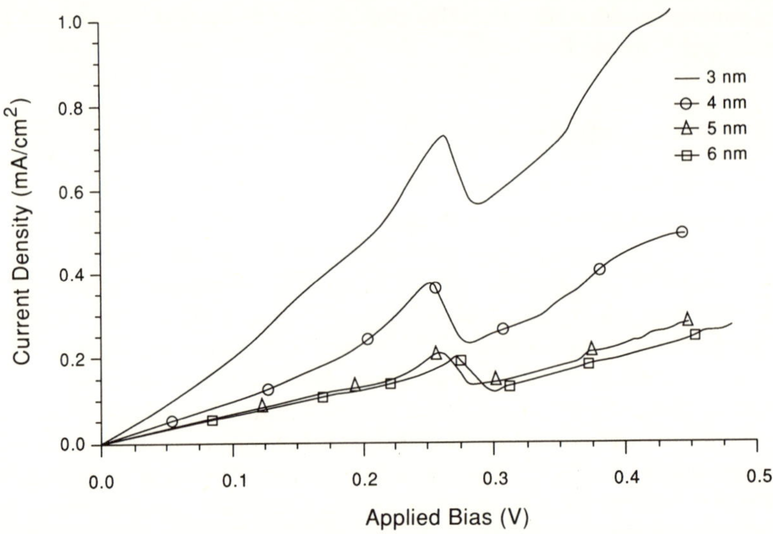

FIGURE 11.10. *I-V* curves for an RTD with 3-nm barriers and a well ranging from 3 to 8 nm.

increases with decreasing barrier thickness, both off- and on-resonance. At biases far away from resonance, the current depends on tunneling essentially through the thickness of two barriers and so decreases exponentially with increasing barrier width. At resonance, the transmission probability is essentially unity, and the peak height is determined by the width of the resonance—the range of energies contributing to the peak. This width decreases as the barriers become thick. At the thickest barriers, the former effect dominates, and the peak-to-valley ratio consequently increases. One way to understand this is by noting that the resonance width cannot decrease indefinitely with increasing barrier width, since the resonance is degenerate with the continuum on the anode side. More precisely, as the barriers lengthen, they begin to be triangular in profile and therefore less effective.

For Figure 11.10 the barriers are held fixed at 3 nm, and the well width is varied from 3 to 8 nm. The peak position, tracking the resonant state energy, varies approximately as the inverse square of the well width at low energy (where the barrier height is much greater than the resonance energy, so the quasibound state is well defined). (In every case, the next resonant energy is above the barrier.) Here the peak-to-valley ratio varies strongly, since the current off-resonance depends only weakly on the well length, but the resonance width—and so the sharpness of the resonance—decreases with decreasing well width (and increasing energy).

In Figure 11.11 we illustrate our fiducial structure (5 nm/5 nm/5 nm) with a GaAs mobility of 300 cm²/V-sec at 300 K. Here there are two curves,

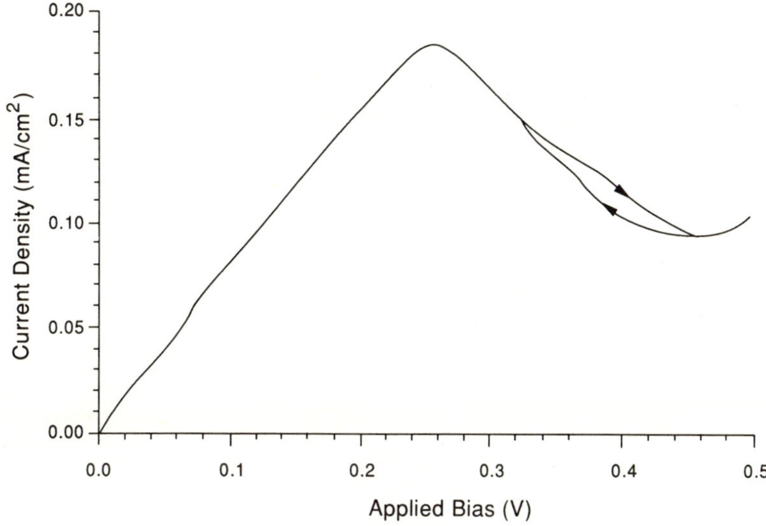

FIGURE 11.11. Self-consistent *I-V* curve for the RTD. The applied bias is increased adiabatically to a maximum of 0.5 V, then decreased. The *I-V* curve shows a hysteresis resulting from redistribution of the potential drop and charge density.

demonstrating hysteresis: after the bias has been increased to a maximum as described above, the potential was decremented toward zero, in like manner, with the self-consistent potential and the current density being calculated along the way. This difference in the two curves demonstrates that the solution of our system of equations is not a unique function of the boundary conditions. This can only arise from nonlinearity. In other words, the hysteresis arises from the rearrangement of charge as the bias is swept.

Simpler models of the RTD, which commonly assume that all potential was dropped across the barrier-well structure,[41,46,59,65] are therefore not capable of detecting the hysteresis. For the fiducial device of Figure 11.11, at the bias which yields the (unique) current maximum, the self-consistent internal potential of the device is illustrated in Figure 11.12. This shows that only about one third of the potential is dropped across the barriers. Of the remaining potential, the greater part is dropped across the cathode side of the device. At this side of the device, the Wigner distribution exhibits heavy depletion, as seen in Figure 11.13. As bias is applied, this region is pulled down, and would be expected to show an accumulation of electrons. In fact, at very low biases, this region does accumulate slightly. As the bias is increased, however, the cathode-barrier interface region is lowered even further in energy until it forms a shallow triangular quantum well that quantizes the electrons, similar to the quantization in an enhancement

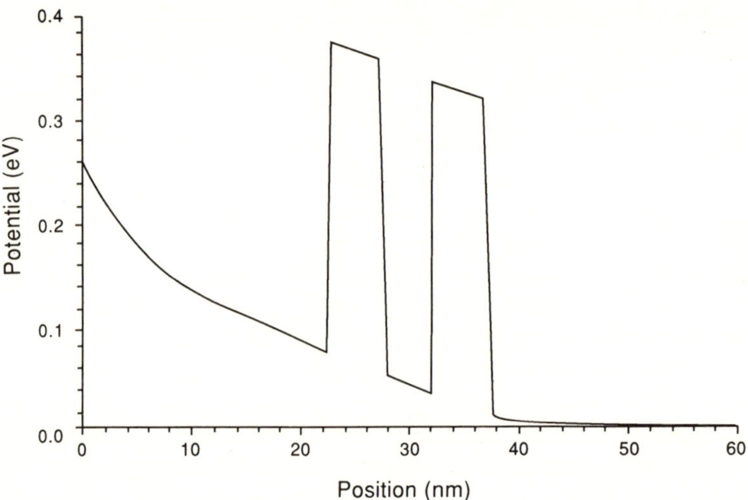

FIGURE 11.12. Potential for the RTD for an applied bias of 0.22 V, the peak of the *I-V* curve (Figure 11.11). Much of the potential is dropped across the cathode side of the device, forming a triangular potential well.

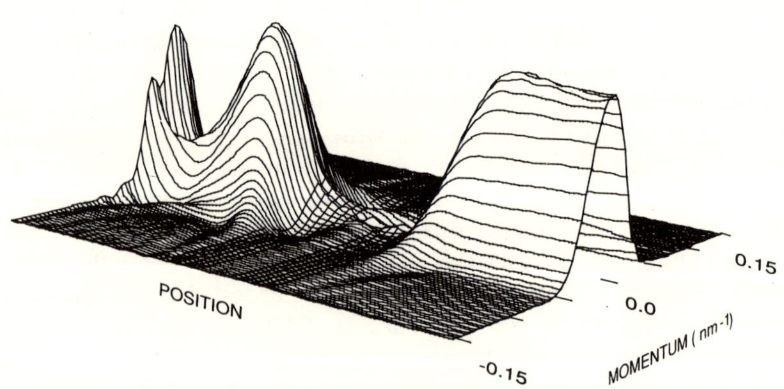

FIGURE 11.13. Steady-state Wigner distribution for bias corresponding to the current peak. Depletion is evident in the cathode region. At the cathode (left) contact, the incoming distribution appears as a shifted Fermi–Dirac distribution.

MOSFET[19,66] or a HEMT.[67] An overall depletion of electrons occurs in the cathode.

11.5.3. Bistability

A controversial aspect of the RTD is the origin of the bistability that has been observed in the *I-V* curve in the NDC regime. An intrinsic origin would correspond to the hysteresis in Figure 11.11. It arises from charge storage within the quantum well, which changes the potential and field near the barriers.[4] This shifts the conduction band edges on either side of the barriers and changes the relative position of the resonant level in the well. Since the current is strongly affected by the positions of these bands and the resonant level, subtle shifts in stored charge can greatly affect the current tunneling through the resonant level between the barriers. Extrinsic bistability, on the other hand, results from the external circuit and may either cause, or be a result of, the device oscillating at very high frequency about the bias point.[2] Such bistability is then a function of the external circuit rather than the quantum well structure.

While experiments have observed bistability in the operation of RTDs, there is general disagreement as to its cause.[2,4,68-71] If all oscillations of the device could be suppressed, any observed bistability would have to be intrinsic. Because the negative resistance R_n is low, however, debate centers around whether all oscillation has in fact been suppressed, and extrinsic bistability observed.

As noted, in many studies the potential is dropped uniformly across the barrier structure.[41,46,59,65] Treatments which specify a fixed electrostatic potential cannot account for the effects of charge storage or self-consistent potential shifts outside the barriers. Since intrinsic bistability is thought to result from changes in internal potentials, it is not surprising that rigid potential models do not show bistability.

The calculation of the *I-V* curve does not include external circuit effects. Any bistability which is found, therefore, must be intrinsic in nature. In fact, one theoretical approach has shown that an intrinsic bistability does exist.[71] Scattering is modeled by a relaxation time approximation, and its primary effect is in determining the potential drop in the resistive regions outside the barriers. This resistive load also influences the slope $-1/R_n$ of the NDC region. Since the mobility at 300 K is density-dependent, the carrier concentration in the device is critical to R_n.

Earlier work, without self-consistent potentials, showed a shoulder in the NDC region of the curve.[41] With the addition of fully self-consistent potentials, the shoulder is not seen, but a soft bistability, similar to ferromagnetic hysteresis, is found in the same region of the *I-V* curve. This model shows one region of bistability with a finite r_n, resulting strictly from differences in the self-consistent potentials and the distributions on either side of the bistable region. Both effects are related to the charge stored in the quantum well. The maximum difference in current in the bistable region is found at an applied bias of 0.36 V. In Figure 11.14, we plot the difference between the steady-state Wigner distribution for increasing potential and the steady-state distribution for decreasing potential in the bistable region. For decreasing potential, less current flows through the resonant structure, and the injected carriers accumulate on the cathode side of the structure, shown as the peak in Figure 11.15. Since less current is flowing in the entire device, the anode side shows a slight depletion of current-carrying distribution.

In an experimental circuit the RTD is loaded, which changes the slope of the measured NDC region as determined by the load line. Experiments have shown two bistable regions of the I-V curve, each of which has a very large R_n. What may be observed experimentally is the intrinsic bistability being heavily modified by the external circuit parameters.

11.5.4. Zero-Bias Anomaly

Some experiments with tunneling devices have observed an increased conductivity near zero bias.[72] It has been argued that this "zero-bias anomaly" arises from impurity-assisted tunneling, in which the spin interaction of the tunneling electron with the impurity leads to several conducting channels.[73,74] In this model, the s-d interaction in third) and higher-order perturbation theory yields a peak in the conductivity at zero bias, which results from the combined effects of spin-flip scattering in the barrier and the exclusion principle in the contacts. Enhanced conductivity is also seen

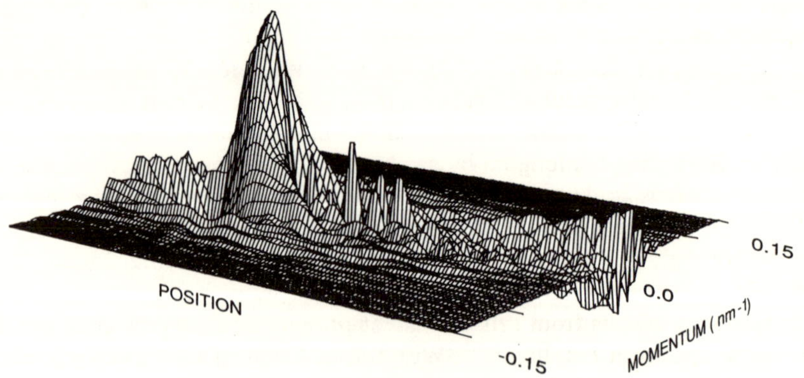

FIGURE 11.14. Difference between the bistable Wigner distributions at a bias of 0.36 V. The quantized state in the cathode well has more carriers. For increasing potential, the current is greater, as indicated by the "ridge" in the distribution.

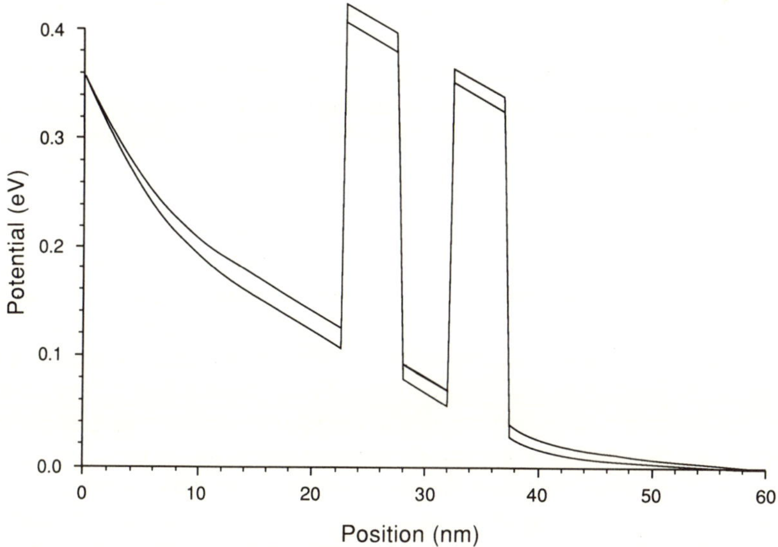

FIGURE 11.15. Potentials corresponding to the two bistable states for an applied bias of 0.36 V. The higher (more repulsive) curve corresponds to the (decreasing voltage) branch of the bistable region.

near zero bias in MBE-grown tunneling structures.[75] Here we describe an alternative mechanism that also produces a high conductivity at low bias, but which results from high-momentum tails of the distribution which depend on the distance to the barrier, and which appear to be nonthermal.

Near a quantum barrier, the electron density is reduced as a result of quantum interaction with the barrier.[17] This is not a depletion layer in the classical sense. It is not caused by band bending; this phenomenon occurs even when there are no electromagnetic fields. The wave function at a particular momentum k is repelled by the tunnel barrier. The first peak in the standing wave occurs at about a distance $\pi/2k$ from the barrier, and so is much closer for high momentum. This leads to a broadened distribution for the density, with a length scale of about λ_{th}. The quantum repulsion is, in a sense, complementary to barrier penetration: just as a nonzero density penetrates a finite distance into a classically forbidden region, a density deficit penetrates a finite distance into a classically allowed region. This was discussed earlier. In Figures 11.4 and 11.5 one can see the momentum spread which results from this quantum repulsion. This momentum spread of the distribution emphasizes a distinction between two components of the current at extremely low biases. The high-momentum tails of the distribution in Figure 11.5 extend above the barrier in energy. A slight perturbation of the potential is sufficient to alter radically the nature of this distribution and to force a current flow. Because the tails of the distribution are characterized by high momentum, the current which results may be characterized by extremely high conductance. After the momentum tails have depleted, the only remaining source of current is tunneling from the main thermal distribution, which is characterized by a much lower conductance. Thus, near zero bias we see a dramatic change in $f(E)$, affecting the current through (11.1.1)

Early models of tunneling were unable to show this effect, since it was assumed that the distribution at the interface between the barriers and the bulk region is essentially thermal. One way to discuss this effect is in terms of an effective temperature which describes the momentum spread at the interface.[76] At zero bias, this effective temperature is much larger than that of the bath. A finite bias, by letting the high-momentum components transmit, lowers this effective temperature. Failure to account for the change in effective temperature at the interface neglects the contribution that these momentum tails make to the current at low biases.

The *I-V* curve calculated for Wigner functions exhibits the zero-bias anomaly below 0.1 mV, as shown in Figure 11.16. Above 0.1 mV, the current begins to show an exponential dependence, but does not extrapolate to zero. Below 20 μV, the current is characterized by a conductivity much higher than that above 0.1 mV. Between these regions, the conductivity changes with the applied bias.

In Figure 11.17a, the difference between the steady-state distribution with 20 μV applied bias and the equilibrium distribution is plotted (the

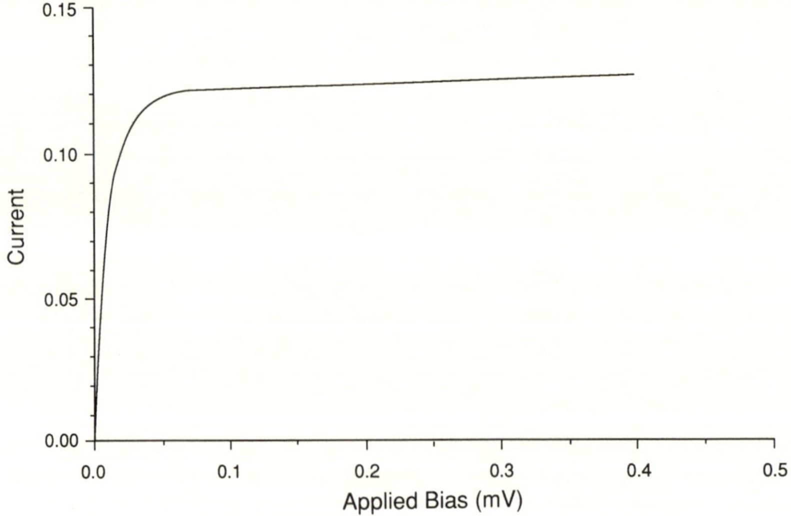

FIGURE 11.16. *I-V* curve of the RTD for low biases. The current displays two different conductances. The exponential dependence above 0.1 mV does not extrapolate to zero.

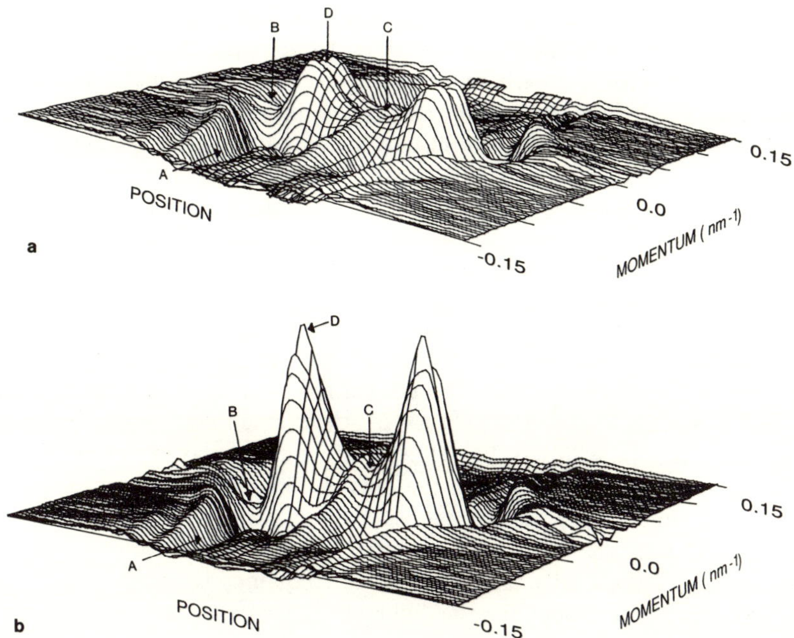

FIGURE 11.17. Difference between the steady-state distribution and the low-bias distributions (magnified 15×). (a) Applied bias is 20 μV. Point A shows the depletion of the high-momentum tails. Points B and D show a shift of the accumulation layer toward the barriers, while C shows a slight enhancement of the distribution within the quantum well. (b) Applied bias is 100 mV. Point A shows the same loss of tails as the 20-μV distribution, while B and D show the accumulation layer shifted clearly into the barriers. The distribution within the well C is greatly enhanced.

amplitude is magnified by a factor of 15). Point A on this figure indicates a loss of carrier density from the high-momentum tails. Quantum repulsion near the barriers is still present, but has been reduced by the small applied bias, leading to the positive peak D. Point B, the location of the accumulation in equilibrium, shows some slight reduction as carriers begin to tunnel through the barriers. Within the quantum well, at point C, the distribution is enhanced by the tunneling carriers. At this small bias, nearly all of the current in the barrier region comes from the high-momentum carriers. These carriers have sufficient energy to tunnel easily through (or over) the quantum barriers. Tunneling from the thermal distribution does not yet contribute significantly to the current.

In Figure 11.17b, the difference between the distribution at a higher applied bias of 100 mV, and the equilibrium one is shown. A comparison of Figures 11.17a and 11.17b reveals that there is little difference at point A, as the high-momentum tails have been completely depleted. The differences between 17a and 17b appear clearly at point *B*. The accumulation layer is moved considerably nearer to the barrier, from point *B* to point *D*. The distribution at point *D* clearly reflects the presence of a thermal distribution right up to the barriers. In fact, it extends well into the barriers, and a large distribution within the well, at point *C*, is evident. Calculation of the current in the barrier region reveals that most of the current now results from carriers at lower energies which tunnel through the barriers.

11.5.5. Transient Behavior

The transient behavior of the RTD may be studied using Wigner functions.[41,63,64] In this study, the system for $t < 0$ is biased at the peak of the *I-V* curve of Figure 11.11. At $t = 0^+$, the bias is raised instantaneously to a value corresponding to the valley of the curve, so as to determine the large-signal transient behavior of the device. The current is recorded as a function of time as the system evolves toward its new steady state. The current and potential transients are then Fourier-transformed, and the conductance of the device as a function of frequency is found.

The current transient for the fiducial (5/5/5) structure is plotted in Figure 11.18. Strong oscillations are evident in the current transient, as well as a very large initial overshoot. A comparison of the transients for different

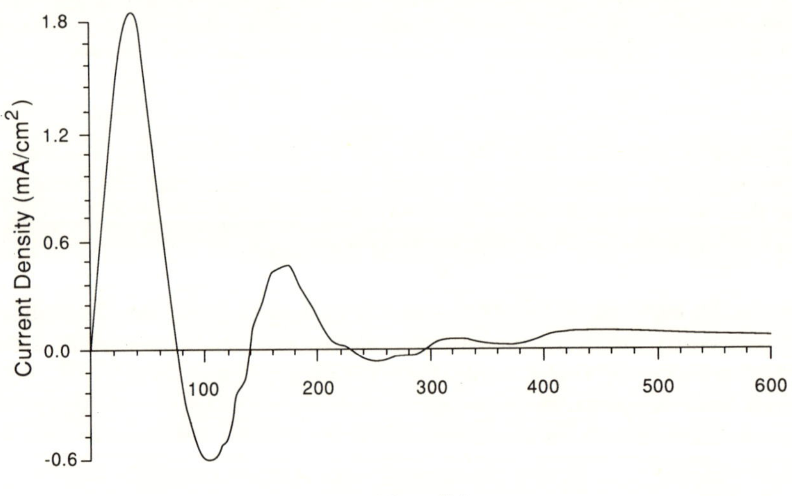

FIGURE 11.18. Current transient of the RTD for the 5/5/5 structure. The device is switched from the peak to the valley of the *I-V* curve. The ballistic inertia of the electrons makes possible plasmalike excitations.

barrier widths (Figure 11.19) shows that the initial overshoot and longer-time oscillations vary differently. The initial overshoot, which is largest for the thinnest tunneling barriers, may be explained as arising from a rapid discharge of the trapped charge in the potential well (compare Figure 11.13). The oscillations are less sensitive to barrier thickness, and appear to be similar in nature to plasma oscillations, although they arise from charge oscillations in the vicinity of the heterostructure. These oscillations are also clearly observable in animations of the full (two-phase-space-dimensional) distribution function. The large overshoot of current is related to an inherent inductive effect of the carriers.

The current transient has been calculated for three values of the relaxation time, corresponding to mobilities of 1500, 3000, and 4200 cm^2/V-sec to determine whether the relaxation time is not the factor limiting the frequency of operation of the device. The absolute magnitude of the conductance for the three mobilities is plotted in Figure 11.20. The conductance shows a peak in all cases at approximately 1.5 THz, with a weak second harmonic peak at 3 THz. The systematic increase in the low-frequency conductance with increasing mobility reflects the normal behavior of the nonresonant component of the current, which is decreased by scattering just as bulk current is decreased.

In Figure 11.21, the conductance is plotted for an RTD with 5-nm well, and a range of barrier thicknesses. It is clear from the insensitivity of the

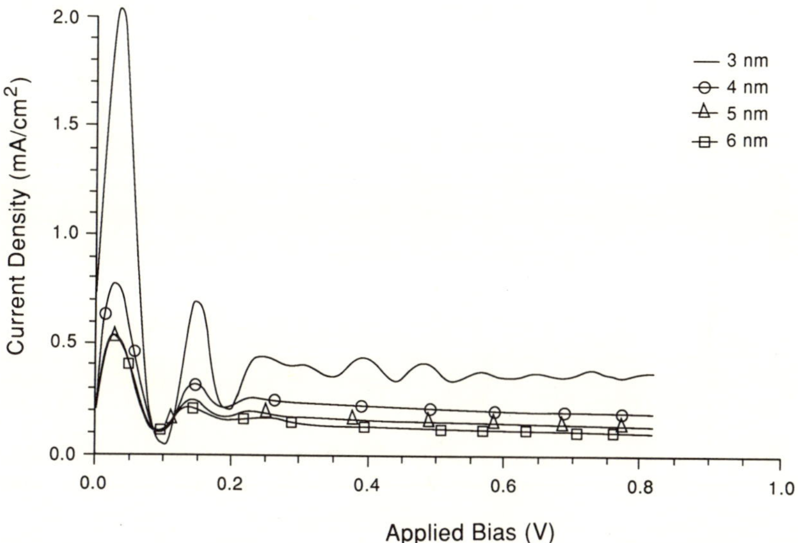

FIGURE 11.19. Current transient of the RTD for a 5-nm well and a range of barrier widths. The device is switched from the peak to the valley of its particular *I-V* curve.

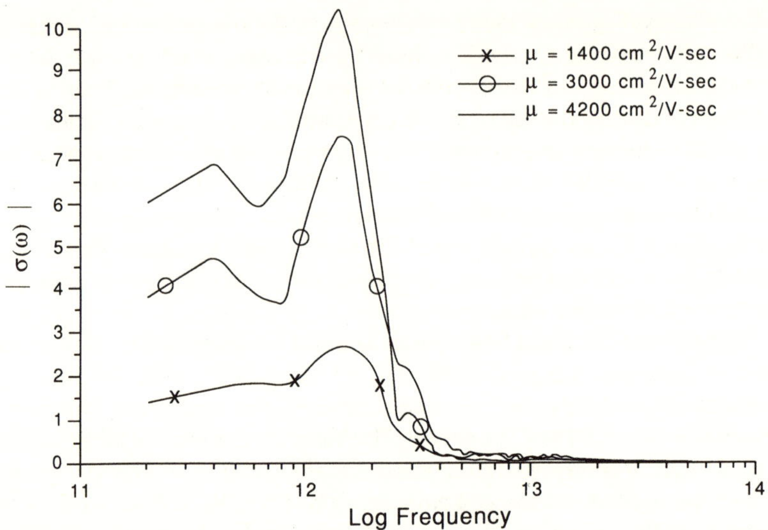

FIGURE 11.20. Magnitude of the conductance of the RTD in the negative differential conductance region, calculated for three values of mobility.

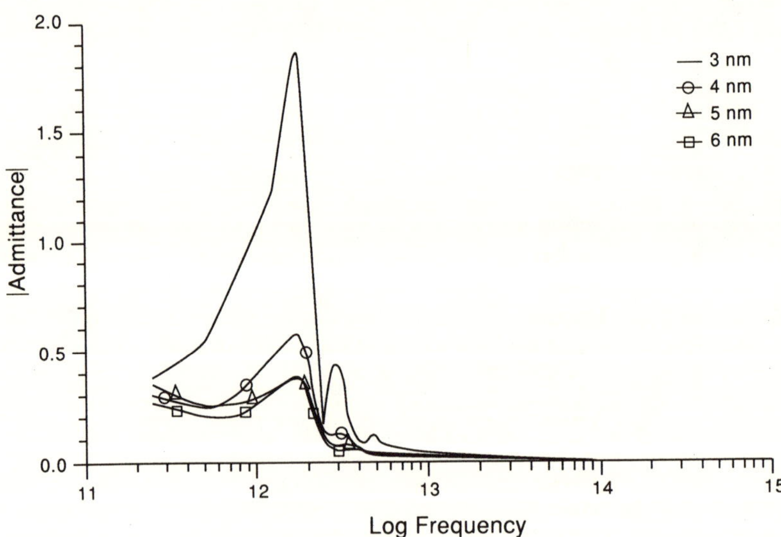

FIGURE 11.21. Magnitude of the conductance of the RTD in the negative differential conductance region, calculated for different barrier thicknesses (well is 5 nm long).

peak position frequency that this frequency does not depend on the time it takes for charge to cross the barrier. (We refer here only to the time that it takes for a quantity of charge, initially one side, to be partially redistributed to the other side of the barrier. We do not refer here to the duration of an individual tunneling event, which is a more controversial issue, and discussed in Chapter 10.)

The imaginary part of the conductance is plotted in Figure 11.22a. For low frequencies, the device is inductive. As the relaxation time and the mobility decrease, the magnitude of the inductive contribution decreases, but the peak position remains unchanged. The electrons move essentially ballistically over a mean free path, and their inertia prevents a sudden reaction to sharp changes in the potential. As the mobility decreases, the electrons are ballistic over a shorter mean free path and less inductive. For frequencies above 2 THz, the device is capacitive, as a result of the successive charging and discharging of the resonant energy level within the quantum well. Experimental evidence supports the buildup of a space-charge layer in the tunneling structure,[2] and earlier studies[59] have shown a charge storage within the quantum well, characterized by a time of the order of 250 fs.

The real part of the conductance is plotted in Figure 22b. For frequencies under 1.5 THz, the real part of the conductance is negative and becomes positive for frequencies over 1.5 THz. The frequency at which the transition between negative and positive conductance occurs is independent of the relaxation time. We note that the peak of the reactance in Figure 11.22a occurs at the point at which the real part is zero, a general behavior that is commonly observed in systems with inertia which satisfy Kramers–Konig relations.

11.5.6. Spacer Layers

Resonant tunneling diodes are often fabricated with undoped spacer layers at the interface between the barriers and the bulklike contact regions. The effects which result from these spacer layers may easily be examined with the Wigner function model. The self-consistent equilibrium Wigner function, potential, and ionized donor concentrations are again calculated, with the spacer layer built into the donor distribution $N_d^+(x)$. While the difference in the equilibrium distributions is only slight, and concentrated at the interface between the barriers and the contacts, the fundamental difference may be easily seen in the potentials $V(x)$. The self-consistent potentials for the basic structure and for the RTD with a spacer layer are plotted in Figure 11.23. Without a spacer layer, the potential at the barrier tends toward a triangular potential well, leading to a slight accumulation of electrons near the barrier. With a spacer layer, the potential is shifted

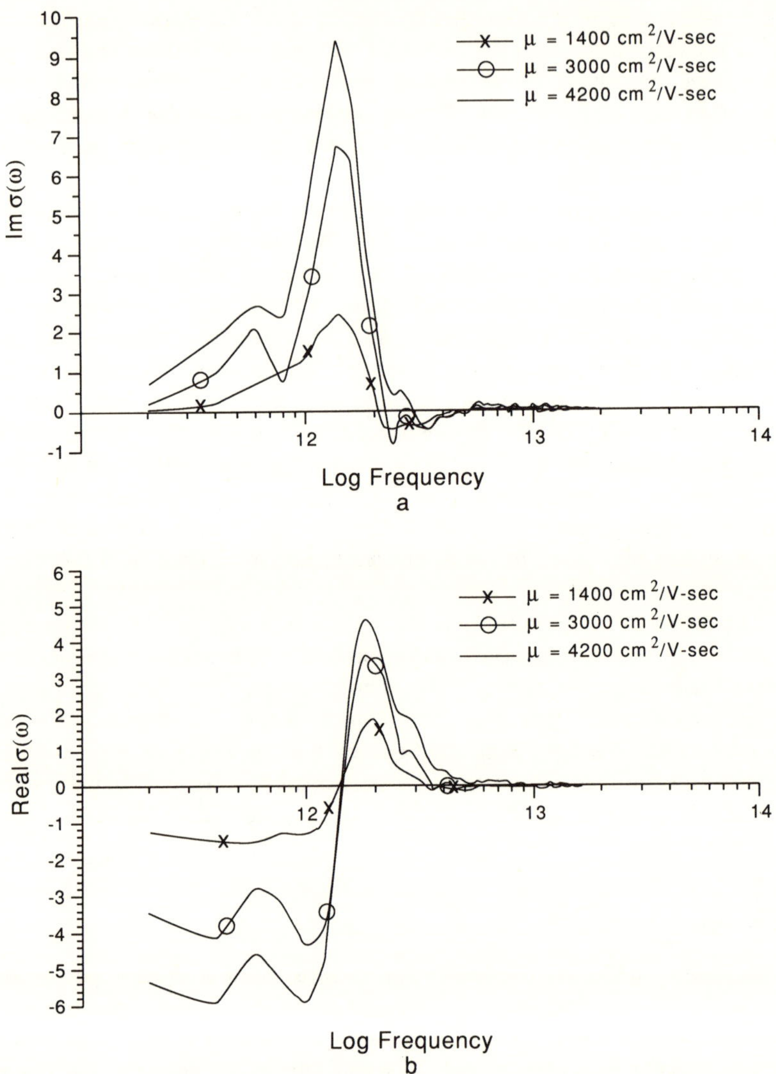

FIGURE 11.22. Real and imaginary components of the conductance of Figure 11.20. (a) Imaginary part of the conductance. Below 2 THz, the system is inductive. (b) Real part of the conductance. Resistance is negative below the resonant frequency of 1.5 THz.

upward, causing a much smaller tendency toward depletion of electrons. In fact, there is a small accumulation in the spacer layer. It is apparent that the spacer layer concentrates the potential near the barrier. This both reduces the contact resistance and reduces the ballistic injection so that the peak-to-valley ratio is improved.

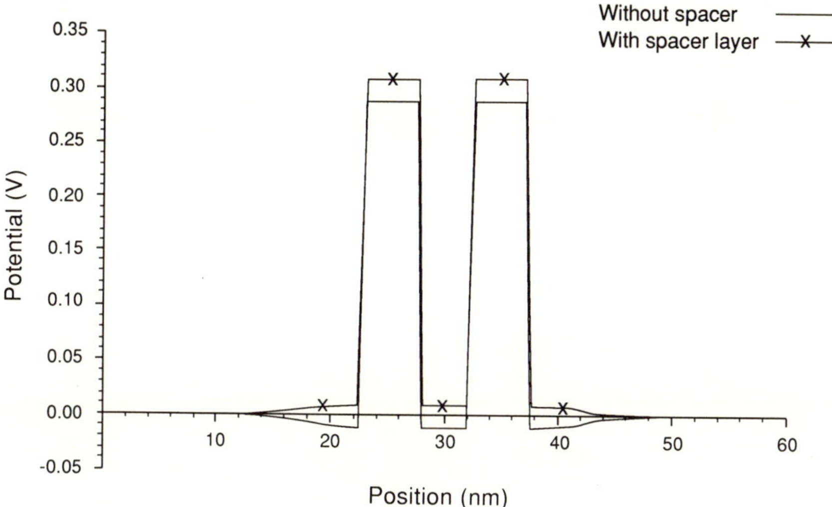

FIGURE 11.23. Self-consistent equilibrium potentials for the RTD with and without an undoped spacer layer.

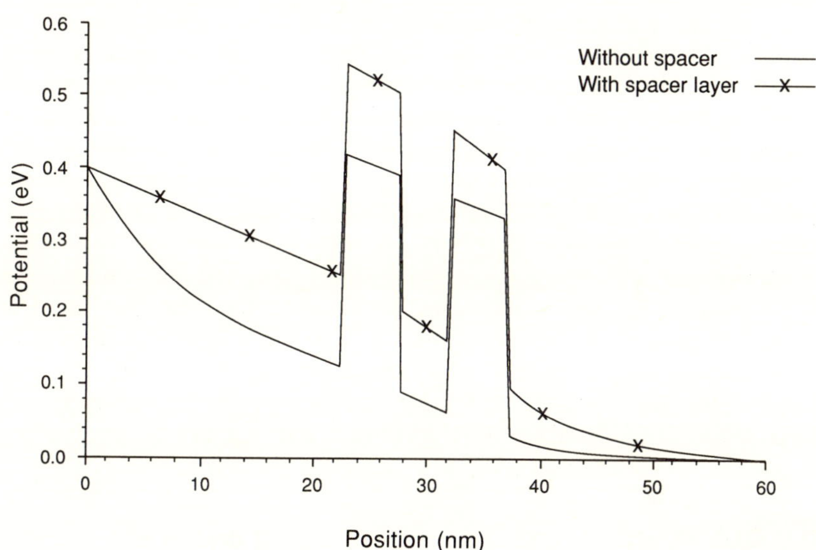

FIGURE 11.24. Self-consistent potentials for an applied potential of 0.4 V for the RTD with and without the spacer. Without the spacer, a large depletion region and triangular potential well form between the cathode contact and the barrier. The spacer layer controls the depth of the well.

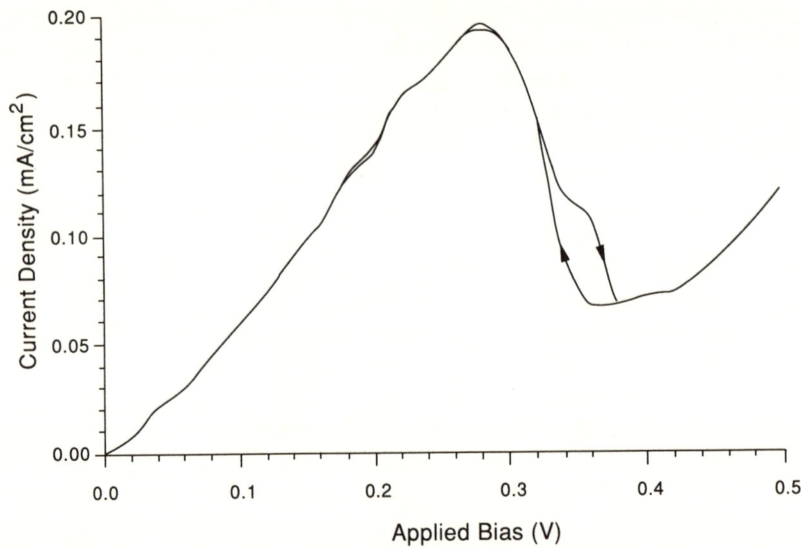

FIGURE 11.25. *I-V* curve of the RTD with the spacer layer. The negative resistance is lower, and the bistable region is more pronounced than without the spacer layer. The peak-to-valley ratio is nearly 3:1.

A second difference results from the effect of the spacer layers on the scattering. Without ionized donors, the electron mean free path is longer, and the relaxation time is correspondingly increased. More of the potential is dropped across the barriers, causing a higher barrier to the ballistic electrons, which reduces the valley current. The self-consistent potential for a bias at the valley is plotted in Figure 11.24, showing how the potential differs from that in the RTD without the buffer layer. The effect upon the valley current is clearly evident in the calculated *I-V* curve of Figure 11.25. Another major difference caused by the buffer layer and reduced scattering is the reduced R_n, which increases the slope of the NDC region of the curve.[4] Finally, the bistability of the NDC region of the *I-V* curve is found to be much sharper, again a consequence of the reduced R_n. We note, however, that the bistability still appears to be intrinsic. The curves are now much closer to the simpler model of Sheard and Toombs.[71]

11.6. SUMMARY

The Wigner function is an effective tool for modeling one-dimensional heterostructures. As we have shown with one- and two-barrier models, it is possible to incorporate accurate self-consistent potentials. In the case of the two-barrier (resonant tunneling diode) structure, modeling based on this function and its explicit time evolution is able to explain a number of experimentally observed phenomena, such as hysteresis and bistability, and the zero-bias anomaly. As a mathematical representation of the physical system, the Wigner function offers a classically oriented description (a "probability" distribution in a space described by phase-space variables) which simultaneously is a faithful representation of the mixed quantum state. Thus, a classical physical intuition allows one to recognize distinct quantum effects, such as phase correlation of scattered packets and nonelectrostatic repulsion at a distance from a barrier.

We have described the successful management of various numerical issues. In particular, Lax–Wendroff explicit time differencing, which incorporates high-order terms in the discretized equation, is used to achieve stability in the temporal evolution. This is unconditionally stable and convergent if the choices of space mesh Δx and the time step Δt satisfy the CFL stability criterion for the entire discretized mesh. Initial conditions for the time evolution are discussed, and we describe a method for finding these which is based on exact solution of the Hamiltonian.

The Wigner distribution function obeys a transport equation similar to the Boltzmann equation. We treat scattering as a local phenomenon, but the Wigner function equation of motion includes nonlocal correlations through the force. This makes it difficult to disentangle the inflow and outflow components at the boundary. This problem is handled through an asymptotic analysis which identifies the fastest components at the boundary. We use this expansion at lowest order to specify incident boundary conditions, so as to eliminate artificial numerical reflections from the boundaries. Success is demonstrated by the absence of spurious reflections in the numerical evolution of isolated wave packet propagation. In general, "nonreflecting boundary conditions" sharply reduce the dependence of modeling results on the size of the simulation region.

In the more physically representative situations, of devices coupled to reservoirs, this makes possible a coupling to realistic thermal reservoirs. These are "ideal" contacts which provide a thermal distribution of carriers to enter the device and which absorb perfectly carriers that leave the device. The distribution within the contact must reflect the physical reality of current flow, so we enforce current continuity by using a drifted Fermi–Dirac distribution within the contacts. In effect, to avoid modeling an arbitrarily long device, one makes the more tractable, yet reasonable, assumption that the mean free paths are extremely short beyond the simulated region.

The electrostatic interaction is included by coupling Poisson's equation to the Wigner equation. This includes the corresponding ionized donor distribution and the Hartree potential of the electrons. Both of these must be evaluated self-consistently. (The ionized donor distribution depends on the Fermi level.) Scattering is included through a single-relaxation-time approximation, with the relaxation time adjusted to fit the bulk mobility.

The self-consistent model is applied to the RTD. The calculated *I-V* curves show a peak-to-valley ratio of 2:1 for a structure with 5-nm well and barriers and, more importantly, show an intrinsic bistability in the NDC region of the curve. This bistability results from charge storage in the quantum well, which changes the internal potentials. Bistability has been observed experimentally, and a recent theoretical model predicted that the bistability is intrinsic.[71] When we model a modified structure with undoped spacer layers adjacent to the quantum barriers, we find a peak-to-valley ratio of nearly 3:1, and a sharp decrease in the negative resistance. The bistable region is correspondingly enhanced. In the absence of the spacer layers, a deep triangular quantum well forms between the cathode contact and the quantum barrier. This well is sufficiently deep to give rise to a quantized state. Approximately half of the applied bias is found in the cathode well. (Essentially similar behavior has been found recently by Frensley.[77]) At the valley of the *I-V* curve, the barriers are pulled down in energy, permitting ballistic electrons from the cathode to travel over the barriers, leading to an enhanced valley current and a degraded peak-to-valley ratio. A net depletion in the cathode region stems from the contact resistance of the "ideal" contact model. Inclusion of spacer layers inhibits formation of the cathode well. The spacer layer causes an upward bending of the conduction band at the barrier interface, which resists the downward bending of the deep quantum well as bias is increased. More of the potential is dropped across the barriers. The peak-to-valley ratio is enhanced by a reduction in the number of ballistic electrons traveling over the barriers.

The transient behavior of the RTD was also analyzed. The device is switched from a bias at the peak of the *I-V* curve to the valley, and the current is monitored during as the system relaxes to its new steady-state. There is a large initial current transient, resulting from the discharging of the quantum well. The current oscillates in response to the adjusting potentials within the device until it decays to its new steady-state value. Fourier analysis of the current transient shows a peak in the conductance near 1.5 THz.

An anomaly in the *I-V* curve at very low bias *I-V* characteristic that has been observed experimentally in tunneling structures is reproduced by the Wigner function model. In contrast to previous explanations of that effect, it is found to result from quantum repulsion from the barrier: this repulsion, complementary to quantum tunneling, leads to high-momentum tails in the electron distribution near the barrier. Since these tails extend to energies above the barrier, they contribute to a current with high conductance. The high-momentum tails are quickly disrupted by a small applied bias, however. At moderately high bias, the tails have vanished and the current comes from the thermal distribution, which is characterized by a much lower conductance.

ACKNOWLEDGMENT

Support for this work was provided by the U.S. Office of Naval Research, and one author (C.R.) was supported by the Air Force Office of Scientific Research.

REFERENCES

1. See, for example, the Les Houches proceedings in *The Physics and Fabrication of Microstructures and Microdevices* (M. J. Kelly and C. Weisbuch, eds.), Springer, Berlin (1986); *Proceedings of the Fourth International Conference on Molecular Beam Epitaxy* (C. T. Foxon and J. J. Harris, eds.), in *J. Cryst. Growth* **81**; M. J. LUDOWISE, *J. Appl. Phys.* **58**, R31 (1985).
2. L. ESAKI and R. TSU, *IBM J. Res. Dev.* **14**, 61 (1970); R. TSU and L. ESAKI, *Appl. Phys. Lett.* **22**, 562 (1973); L. L. CHANG, L. ESAKI, and R. TSU, *Appl. Phys. Lett.* **24**, 593 (1974).
3. T. C. L. G. SOLLNER, W. D. GOODHUE, P. E. TANNENWALD, C. D. PARKER, and D. D. PECK, *Appl. Phys. Lett.* **43**, 588 (1983); T. C. L. G. SOLLNER, P. E. TANNENWALD, D. D. PECK, and W. D. GOODHUE, *Appl. Phys. Lett.* **45**, 1319 (1984).
4. V. J. GOLDMAN, D. C. TSUI, and J. E. CUNNINGHAM, *Phys. Rev. Lett.* **58** 1256 (1987); V. J. GOLDMAN, D. C. TSUI, and J. E. CUNNINGHAM, *Phys. Rev. Lett.* **59**, 1623 (1987).
5. G. BERNSTEIN and D. K. FERRY, *IEEE Trans. Elec. Dev.* **ED-35**, 887 (1988); G. BERNSTEIN and D. K. FERRY, *Z. Phys. B* **67**, 449 (1987).
6. R. A. PUECHNER, JUN MA, R. MEZENNER, W.-P. LIU, A. M. KRIMAN, G. N. MARACAS, G. BERNSTEIN, D. K. FERRY, P. CHU, H. H. WIEDER, and P. NEWMAN, *Surf. Sci.* **228**, 520–523 (1990).
7. See *IBM J. Res. Develop.* (Mar. 1988).
8. B. RICOH and M. AZBEL, *Phys. Rev. B* **29**, 1970 (1984).
9. R. LANDAUER, *IBM J. Res. Dev.* **1**, 223 (1957).
10. W. W. LUI and M. FUKUMA, *J. Appl. Phys.* **60**, 1555 (1986).
11. E. MERZBACHER, *Quantum Mechanics*, Wiley, New York (1970).
12. J. HEADING, *An Introduction to Phase Integral Methods*, Wiley, New York (1962).
13. E. L. HILL, *Phys. Rev.* **38**, 1258 (1931).
14. R. PÉREZ-ALVAREZ and H. RODRIGUEZ-COPPOLA, *Phys. Stat. Sol.* (*b*) **145**, 493 (1988), has extensive references.
15. R. LANDAUER, *Z. Phys. B* **21**, 247 (1975).
16. M. BÜTTIKER, *Phys. Rev. Lett.* **57**, 1761 (1986).
17. A. M. KRIMAN, N. C. KLUKSDAHL, and D. K. FERRY, *Phys. Rev. B* **36**, 5953 (1987).
18. U. RAVAIOLI, M. A. OSMAN, W. PÖTZ, N. C. KLUKSDAHL, and D. K. FERRY, *Physica B + C* **134B**, 36 (1985).
19. S. M. GOODNICK and D. K. FERRY, in *Physics and Chemistry III-V Compound Semiconductor Interfaces* (C. W. Wilunson, ed.), Plenum, New York (1985).
20. E. WIGNER, *Phys. Rev.* **40**, 749 (1932).
21. J. E. MOYAL, *Proc. Cambridge Phil. Soc.* **45**, 99 (1949).

22. I. B. LEVINSON, *Zh. Eksp. Teor. Fiz.* **57**, 660 (1969) [*Sov. Phys.–JETP* **30**, 362 (1970)]; P. CARRUTHERS and F. ZACHARIASON, *Rev. Mod. Phys.* **55**, 245 (1983).

23. G. J. IAFRATE, H. L. GRUBIN, and D. K. FERRY, *Phys. Lett.* **87A**, 145 (1982).

24. L. MANDEL and E. WOLF, *Rev. Mod. Phys.* **37**, 231 (1965).

25. H. HAKEN, H. RISKEN, and W. WEIDLICH, *Z. Phys.* **206**, 355 (1967).

26. M. LAX and W. H. LOUISELL, *IEEE J. Quantum Electron.* **QE-3**, 47 (1967).

27. M. J. BASTIAANS, *Opt. Comm.* **25**, 26 (1978).

28. K. H. BRENNER and J. OJEDA-CASTANEDA, *Opt. Acta* **31**, 213 (1984).

29. M. J. BASTIAANS, *Appl. Opt.* **19**, 192 (1980).

30. M. CONNER and Y. LI, *Appl. Opt.* **24**, 3825 (1985).

31. B. V. K. VIJAYA KUMAR and C. W. CARROLL, *Opt. Eng.* **23**, 732 (1984).

32. J. R. BARKER, *J. Phys. C* **6**, 2663 (1973).

33. J. R. BARKER and S. MURRAY, *Phys. Lett.* **93A**, 271 (1983).

34. N. C. KLUKSDAHL, W. PÖTZ, U. RAVAIOLO, and D. K. FERRY, *Superlatt. Microstruct.* **3**, 41 (1987).

35. W. R. FRENSLEY, *J. Vac. Sci. Tech. B* **3**, 1261 (1985).

36. R. BALESCU, *Equilibrium and Nonequilibrium Statistical Mechanics*, Wiley, New York (1975).

37. L. P. KADANOFF and G. BAYM, *Quantum Statistical Mechanics*, Benjamin, New York (1962).

38. W. HÄNSCH and G. D. MAHAN, *Phys. Rev. B* **28**, 1902 (1983); G. MAHAN, *J. Phys. F* **13**, L257 (1983); G. MAHAN and W. HÄNSCH, *J. Phys. F* **13**, L47 (1983).

39. A. P. JAUHO and J. W. WILKINS, *Phys. Rev. B* **29**, 1919 (1984).

40. J. R. BARKER, in *Physics of Nonlinear Transport in Semiconductors* (D. K. Ferry, J. R. Barker, and C. Jacoboni, eds.), Plenum, New York (1979).

41. N. C. KLUKSDAHL, A. M. KRIMAN, C. RINGHOFER, and D. K. FERRY, *Solid-State Electron.* **31**, 743 (1988).

42. A. M. KRIMAN, J. ZHOU, N. C. KLUKSDAHL, H. H. CHOI, and D. K. FERRY, *Solid-State Electron.* **32**, 1603 (1989).

43. A. M. KRIMAN, A. SZAFER, A. D. STONE, and D. K. FERRY, submitted for publication.

44. A. M. KRIMAN and D. K. FERRY, *Superlatt. Microstruct.* **3**, 503 (1987).

45. E. WIGNER, *Phys. Rev.* **98**, 145 (1955).

46. W. R. FRENSLEY, *Phys. Rev. B* **36**, 1570 (1987).

47. R. COURANT, K. FRIEDRICHS, and H. LEWY, *Math. Ann.* **100**, 32 (1928).

48. Y. I. SHOKIN, *The Method of Differential Approximation*, Springer-Verlag, Berlin (1983).

49. J. R. BARKER, private communication.

50. M. BÜTTIKER, *Phys. Rev. B* **33**, 3020 (1986).

51. R. LANDAUER, *Philos. Mag.* **21**, 863 (1970).

52. R. LANDAUER and M. BÜTTIKER, *Phys. Rev. Lett.* **54**, 2049 (1985).

53. M. BÜTTIKER, *Phys. Rev. B* **32**, 1846 (1985).

54. M. BÜTTIKER, Y. IMRY, R. LANDAUER, and S. PINHAS, *Phys. Rev. B* **31**, 6207 (1985).

55. M. BÜTTIKER, *Phys. Rev. B* **35**, 4123 (1987).

56. B. ENGQUIST and A. MAJDA, *Math. Comp.* **31**, 629 (1977).

57. C. RINGHOFER, D. FERRY, and N. KLUKSDAHL, *Transp. Thry. Stat. Phys.* **18**, 331 (1989).

58. L. NIRENBERG, *Lectures on Linear Partial Differential Equations*, CBMS Regular Conference Series in Mathematics, vol. 17, American Mathematical Society, Providence, RI (1973).

59. N. C. KLUKSDAHL, A. M. KRIMAN, and D. K. FERRY, *Superlatt. Microstruct.* **4**, 127 (1988).

60. N. C. KLUKSDAHL, C. RINGHOFER and D. K. FERRY, unpublished.

61. J. S. BLAKEMORE, *Semiconductor Statistics*, Pergamon Press, New York (1962).

62. See, for example, K. SEEGER, *Semiconductor Physics*, Springer-Verlag, Berlin (1985).
63. W. R. FRENSLEY, *Solid-State Electron.* **31**, 739 (1988).
64. N. C. KLUKSDAHL, A. M. KRIMAN, and D. K. FERRY, *IEEE Elec. Dev. Lett.* **EDL-9**, 457 (1988).
65. W. R. FRENSLEY, *Phys. Rev. Lett.* **57**, 2853 (1986).
66. A. MANY, Y. GOLDSTEIN and N. B. GROVER, *Semiconductor Surfaces*, Wiley, New York (1965).
67. D. DELAGEBEAUDEUF and N. T. LINH, *IEEE Trans. Electron Devices* **ED-29**, 955 (1982).
68. V. J. GOLDMAN, D. C. TSUI, and J. E. CUNNINGHAM, *Phys. Rev. B* **35**, 9387 (1987).
69. T. C. L. G. SOLLNER, *Phys. Rev. Lett.* **59**, 1622 (1987).
70. L. EAVES, E. S. ALVES, T. J. FOSTER, M. HENINI, O. H. HUGHES, M. L. LEADBEATER, F. W. SHEARD, G. A. TOOMBS, K. CHAN, A. CELESTE, J. C. PORTAL, G. HILL, and M. A. PATE, *Proceedings of the 1988 Mounterndorf Conference on Physics and Technology of Submicron Structures*, to be published.
71. F. W. SHEARD and G. A. TOOMBS, *Appl. Phys. Lett.* **52**, 1228 (1988).
72. L. Y. L. SHEN AND J. M. ROWELL, *Phys. Rev.* **165**, 566 (1968).
73. C. B. DUKE, *Tunneling in Solids, Solid State Physics* Suppl. 10, Academic Press, New York (1969).
74. J. A. APPELBAUM, *Phys. Rev.* **154**, 633 (1967).
75. D. E. LACKLISON, B. DUGGAN, J. J. HARRIS, C. T. B. FOXON, D. HILGON, C. ROBERTS, and C. M. HELLON, *Appl. Phys. Lett.* **52**, 305 (1988).
76. N. C. KLUKSDAHL and D. K. FERRY, unpublished.
77. W. FRENSLEY, in *Proc. NCCE Workshop on Semiconductor Device Modeling* (in press).

Index